# ETHICS IN ENGINEERING PRACTICE AND RESEARCH

## Second Edition

The first edition of Caroline Whitbeck's *Ethics in Engineering Practice and Research* focused on the difficult ethical problems engineers encounter in their practice and in research. In many ways, these problems are like design problems: they are complex, and often ill defined; resolving them involves an iterative process of analysis and synthesis; and there can be more than one acceptable solution. In the second edition of this text, Dr. Whitbeck goes above and beyond by featuring more real-life problems, stating recent scenarios, and laying the foundation of ethical concepts and reasoning. This book offers a real-world, problem-centered approach to engineering ethics, using a rich collection of open-ended case studies to develop skill in recognizing and addressing ethical issues.

Caroline Whitbeck is the Emerita Elmer G. Beamer–Hubert H. Schneider Professor in Ethics at Case Western Reserve University. Dr. Whitbeck teaches in both the Philosophy and the Mechanical and Aerospace Engineering departments. Her research spans numerous fields, such as philosophy, engineering, technology, medicine, and feminist philosophy. Dr. Whitbeck is currently the Director of The Online Ethics Center for Engineering and Science at the National Academy of Engineering. Dr. Whitbeck has published numerous articles on bioethics and is the author of the first edition of *Ethics in Engineering Practice and Research* (1998, Cambridge University Press).

T0211257

# ETHICS IN ENGINEERING PRACTICE AND RESEARCH

## Second Edition

**Caroline Whitbeck**

Case Western Reserve University

# CAMBRIDGE
## UNIVERSITY PRESS

32 Avenue of the Americas, New York NY 10013-2473, USA

Cambridge University Press is part of the University of Cambridge.

It furthers the University's mission by disseminating knowledge in the pursuit of education, learning and research at the highest international levels of excellence.

www.cambridge.org
Information on this title: www.cambridge.org/9780521723985

© Caroline Whitbeck 1998, 2011

First edition published 1998
Second edition published 2011

*A catalogue record for this publication is available from the British Library*

*Library of Congress Cataloguing in Publication data*

Whitbeck, Caroline.
Ethics in Engineering Practice and Research / Caroline Whitbeck. – Second edition.
    p.   cm
Includes bibliographical references and index.
ISBN 978-0-521-89797-6 (hardback) – ISBN 978-0-521-72398-5 (paperback)
1. Engineering ethics.   I. Title.
TA157.W47   2011
174'.962–dc22        2010044633

ISBN  978-0-521-89797-6  Hardback
ISBN  978-0-521-72398-5  Paperback

*To the memory of James R. Melcher (1936–1991)*

# Contents

## 9. Responsible Authorship and Credit in Engineering and Scientific Research    324

## PART 4: THE FUTURE OF ENGINEERING

## 10. Responsibility for the Environment    351

# Note to Students

The educational goal of this book is to help prepare you for your professional responsibilities as engineers. It is designed to help you recognize and think through ethically significant problem situations that are common in engineering and to evaluate the existing ethical standards for engineering practice.

The central subjects that guide this work are the ethically significant problems that arise in engineering, the ethical and other values at stake in responding to them, and the concepts necessary to clearly understand those problems and respond to them. As a philosopher (and former mathematician) I have contributed conceptual clarifications necessary to understanding the ethically significant problems that commonly arise in engineering and I cite useful clarifications by others. The problems themselves are ones I have gathered from engineering societies and individual engineers who have generously shared their experience with me. These engineers are thanked in the preface to the first edition. Although some concepts, such as a conflict of interest, are common to many areas of professional ethics, engineering ethics differs from medical ethics or legal ethics in that the ethically significant problems encountered in engineering practice are different from those problems commonly encountered in medical or legal practice.

As aids to learning I have added sidebars that emphasize main points, and at the beginning of each section is a query to raise issues that are helpful for you to think through. At the end of each section is an exercise question similar if not identical to the opening query. The section will have given you help in formulating at least one good answer to the question, but there may be other good answers, and *you should feel free to give the best answer you can.*

Throughout the text I have added sidebars like the one here to emphasize main points because I have found some of my students benefited from such emphasis.

> The goal of this book is to prepare you to recognize and think through the problems you will encounter as an engineer and to evaluate the existing ethical standards for engineering practice.

The definitions offered in this book are primarily philosophical or **conceptual definitions**; that is, they show how one concept or idea is related to or composed of others. When a definition is offered in this book, it appears in **bold type**. You have encountered conceptual definitions in geometry, which define a concept in terms of other, simpler concepts. For example: A straight line is the shortest distance between two points. A dictionary occasionally gives a conceptual definition but more often will just tell you how a word is *used*, perhaps simply by giving a synonym for it. The conceptual definitions given here are like those given for concepts in physics and unlike the definitions in, say, a Spanish–English–Spanish dictionary.

Throughout this book, you will find boxes with thick borders, which contain brief but complete accounts of situations that illustrate or expand on some point discussed in that section of the text. Other boxes with thin borders contain open-ended problem situations to which you are asked to respond. These open-ended problems are often based on problems that engineers and my previous engineering students have found salient, perhaps because they experienced them on the job. You are asked to think about how best to respond to them both to build your problem-solving abilities and to help you recognize and anticipate problems that actually do arise in the engineering workplace. Occasionally, when some background information about the problem is useful but not explicitly discussed in the text, I have added that information in a section at the end titled "Getting Started." However, what you learn from the text is often not enough to construct a complete response to these open-ended problem situations. When actually faced with such a problem, you would need to interrogate both the problem and the resources available to you in that circumstance. (Such interrogation is discussed in Chapter 3, Ethics as Design.)

# Foreword to the First Edition

I want to die proud of having been an engineer. Since that can happen only if we engineers behave ethically, and since I see a connection between this book and gracious professionalism, I am very enthusiastic about Dr. Whitbeck's effort to help us think effectively and somewhat pragmatically about professional ethics. Everyone, professionals in particular, must expect ethically complex situations to arise. When that happens, each of us badly needs a self-image that includes conviction that our intellect and heart can help make choices that will dramatically affect the course of events. That point of view will not materialize out of the ether. It must be nurtured and encouraged. This book will help seasoned professionals clarify their approach to their own behaviors, and this book can profoundly affect those who face a messy situation for the first time.

Caroline's arguments penetrate some of the fog around ethics. Most people think of it as an obscure topic belonging to an elite few who can spend their lives in deep and abstract thought. Even many professors of engineering regard ethics as a somewhat untouchable topic. "Students will never listen! Why waste our time and theirs?" Several have argued that post–high school is too late to influence students' proclivity to behave in society's best interest. I strongly disagree. Since I have spent most of my teaching career encouraging students to trust their own creative abilities, I have developed a thick skin about comments like "You cannot teach creativity!" I do not debate that assertion. I think I know that one can unleash creative behavior by ensuring that it is overtly rewarded and by providing people with an assortment of "tools" that facilitate creativity. Likewise, after ten years of knowing Dr. Whitbeck and listening to her discussions, I am convinced that one can develop a self-image that includes self-confidence in dealing with ethically complex situations. I think that self-image is part of the foundation for a role as one of the protectors of society. It is essential to one who derives satisfaction from doing the thing that is right rather than easy or lucrative in the short term.

If students are told about an ethically complex situation and asked what course they would take if they found themselves in such a plight, they are quite likely to argue that they should call a press conference and blow the whistle on the bastards! Only after some discussion do they start to visualize the many scenarios that might accompany the choices made by the players. In a successfully guided discussion, they see that their creative and problem-solving talents are important resources and start to propose actions that minimize damage without "selling out." They start to synthesize solutions rather than judge the players. Thus, as Caroline argues, there is a strong parallel between the process of design and the process that should be used to guide one through fate's hammer-locks. The "problem" is ill-defined and resplendent with ambiguity and untruth; creativity and the wisdom to recognize what is important are critical, the iterative process of synthesis and analysis applies, and the solutions are not likely to be perfect, especially as judged by many stakeholders. One is not born comfortable with such a fuzzy and emotional process. Like design, one best learns it through a supervised opportunity to practice. This book provides such a guided opportunity.

The case studies provide very rich examples of successful and unsuccessful attempts to deal with ethical complexity. They illustrate that the right path is sometimes frightening and very rough. In the case of Roger Boisjoly and the *Challenger* disaster, he was forced to endure personal and professional persecution before being recognized as a most exemplary advocate of the "right thing" in an industry obsessed with the "right stuff." Mental experiments, classroom exercises, and personal introspection founded on Mr. Boisjoly's incredible story can be very productive. To borrow from the late Senator Everett Dirkson, an epiphany here, an epiphany there, and before long, we are talking real understanding. Interaction with Caroline has helped me understand what I think about when forced to confront ethical complexity. Thankfully, my ethics and religion are very simple and grow from the notion that we should all behave in a way that enhances the community good. I struggle with deciding the proper scope of "community." To me it includes animals, but what about plants? My most robust observation about "good" is that it is only a function of time until reelection, or it depends only on the time period over which the evaluation is performed. But given those vagaries, I find that Dr. Whitbeck has given me a nice road map for thinking about my actions. I recommend that you enjoy this book and allow it to make your brain hurt a bit to ensure that the message sticks. Many times, we can do well while doing good.

Woodie Flowers
Papalardo Professor of
Mechanical Engineering, MIT

# Preface to the First Edition

*Ethics in Engineering Practice and Research* is about professional responsibilities of engineers and applied scientists. It is about professional responsibilities: the character of problem situations in which those responsibilities must be fulfilled and the moral skills for fulfilling them. Interspersed throughout the text are open-ended scenarios that present ethically significant situations of the sort engineers and applied scientists commonly encounter. These have been set apart in centered boxes to aid the use of them in group discussion and for homework assignments. Also set apart from the text, in boxes, are fine points, which may enhance the reader's understanding but are not essential to the main argument. Most of these fine points concern philosophical issues.

## Outline and Summary

The introduction on concepts provides a clarification of many general ethical terms and provides a general framework for considering ethical questions. This framework draws on readers' prior experience of moral life and of moral reflection. Other more specialized ethical concepts are introduced as needed throughout the book.

Chapter 1 discusses what moral problems look like to a person in the situation who must respond to them. The frequent need to cope with an ambiguous situation and to formulate responses to the problem situation shows that addressing ethically significant problems is more demanding than simply evaluating the relative merits of preestablished responses. In many respects challenging ethical problems resemble challenging design problems.

Chapter 2 discusses professional responsibility and its basis and scope, and provides comparison of engineering with other professions. (Beginning with this one, the order of the chapters roughly corresponds to the sophistication of their subject matter.)

The Central Professional Responsibilities of Engineers and applied scientists, especially the responsibility for safety, is the subject of Chapter 3. Public safety, consumer safety, operator safety, occupational safety, and laboratory safety are considered.

Chapter 4 recounts the stories of two engineers who discharged their responsibility for safety in exemplary ways. Their stories are told in detail to show the development of the problem situation they faced and the appropriate responses that they made at different stages.

Chapter 5 treats workplace rights and responsibilities, focusing on engineers in corporations or governmental organizations.

Chapter 6 on the responsibility for research integrity and later chapters on research ethics carry over the discussion of complaint handling in Chapter 5 to universities dealing with charges of research misconduct.

Chapter 7 examines investigators' responsibilities for the subjects of their research experiments.

Responsibility for the environment, which is the subject of Chapter 8, is found to have a more complex basis than the responsibility for research subjects.

Chapter 9 deals with fair credit in research and scientific publication, and Chapter 10 examines credit and intellectual property issues arising in engineering practice.

The epilog presents two stories of engineers who went beyond fulfilling their professional responsibilities to incorporating their values and aspirations into their work as engineers.

## Order of Topics and Use in Courses

The interested engineer, scientist, or scholar may wish to begin by reading the entire Introduction or by simply skipping it. A detailed table of contents is provided as an aid for the general reader who wishes to read selectively, although each chapter does build on previous chapters.

If this book is to be used as a course text, the sections of the Introduction are best considered in concert with the early chapters. For example, Part 3 of the Introduction, on moral character and moral responsibility, is well considered in conjunction with the substantive discussion of professional responsibility and the engineer's responsibility for safety in Chapter 2 or 3. (A scheme for using the book in a single course is provided in the syllabus for *Real World Ethics*, one of the courses in engineering ethics available through the WWW Ethics Center for Engineering and Science (http://ethics.cwru.edu).) Cases and materials marked with "www" may be found in the WWW Ethics Center. The book does not presuppose any particular prior course of study, and its early chapters are accessible to all undergraduates.

Because the book provides a coherent guide to many topics within engineering and research ethics, it is suited to unifying the educational experience of engineering and science students who are learning engineering ethics by the "pervasive method," that is, having topics in engineering ethics and research ethics included in their science and engineering courses. Used for pervasive ethics education, Chapters 2 through 4, together with related case materials on the worldwide web (WWW) and available on videotape, are suitable for use with first- and second-year students. The remainder of the book is best used in upper-level undergraduate and graduate courses.

When the book is to be used as a primary text in a freestanding course in engineering ethics or research ethics, it should be a course for upper-level or graduate students. Students will best understand the issues if they have some experience handling complex responsibilities. Many students enter college with such experience, but not all do. Summer work experience often provides a very useful experience on which to draw in class discussion.

The only topic I regularly address in my own undergraduate course that I have omitted from the book is the topic of academic honesty. I have omitted it because of my commitment to active learning and the realization that the most effective approach to active learning about academic honesty requires linking problems and cases to specific policies and issues on one's campus. For example, where there is an honor code, it will be important to examine how that functions. If there is a student court, then it may be appropriate to spend some time on questions of procedural justice. Academic honesty is one of the first topics to cover in the pervasive method of teaching professional ethics. I find that the subject of research ethics provides a useful reprise for upper-class undergraduates and graduate students on such topics as plagiarism.

An appendix to this book discusses several major trends in philosophical ethics since 1980. To spare student readers the added expense of a larger book with an appendix that few of them would actually read, I have placed the appendix on the WWW. Engineers, scientists, philosophers, and social scientists who are interested in an explicit discussion of the philosophical position underlying this book will find it there. Here I will simply say that active learning in professional ethics should involve students in hands-on/minds-on learning. Students should learn how to reflectively consider moral problems and moral standards and examine such standards with others of diverse backgrounds. Philosophical work on topics such as trust, responsibility, and harm is useful in such reflection, but theories about how one might found ethics on reason alone are best reserved for courses in the history of philosophical ethics. [In practice, what often happens when terms such as "utilitarian" or "rights theory" are introduced in courses in professional ethics is that students get the ludicrous impression that they are expected to choose between considering consequences and rights (or duties or considerations of virtue) in making ethical assessments.] The view that the reflection that differentiates ethics from mere custom is *social* reflection, and that it is carried out with respect to particular problems and issues, rather than being the reflection of a lone scholar who considers action in the abstract, finds support in the challenges that many of the most distinguished philosophers recently have offered to the abstract and detached model of philosophical reflection. Annette Baier summarizes some of those challenges in the following terms:

> Bernard Williams and Thomas Nagel have both in their recent books[1] raised the question of what philosophical reflection [that is, detached, abstract consideration], especially that which Hume called "a distant view or reflexion" (T 538), does to what Williams calls our "confidence" in ourselves and our mores, and our personal projects and commitments. Is what Nagel calls "objective engagement" a real possibility for us, or will the attempt to be detached and reflective have the effect

[1] Williams, Bernard. 1985. *Ethics and the Limits of Philosophy*. Cambridge, MA: Harvard University Press; Nagel, Thomas. 1986. *The View from Nowhere*. New York: Oxford University Press.

of detaching us from all engagements, destroying our confidence in any project, making all our concerns seem "absurd"? Will the philosophically examined life be found to remain worth living? Williams says "the ideal of transparency and the demand that our ethical practice should be able to stand up to reflection do not demand total explicitness, or a reflection that aims to lay everything bare at once . . . I must deliberate from what I am. Truthfulness requires trust in that as well, and not the obsessional and doomed drive to eliminate it" (p. 200). Though I welcome Williams's emphasis on the importance and fragility of confidence, and his reminder of the close link between the trusty and the true, I would amend his statement to "we must deliberate from where we are"; for, as he himself emphasizes, confidence and trust are social achievements. We may be able more successfully to combine self-trust with explicitness and reflectiveness if we can abandon the "forelorn solitude" of that singular philosophical thought which turns each of us into "a strange uncouth monster" (T 264) and incorporate into our philosophical reflections on morality more of the social and motivational resources of morality itself. For our form of life to be able to "bear its own survey" (T 620), maybe both the life and the method of surveying will have to change.[2]

---

[2] Baier, Annette. 1986. "Extending the Limits of Moral Theory," *Journal of Philosophy* 77: 538–545.

# Acknowledgment

This second edition includes material supported by the National Science Foundation under Grant No. 0428597.

# Acknowledgments to the First Edition

Many people have contributed to my understanding of engineering ethics and research ethics and to the writing of this book. First are the faculty with whom I have taught, especially Stephanie Bird, Larry Bucciarelli, Peter Elias, Woodie Flowers, Nelson Kiang, Albert Meyer, Igor Paul, Steve Senturia, Tom Sheridan, Leon Trilling, and guest lecturers in several courses: Stephen Chen, Randall Davis, Stephen Fairfax, Yolanda Harris, J. J. Jackson, Vera Kistiakowsky, Freada Klein, Mark Kramer, Elizabeth Krodel, Judith Lachman, Jenny Lee McFarland, Richard Petrasso, Steve Robbins, Andrew Rowan, Mary Rowe, Susan Santos, Gerald Schneider, David G. Wilson, and the students in my courses, some of whose scenarios appear in this book and whose projects are available on the WWW.

I also thank the many people who aided the effort to teach engineering ethics at MIT and in the larger community: Mildred Dresselhaus, Hermann Haus, Jack Kerrebrock, Robert W. Mann, Paul Penfield, Sheila Widnall, and David Wormley.

Thanks to Kathryn Addelson, Pamela Banks, Stanley Hauerwas, Jennifer Marshall, David Neelon, Helen Nissenbaum, Aarne Vesilind, and David Gordon Wilson, who criticized parts of the manuscript; to Djuna Copley-Woods who drew the diagrams; and to the students in my Real World Ethics class in the spring terms of 1995, 1996, and 1997 who gave responses to it as a text.

I especially thank my wise and tender husband, David Neelon, for the sure pleasure of his company throughout this work.

# PART 1

# VALUES AND THE EVALUATION
# OF ACTS IN ENGINEERING

# Introduction to Ethical Reasoning and Engineer Ethics

**Ethics, Values, and Reason**

### Values and Engineering

What makes a good engineer and good engineering? What values underlie engineering practice today? Which of those values are specifically ethical values? What is the experience of living by those values and working in a society and in organizations that trust you to practice those values? How do these values reflect and affect the person you are and the person you become by practicing them?

This book will help you answer those questions. To answer them requires an understanding of values and value judgments in general and ethical values and ethical judgments in particular.

Societies, especially technologically developed democracies, place trust in professions and the members of professions, such as engineers (including computer professionals). In this book, we will examine what is entrusted to engineers (and computer professionals), together with the factors that created and continue to mold the expectations ingredient in that trust, and what is necessary for engineers and computer scientists to be worthy of that trust. We will consider morally significant problems that arise in engineering and computer fields, and what constitutes fulfilling the trust placed in those professionals. We will also examine the features of work environments that support the fulfillment of that trust.

> The engineering examples chosen for this book reflect *actual engineering experience* so that the discussion of engineering ethics will help introduce engineering students to the realities of the profession for which you are being educated. Therefore, they can help you understand the sort of professional life you will be entering, if you become an engineer.

Most of the readers of this book will be engineers or student engineers. The engineering examples chosen for this book reflect *actual engineering experience* so that the discussion of engineering ethics here will help introduce engineering students to the realities of the profession for which you are being educated. Therefore, they can help you understand the sort of professional life you will be entering, if you become an engineer, and help you find an environment in which you can work with integrity and in an atmosphere of mutual trust (or help you decide at an early date to seek a career elsewhere). An engineering education provides excellent intellectual preparation for many fields in addition to engineering fields, so deciding on a different career, say one in medicine, law, or business, need not mean that you should stop studying engineering, and this book will shed some light on ethical issues in other professions, especially the science-based professions.

This introduction examines basic ideas in ethics and draws illustrations from daily life, especially college life, as well as engineering practice and research. Illustrations will sometimes be drawn from other professions, too, especially medicine. Not only will some readers be studying engineering to prepare themselves for a career in the technologically sophisticated world of medicine, but medicine is a profession that engineering students and their families are likely to have experienced from the *client side*. That client experience gives you a second perspective on the importance of trustworthiness of professionals, to complement your perspective as professionals in training. In Chapter 1, we will turn attention to the specific context of engineering practice, the moral problems – by which I mean the ethically significant practical problems – that are likely to arise in that context, and the guidance that the profession offers to new entrants.

Understanding the ethical significance of problems is the first step in responding well to them, so preparing you to both recognize and understand the ethical significance of problems that commonly face engineers is one purpose of this book. Clear concepts and distinctions will aid your understanding and are necessary for the reflective examination of the ethical validity and soundness of conduct, practices, and customs. *The ability to withstand such examination is what distinguishes a **rationally based ethical conviction** from a mere opinion, an opinion that has no rational basis.* Such opinions with no rational basis may be firmly established in popular culture or a particular subculture even if they are not well supported with reasons and evidence.

> Understanding the ethical significance of problems is the first step in responding well to them, so preparing you to both recognize and understand the ethical significance of problems that commonly face engineers is one purpose of this book. Clear concepts and distinctions will aid your understanding and are necessary for the reflective examination of the ethical validity and soundness of conduct, practices, and customs. *The ability to withstand such examination is what distinguishes a **rationally based ethical conviction** from a mere opinion, an opinion that has no rational basis.*

The tendency to avoid ethical language is widespread in today's society, so that even common terms for describing ethical situations may seem unfamiliar. Although, in some circumstances, avoiding ethical language may reduce the defensiveness of those whose actions or policies are being questioned, such avoidance inhibits the understanding of ethical problems that commonly occur and obscures the ethical notions and distinctions that are marked by ethical terms. As was noted in the Foreword to Students, the precise use of concepts is essential for careful reasoning in any field from physics to ethics.

The consistent use of terms, although a separate matter from the clarification of concepts, is also important in engineering and ethics. You may notice that the government's reports on the failings that led to the 2003 explosion of the shuttle *Columbia* highlighted *miscommunication due to vague and inconsistent use of terms*. A consistent use of terms is also important in discussing ethics *so that parties will be able to recognize when they are agreeing, disagreeing, or addressing different subjects.*

The purpose of this introduction is to clarify ethical concepts and distinctions needed to understand many of the widely accepted ethical standards for the practice of engineering and to introduce a model of ethical life (one that centers on the moral evaluation of the acts that people or institutions perform). *Acts* are judged as right and wrong, morally good or bad, according to several sorts of criteria*:

1. The nature of the *acts* and/or whether they respect others' rights or fulfill one's own duties – for example, killing is wrong.
2. The specific *circumstances* surrounding a particular act – for example, Arthur's *unprovoked* assault on Burt was wrong.
3. The *motives* with which the agent committed the act – for example, Cedilla's criticism was *motivated by hostility* rather than a sincere attempt to improve performance and, therefore, wrong.

In the contemporary United States, adversarial disputes tend to dominate media treatment of ethics and values disputes. Popular culture tends to regard ethical questions and value questions more generally, as a matter of deciding on which of two opposing sides to stand on a variety of controversial questions. The goal of this book is not to argue for any particular side in these two-sided debates, but *to help you think critically about ethics and values questions and those that arise in engineering ethics in particular.*

If discussion of ethical concepts and terms is new to you, you may want to initially read only the main text in this book and skip over the "fine points" that are set off in gray in smaller type. Those "fine points" are primarily philosophical and conceptual points that are not necessary for understanding the principal issues. In the first part of this book, we will consider the moral evaluation of acts and in this introduction examine the concepts needed for that examination. In the second part of this book, we will examine aspects of moral responsibility that go beyond the ethical evaluations of acts, along with related concepts of character. You will find other specialized ethical, legal, and technical notions introduced as needed throughout the book.

*What makes a good engineer and good engineering? To which of those characteristics do you aspire and why?*

### Ethics in Popular Culture and in Reality
Is this book intended to help you choose the right side in ethical struggles?

In the contemporary United States, adversarial disputes tend to dominate media treatment of ethics and values disputes. Popular culture tends to regard ethical questions and value questions more generally, as a matter of deciding on which of two opposing sides to stand on a variety of controversial questions. The goal of this book is not to argue for any particular side in these two-sided debates, but *to help you think critically about ethics and values questions and those that arise in engineering ethics in particular.* Critical thinking skills will often reveal

---

*A **criterion** is a standard upon which judgments can be based. (*The plural is* "criteria.") Example: In addition to having driving skills, one criterion for being a *qualified driver of some specific type of automotive vehicle* is that when the driver is sitting in the driver's seat, she can operate all of the controls. If some person could not operate all of the controls when seated in the driver's seat of some specific vehicle, that person would not be a qualified driver for *that* vehicle.

a greater complexity than *either* "side" in the well-known two-sided debates considers. Truth is often complex. Your college education is meant not merely to help you get a good job, but to prepare you to think through all the problems you encounter in life.

---

### Engineers and Obsolescence

When Corey was in college there was a lot of discussion in the Big Tech student newspaper about whether the engineering students were being educated to be more than "tools." Indeed, the engineering students in Corey's living group had taken to using "wedge" (the simplest tool) as a joking insult to one another. Working hard on a problem set had come to be called "tooling." The discussion in the student paper centered around whether the engineers could think about the larger goals that they served by doing their technical work. It also discussed the growing evidence that many engineers at mid-life were finding themselves without jobs because their employers found that to keep abreast of technological advancements, the easiest course was to replace their mid-life engineers with recent graduates, much as one might replace an obsolete tool.

Corey had been too busy to take much part in the discussion but recalled it after being hired after graduation at the Major Widget Company where Corey had worked for several summers. Corey was hired to adapt a new technology to make widgets, and heard that the engineers who had been working with the technology previously used to make widgets had been let go except for the one who had gone into management. Corey was never attracted to management while in school. Indeed Corey had been among the engineering students who snickered that one majored in management if one couldn't hack an engineering major.

What can/should Corey do now?

#### Getting Started

Reflecting on what you want and expect from your career is a good idea as your career develops. As it does, you will acquire more experience but may find yourself no longer up on the latest technology. How do you expect to grow and benefit from your experience? Does further education appeal to you? Do you like managing people? Are there unusual ways to use your special engineering expertise?

---

*Many ethical problems are discussed in this book. How is this book intended to help you in thinking about them?*

### The Perspective of This Book

What do you need to understand about:

- Values and ethics,
- Ethical arguments,
- Media stories about government policies,
- Court decisions,
- The alternatives that physicians and other professionals present for you and your family members,
- The questions about your priorities that financial advisors, lawyers, and other advisors ask you,
- How to evaluate the likelihood that a course of action will actually achieve your most important goals?

## Why This Book Contains Few, If Any, Coined Terms

For the first half of the twentieth century, it was common in analytic philosophy as well as continental philosophy to coin technical and philosophical terms and to stipulate special senses of familiar words and phrases. This tendency reached an extreme in the work of the Vienna Circle and the school of thought called "logical positivism" and its successor, "logical empiricism."

Logical positivism especially looked upon natural languages as too confused to be useful for clear thinking and, at least in its early stages, often regarded sentences about ethical and other values as simply nonsense and unworthy of attention. Later adherents took the position that what appeared to be ethical statements actually expressed emotion or were recommendations. (C. L. Stevenson provided some of the most nuanced arguments for this view.[a])

Partially in reaction to this trend, a philosophical movement called "ordinary language philosophy" was born and championed by a variety of major philosophical figures from John L. Austin to Ludwig Wittgenstein in his later life. These philosophers renewed respect for and interest in the myriad functions of natural languages and the distinctions they express, although they recognized that language developed for ordinary life may occasionally need to be augmented with new terms, including ethical terms, to capture novel insights or for specialized purposes.

---

[a] See especially his 1941 essay, the "Nature of Ethical Disagreement," which may be found in his *Facts and Values: Studies in Ethical Analysis* (New Haven, CT: Yale University Press: 1963).

The definitions of ethical terms in this book follow accepted English usage closely. Sometimes, when a word has several senses, I chose one for the sake of clarity. I avoid stipulating *new* technical senses of words, however, for three reasons. First, the ethical distinctions marked in language express many important and subtle distinctions that will often remind readers of distinctions they have been using all their lives even if they have not reflected on that use before. Second, part of my purpose is to prepare readers to discuss ethical problems, concerns, and questions with others who have never read this book, a goal that would be undermined by introducing new jargon. Third, I share the philosophical view that it is pompous and unhelpful to stipulate special senses of terms except when necessary to clearly present major philosophical points.

Therefore, I do not stipulate any distinction between the terms "moral" and "ethical." The latest edition (the eleventh [2005]) of *Merriam-Webster's Eleventh Edition Collegiate Dictionary*[1] lists "ethical" as a synonym for "moral." Many different distinctions have been drawn between the terms "moral" and "ethical." For example, philosophers often reserve the term "ethics" for the *study* of morality. Others, including many engineers, take "moral" to apply to private as contrasted with professional life. To use any one of the distinctions would invite confusion with a host of others. Therefore, I use the terms interchangeably in this book. My goal is to prepare you to understand, discuss, and advance the ethics of engineering and present only as many distinctions as you will need to do that. The ethical concepts and distinctions I discuss are those that are directly applicable to ethical problems in engineering and science. They are usually concepts for which English has adequate terms. These distinctions are not precisely the same as those found in other languages, however. This book does use distinctions expressible in contemporary (American) English. To that extent this book does embody a cultural perspective, although I try to show some ways of expressing a variety of cultural and religious views on ethical matters.

---

[1] The *Third College Edition of the American Heritage Dictionary* also lists ethical as a synonym for "moral" and presumably the latest edition (fourth) does as well.

## "Ethical" and "Moral"

Beginning with H.A. Pritchard in the early 1900s, many distinguished philosophers, especially those philosophizing about moral life rather than "meta-ethics," have referred to their work as "moral philosophy."

Although I make no distinction between **"moral"** and **"ethical,"** I follow the common practice of tending to use "moral" for topics that are more concrete and "ethical" for ones that are more abstract. Thus, I usually speak of moral problems and ethical theory.

Some notions carry built-in cultural or political assumptions. "Privacy" is sometimes claimed to be a notion that is used only in relatively individualistic societies. Languages such as Japanese have no term for it. Even if only relatively individualistic societies emphasize the privacy of the individual, actions that many Americans would see as violations of individual privacy may be seen in other cultural settings as rudeness or unwarranted invasions of family or group life. Therefore, discussions of subjects such as the influence of technology on privacy may have relevance for societies that see those influences in other terms than their effects on individual privacy.

This introduction is intended to provide a vocabulary that is rich enough to express ethical problems and make ethical judgments. It is not intended to establish whether some act, motive, or character trait is ethically acceptable. I have tried to choose illustrations of ethical concepts that are relatively noncontroversial. If you disagree – for example, if you think one of my examples of a human right is not a human right at all – understand that *such questions are not supposed to be settled by my discussion.* The examples are simply intended to make the concepts easier to grasp.

> This introduction is about concepts. It is intended to provide a vocabulary that is rich enough to express ethical problems and make ethical judgments. It is not intended to establish whether some act, motive, or character trait is ethically acceptable.

The problems addressed here arise primarily in engineering as it is practiced in technologically developed democracies, especially signatories to the so-called Washington Accord. This accord or agreement specifies the education and proficiency that may be assumed of persons with degrees in engineering. Common expectations are needed for engineering in the global marketplace. Australia, Canada, Hong Kong, Ireland, New Zealand, South Africa, United Kingdom, United States, and more recently, Japan, agree on these common standards for engineering. The point, however, is to understand ethical notions, whether or not English or some other language has ready terms for them. Ethical terminology changes over time. For example, although the notion of an ethical right, especially an ethical right of an individual, arose only with the individualism that marks modern thinking in Western European cultures, the notion of moral rights of individuals, and more specifically of human rights, now finds widespread international acceptance.

The same general conditions of engineering and scientific practice hold for most technologically developed democracies. Some specific conditions of practice vary among them, however, and even vary among signatories to the Washington Accord. For example, although some states in the United States are now moving to require engineers practicing within their borders to become licensed, U.S. engineers employed in industry are currently exempt from the requirement that they be licensed. As a result, the majority of employee engineers in the

United States are not licensed. In Canada, all engineers must become licensed. There the engineering society in each province possesses the legal authority to revoke licenses. In Australia, there is at present no general requirement of licensure for engineers, although engineers must fulfill special requirements to be able to certify drawings. Australia is moving toward licensure, but because government is more centralized in Australia than the United States or Canada, licensure will be administered rather differently from either the United States or Canada, which license professional practice through the states or provinces.

> *What concepts do you need to understand questions of values and ethics, ethical arguments, media stories about government policies and court decisions, not to mention the alternatives that physicians present for you and your family members, the questions about your priorities that financial advisors, lawyers and your other advisors will ask you, or how to evaluate the likelihood that a course of action will actually achieve your most important goals?*

### One Model of Ethics

How much of engineering ethics or professional ethics can be expressed in terms of what acts are required, which are forbidden, and which are permitted?

There are a variety of models of moral life and moral learning, some more complicated than others. One of the simplest is the supposition that humanity is divided into heroes ("good guys") and villains ("bad guys"), that moral life is a struggle between them, and that the good guys always win. (The ethical and prudential task is then seen as one of being a hero rather than a villain.) Such a model is too simple to help in thinking about engineering ethics, however.

Because these opening chapters are designed to be accessible to beginning students, this first part starts with a simple model of ethics that can express some important judgments and arguments in professional ethics. This simple model focuses on the ethical evaluation of various *acts* and types of acts. The ethical code and guidelines of engineering societies are mostly written in terms of acts, and in terms of the moral rules and obligations that specify what acts are forbidden or required. (Occasionally they also express rights that specify what acts are permitted to the holder of the right. Therefore, our initial model will allow us to examine those codes and guidelines to see how they view engineering ethics.)

Moral obligations specify acts that are required (must be performed) or forbidden (must not be performed). Rights specify acts that it is permissible for the rights-holder to perform, for example, the moral right to vote or prohibitions of interference with the rights-holder in some general area of life. The rights to be free of interference imply obligations upon others to refrain from interfering. Thus the right to freely exercise religion restrains others from interfering with one's religious practices, *whatever they are*, so long as they do not violate other moral rules. (Human sacrifice might be a religiously significant act but would violate other moral rights.)

Moral rules give an alternative way of specifying the acts that are ethically required, permitted, or prohibited. For example, one such *moral rule* recognized throughout engineering is the rule against offering or accepting bribes. This rule expresses the *moral obligation* to refrain from two types of acts: offering bribes and accepting bribes. It appears in one form or another in the codes of ethics of most engineering societies.

Noticing what engineering societies choose to include in their codes of ethics and how engineering codes of ethics differ from the codes of ethics of other professions will also draw our attention to some of the special features of professional practice in engineering and the features of the practice to which some of you will devote your lives. In Part 2, we will augment this initial model of ethics that focuses on acts with attention to responsibility for future states of affairs. Other specialized ethical, legal, and technical notions needed to understand special issues will be introduced as necessary throughout the book.

*How much of what you know of engineering ethics or professional ethics can be expressed in terms of what acts are required, which are forbidden, and which are permitted?*

### Moral and Amoral Agents

You have probably heard people say to their dogs: "Bad dog." How do you interpret what they intend to say? For example, do they think that dogs (or at least *their* dogs) are moral agents and that the dogs have done something that is morally bad? If not, do you explain their behavior in some other way? Wherever you draw the line between moral agents and amoral beings, discuss your reasons for counting some beings as moral agents and others as not.

Acts, agents, and the character and motives of agents are all objects of moral evaluation. However, it makes sense to morally evaluate only agents who can act for moral reasons. Such agents are called "***moral* agents**." The statement "the storm was responsible for three deaths and heavy property damage" means that the storm caused these outcomes. Although the storm was the agent of destruction, the actions of the storm are not subject to *moral* evaluation. The storm is not guilty of murder or even manslaughter.

Moral agents are not necessarily morally good individuals. They are just those who can and should take account of ethical considerations. *Moral agents are those of whom one may sensibly say that they are moral or immoral, ethical or unethical.* A competent and reasonably mature human being is the most familiar example of a moral agent. In contrast, most nonhuman animals are generally taken to be **amoral**. Saying they are amoral is to say that they are not capable of acting for moral reasons, and, therefore, questions of morality are not appropriate in evaluating them and their acts. It does not imply that they are not entitled to ethical consideration. We will take up the question of who or what is entitled to moral consideration, the question of "moral standing," in Section 4 of this introduction.

Highly intelligent and social beings such as mammalian dolphins are sometimes argued to qualify as moral agents because of their intelligence and ability to live in a complex social system. Various religious traditions speak of beings, such as angels, whose actions are subject to moral evaluation and thus are moral agents. Examples of nonhuman moral agents are also found in fiction. Boulle's book *The Planet of the Apes* portrays apes as moral agents. Science fiction often describes nonhuman extraterrestrials as persons and moral agents. These examples show that it is not self-contradictory to think that some nonhumans could be moral agents.

> *You have probably heard people say to their dogs: "Bad dog." How do you interpret what they intend to say? For example, do they think that dogs (or at least their dogs) are moral agents and that the dogs have done something that is morally bad? If not, do you explain their behavior in some other way? Wherever you draw the line between moral agents and amoral beings, discuss your reasons for counting some beings as moral agents and others as not.*

## Section 2. Values and Value Judgments

### The Difference between Values and Preferences

In deciding to enter engineering, what value judgments did you make (or others, such as parents and guidance counselors, make for you)? Such value judgments might vary from ones about the material comforts obtainable with a good starting salary to relationships with friends and relatives who are engineers. Have those values changed as you have learned more about engineering?

What makes a good engineer and good engineering? What reasons can you give to support your value judgments about engineers and engineering?

One consideration used in determining the goodness or rightness of an act is the consequences produced by said act. Thus, the invention and dissemination of technologies that benefit humankind are often judged to have been ethically good acts. Examples of such beneficial technologies include the technologies introduced by civil engineering to provide clean water and improve sanitation. These innovations were introduced in many technologically developed countries in the late 1800s and early 1900s. They produced a greater reduction of infant mortality rates than even vaccination and other medical innovations of that period. To evaluate consequences, we will need some understanding of value judgments in general and the relationship of other types of value judgments to those that are specifically *ethical judgments*. *Ethical judgments are only one type of value judgment.* Furthermore, value judgments are only one type of judgment. Judgments are one type of statement. Sentences express statements, but also many other sorts of things (see Figure I.1).

The question of what is *good or bad, better or worse, desirable or undesirable* is a question of merit or worth. It calls for a **value judgment**. *A value judgment is any judgment that can be expressed in the form "X is good/superior/ meritorious/worthy/desirable" or "X is bad/inferior/without merit/worthless/ undesirable," at least in some respects.* The judgment that *some knife is a good knife* is a value judgment. Any judgment, including any value judgment, that is to stand up to critical evaluation must be based on relevant criteria, that is, there must be *good reasons* for making that judgment. In the case of a knife, relevant criteria would be having a sharp blade, being well balanced, and having a comfortable grip. Being bright blue would not be a relevant criterion for being a good knife *per se* even if under some *special circumstances* one might want one's knife to be bright blue. Saying that *value judgments are **objective** in the sense that*

Sentences may express any of the following:

QUESTIONS—Example: Do you know what time it is?

EXCLAMATIONS—Example: Wow!

. . .

COMMANDS—Example: Stay in line.

STATEMENTS—(Statements have truth value, unlike questions, exclamations, and so on, that is, *they can be true or false*.) Below are three of the many sorts of statements.

   I. Simple descriptions of things and situations. Example: It is twilight.

   II. Statements of preference—are **about the person** or persons whose preferences are stated, rather than **about the thing preferred**. Examples: I detest licorice. She prefers a Macintosh computer.

  III. Judgments—implicitly or explicitly use, or refer to, standards (beyond the meaning of the words in the statement of the judgments). Some but not all judgments are value judgments.

     A. Judgments without value implications. Examples: "The book is 9 inches long." "The book is of medium size." "Diamonds have become cheaper."

Fine Point:   B. Intermediate cases that judge items in relation to human purposes. Examples: "This is a medicinal plant." "That is food." "That pile is just trash."

     C. **Value** judgments—say that something is good or bad in some respect. Examples: "*Monsters Inc.* was a great family movie." "That is a poorly written article." "This song has a beautiful melody." "That would be a good car for you, because it would fit the driving you do most." ["Good" is defined by Aristotle (and many other philosophers) as *what it is rational to want*. John Dewey characterizes the good as the desir*able* as contrasted with what is merely desired.] Value judgments can be of several types depending on the type of value to which they refer.

       • Aesthetic judgments (beauty or ugliness)

       • Epistemic judgments (knowledge value)

       • Religious judgments (sacred and profane)

       • . . .

       • Prudential judgments      Aristotle, among others, does not distinguish between ethical and (long-range) prudential considerations.

       • **Ethical/moral judgments are judgments of:**

         1 People. Example: "She is a fine person."

         2 Character and character traits. Examples: "Honesty is a central virtue." "Lying shows a cowardly nature."

         3 About motives (emotions) and intentions (plans). Example: "She meant well."

         4 About acts, in which case judgments may focus on:

           a Consequences of the act or kind of act (e.g., harms, benefits, damage, improvement, costs)

           b Whether the act is of a kind that is ethically required, permissible, or prohibited

**Figure I.1**    A Typology of Value Judgments and Their Relationship to Other Judgments and Statements

*they are based on relevant reasons and evidence* does not guarantee that everyone, or even every reasonable person, will agree on a particular judgment. Disagreements are especially likely when many factors must be weighed in making an evaluation. People are unlikely to disagree for long about whether one board is longer than another, but competent engineering or software designers may disagree on the best approach to fulfilling a design assignment, even when all have made explicit the reasons for their approach. Similarly, competent physicians may disagree on the diagnosis of a particular patient, even when all have articulated the reasons for their diagnostic judgments. Competent research investigators may argue for decades about the correct interpretation of some experiment in a cutting-edge area of research.

When we consider *value* judgments, the first point to consider is the difference between being desir*able* or worthy in some respect, and simply being desired, liked, or preferred by some person or group. This distinction is crucial to our later discussion of ethical judgments and standards for engineering practice. Consider these statements:

"I like fried peppers." "John likes them, too."
"I am unalterably opposed to having cats in the neighborhood."

These are **statements of preference**. Statements of preference are not judgments about whether something is good or bad, but are expressions of someone's likes, dislikes, or habitual attitudes.

---

### Expressions and Statements

Some expressions of dislike are not statements at all. For example, "Cough syrup, yuck" is not a statement. A statement has truth-value; that is, it is true or false, perhaps to varying degrees of accuracy. However, if someone said, "Cough syrup, yuck," it would be reasonable to surmise that the speaker dislikes the taste of cough syrup.

---

Unlike a value judgment, such as *fried peppers make a good side dish*, a statement of preference, such as "I like fried peppers," is an assertion about the *speaker's likes* rather than about the characteristics of fried peppers. Statements of preference are false only if they misrepresent *the subject's* feelings, views, or attitudes. They are **subjective** in the straightforward sense that their truth-value depends only on characteristics of the *subject* whose preferences are under discussion and not on characteristics of the object that the subject does or does not prefer. If one offers the judgment that fried peppers make a good side dish, one would be expected to back it up with reasons, such as characteristics of the flavor, texture, nutritional, or other *properties of fried peppers* that make them complement other foods. One could question whether fried peppers fulfilled the *criteria* mentioned or even whether such criteria were relevant. If it were asserted, for example, that peppers make a good side dish because they are colorful, the hearer might dispute whether color is a *relevant* characteristic of side dishes.

In what follows, we will examine some controversies in science and engineering ethics. When examining an ethical controversy, it is important to identify the points of agreement and disagreement. Controversies and disagreements do not show that the judgments are subjective, in the sense of depending only on the party engaged in the controversy who holds a certain view, because disagreements are *about the topic in question*, not about the people who make the disputed judgments. As we shall see, disagreements are sometimes about some appropriate limits on action; for example, the acceptable limits on the value of gifts that one may accept from a business associate. Often controversies come down to disagreements about which of several values (or "goods") is more important or which of two evils is the lesser.

It is normal to feel repugnance at wrongdoing, but the strength of one's feelings often fails to be a reliable guide to the gravity of an offense. As people mature, they learn to distinguish between their feelings on a subject and their moral judgments. For example, someone may believe that, ethically speaking, shooting a person is much worse than shooting a dog. If that person's own beloved dog had been shot recently, that person might well experience stronger revulsion when hearing about the shooting of a dog than when hearing about the shooting of an innocent person. Like this speaker, persons may know the origins of their preferences and attitudes and may give causal explanations in terms of psychological factors that have contributed to their development. For example:

"I like fried peppers. We always served fried peppers at celebrations when I was growing up."

"I am opposed to having cats in the neighborhood. When I was a young child, my closest friend was attacked by a cat."

Alternatively, she may analyze her preferences to identify more precisely what it is she likes or dislikes:

"I am opposed to having cats in the neighborhood. I can't stand the sound of cats fighting."

Such a person *may* even give you *reasons* for thinking that what she prefers is desirable or at least desirable for him, such as:

"Cats carry disease."

"I am extremely allergic to cats."

However, the speaker *need not* give any reasons for a preference. For some matters, such as preferring one flavor of ice cream to another, people usually do *not* have reasons for their preference. When you state your preference, you are stating your attitudes or feelings, not giving a reasoned judgment. A person may have a strong preference for something while believing neither that it fulfills some criteria or standards for goodness of that kind of thing, nor that it will bring about some good. He may not even know how he came to prefer what he does.

If one claims that something is *good* or desir*able*, one makes a statement about *the thing* that is claimed to be good, rather than about *the person* who likes it. As Aristotle first observed, to say that something is **good** or **desirable** is to say that it has qualities that it is *rational* to want (in a thing of that sort). As we saw, a *good* knife is one *with the properties it is rational to want* in a knife, that is, a tool with one blade used for cutting. Such properties might include being sharp, well balanced, and having a comfortable grip. A good chair would have properties that it would be rational to want in a seat with a back for one person, such as being comfortable, sturdy, and stable. To claim that something has the qualities that it is *rational* to want in that sort of thing is to claim that there are *reasons* for wanting it.

*What makes a good engineer and good engineering? What reasons can you give to support your value judgments about engineers and engineering?*

### Opinions and Judgments

What criteria must an opinion meet to count as an expert opinion? What criteria must an opinion meet to count as a rationally based judgment?

If someone makes what looks like a value judgment, "X is good," but does not give reasons, that person likely will be met with the retort, "That's only your opinion." **Opinions** may be reasoned judgments. "Mere opinions" are judgments for which reasons are not or cannot be articulated. If someone is a recognized

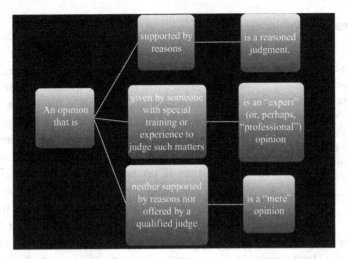

Figure I.2    Opinions and Judgments

expert in the field then that person's **expert opinion** may be accepted without the person giving reasons, but then the *criteria for regarding that person as an expert* must be satisfied. Figure I.2 summarizes these distinctions. Typically those criteria are the person's education or experience or a track record of opinions that proved accurate in the past. As an example of expert opinion: many experienced clinicians can diagnose a disease accurately without being able to say exactly what about the patient's signs or symptoms leads them to a particular diagnosis.

If a person's special ability to recognize something is a result of that person's experience (and perhaps caring and attention), rather than formal education and training, that ability is called "intuition" rather than "expert opinion." **Intuition** is the ability to immediately recognize something that is not evident to most people. (If it *were* evident to people with typical sensory faculties, it would be *perceived* rather than *intuited*.) There is nothing mysterious about intuition that is developed through experience. The ability to recognize something without being able to articulate the basis for one's recognition is familiar in everyday life. One may recognize an acquaintance at a great distance just from the person's walk, without being able to say what it is that is distinctive about that walk. Many parents can distinguish the cry of their own child from that of other children, although few are able to describe what is distinctive about the cry of *their* child.

Opinions are judgments. "Mere opinions" are judgments for which reasons are not or cannot be articulated. If someone is a recognized expert in the field then that person's expert opinion may be accepted without the person giving reasons, but then the *criteria for regarding that person as an expert* must be met.

You may be able to correctly identify an acquaintance at a distance, but unless you are able to give *reasons* for thinking that the person in the distance is who you think it is, or you are known to be specially qualified to identify this person, there is *no reason* for others to regard your *opinion* as a *judgment*, or *as accurate*, either. Being correct is not enough; to convince others to accept your view, you must either give reasons or

## Values, Preferences, and "Willingness to Pay"

Some thinkers, such as Alasdair MacIntyre, argue against the use of the term "values" because it may suggest that all values are somehow on a par. Worse yet, it may suggest that all values are reducible to monetary value or the measure of "willingness to pay," a measure that is commonly used in discussions of economic evaluation.

A person's "willingness to pay" is sometimes used as a measure of that person's degree of preference (notwithstanding people's different access to money) and, hence, is subjective. Some like MacIntyre avoid the term "values" and speak instead of "goods" or "types of flourishing" when discussing desirable outcomes. The term "values" is so widely used, however, that it would only invite confusion to try to avoid using it. If the term "value" is used carefully as a noun or an adjective and not as a verb, the distinction between values and preferences will not be blurred.

The use of the term "value" as a verb, "to value," is confusing because it is often used synonymously with either "to like" or "to assign a monetary value to." Because using "value" as a verb obscures the distinction between value judgments and mere statements of preference, I never use "value" as a verb in this book. If a verb is needed, I use "**evaluate**," meaning assess according to standards or criteria. This preserves the connection of value with reasons or standards and the distinction between value commitments and preferences.

evidence for the view, or give others grounds for thinking that you are an expert on the subject.

Given the differences between value judgments and statements of preferences, others will expect you to be able to back up your judgments in a way not demanded for your statements of preference. If you make a value judgment, others are likely to ask you for the reasons you judge it rational to want (or reject) the thing in question. If, on the other hand, you merely state your preference, you need give no further reasons for your liking or disliking. You may or may not *have* reasons underlying your preference. Value judgments on very major questions – such as "this is a good (or "the right") career/vocation for me" – are not likely to rest on *a few simply stated criteria*, as is the judgment that something is a good chair or a good knife, however. It may take much time and thought to make explicit the criteria for a major value judgment.

Early in life, people develop habits, ways of thinking and acting, that reflect the value judgments of the adults who raised them, the culture in which they were raised, and their own particular life experiences. Part of the work of adolescence is to begin the examination of those habits and see which are justified, morally or otherwise. Universities and high-tech workplaces are environments in which people typically encounter those with habits and values very different from their own. Contrast with the habits of others may stimulate examination of one's own habitual ways of thinking. It requires maturity, however, to simultaneously show tolerance for others with very different habits *and* to critically examine one's own and other's actions and values.

*What criteria must an opinion meet to count as an expert opinion? What criteria must an opinion meet to count as a rationally based judgment?*

### Types of Value and Value Judgments

What values underlie engineering practice today? Which of those values are specifically ethical values?

There are different types of value and value judgments. Both works of art and naturally occurring objects and events may be judged in terms of **aesthetic value**.

**Figure I.3** Strip Mining of Coal (Photo © Stephen Codrington, 2005)

Words like "beautiful," "harmonious," "elegant," and "engaging" are terms of aesthetic praise. Words like "ugly," "banal," "dull," and "lopsided" are terms of aesthetic scorn. Interventions involving engineering are sometimes evaluated in aesthetic terms. For example, surface or "strip" mining is blamed for defacing the natural environment. An example of strip mining is shown in Figure I.3. Are engineers who develop methods for surface mining morally blameworthy for the ugliness (distinct from any environmental damage) that results?

---

### An Assignment to Plan a Strip Mining Project

You are a mining engineer asked by your company to advise on a surface coal mining project. You are to develop a plan to

- *Confirm identification of the layers or seams of coal in the area*
- *Remove trees and vegetation and then soil and rock above the coal*
- *Drill and blast the hard strata over the coal to give access to it*
- *Remove the blasted material and clean the top of the coal layer*
- *Fragment the coal layer by drilling and blasting*
- *Remove and transport the coal*
- *Reclaim the land affected by the surface mining*

A lifelong resident of the area tells you that strip mining will destroy the beauty of the area and so you ought to make a case to your company for using methods other than strip mining, and if your company will not change its plan, refuse to take part. Evaluate (that is, identify the strengths and weaknesses of) the resident's argument that the actions she suggests are morally required of you as a professional engineer. In your answer, avoid confusing the claim that the project will *harm the environment or the beauty of the environment* with the claim that the project will *harm the residents* who enjoy the beauty of the land.

Statements along with hypotheses, research studies, theories, and designs for experiments are also judged to be good or bad in terms of what are sometimes called **knowledge values** or **epistemic values**. These include truth, informativeness, precision, accuracy, and significance. Research is judged by multiple criteria including:

- Whether the results reveal a relationship that is unlikely to have occurred by chance (is "statistically significant")
- Whether it used adequate "controls" in the study to eliminate the possibility that the observed effects were due to factors other than the one(s) under study
- The importance of the research results for the research questions under examination
- The fruitfulness of research conclusions in suggesting further lines of inquiry

Hypotheses are judged in terms of

- Their plausibility
- The scope of the phenomena they explain
- Their testability

### The Values "Internal" to Engineering

Some values are said to be "internal" to engineering, meaning that they are values necessary to the flourishing of engineering itself. These values are independent of purposes for which engineering knowledge is used. Although most values internal to engineering are epistemic values, the ethical value of honesty is also internal to engineering. Without a commitment to honesty at least about methods and results, no one would have reason to trust engineering findings and so the practice of engineering would wither away.

### Must Lies Be Statements?

Sissela Bok's definition of a lie in her famous 1978 book, *Lying,*[a] requires a lie to involve making a statement. This definition gives a criterion that distinguishes lies from *other forms of deception*, namely, that statements are required for lies. Hiding something might be a deception, but it would not be a lie by this definition.

[a] Bok, Sissela. 1978. *Lying, Moral Choice in Public and Private Life.* New York: Vintage Books.

Knowledge values of truth and accuracy often have a place in ethical codes and guidelines for engineering. These values are distinct from the ethical value of honesty. One may say something false although honestly believing it to be true. The opposite of honest behavior is deceptive behavior. Deception may take the form of saying what one believes to be false – usually with the further aim of inducing the deceived party to act differently because of the mistaken belief. If the statement someone made intending to deceive were to turn out to be *true*, the statement would still be a lie by the commonly used definition of a **lie** as a statement made with the intention to deceive, because the true statement was made with the *intent to deceive*. As an example, suppose you were looking for a person named "Chris" and asked someone if she knew whether Chris was in. If she said, "Well, I saw Chris yesterday" while knowing that Chris was in the building *today*, she would be lying even if it were *true* that she had seen Chris yesterday, because the true statement *was meant to mislead* you about Chris's whereabouts, or at least her knowledge of it.

The knowledge value of truth is distinct from ethical questions of honesty and dishonesty. They are related only in that *by* misleading people (which would be depriving them of something that has knowledge value) one may injure them, which is an ethical consideration.

One sort of deception that seems to present a paradox is self-deception. At the very least, one does not deceive oneself in the same way that one may deceive others, because one cannot both be deceived and know about the deception at the same time, although one can know about one's deception of another at the same time that *the other* is deceived. The most helpful conceptual definition for understanding self-deception as a factor in engineering ethics is that of Herbert Fingarette: **self-deception** is the failure to spell out (or make explicit), even to oneself, what one is doing, in circumstances under which it would be normal to do so.[2] This definition makes self-deception close to avoidance of the realization of some truth.

> **Self-deception** is the failure to spell out (or make explicit), even to oneself, what one is doing, in circumstances under which it would be normal to do so.

Engineering societies emphasize the importance of honesty for engineers. The American Council of Engineering Companies' Ethical Guidelines, and the ethical codes of the American Society of Civil Engineers (ASCE), the National Society of Professional Engineers (NSPE), and the American Society of Mechanical Engineers (ASME) all agree in saying that engineers should "[i]ssue public statements only in an objective and truthful manner." (This agreement in wording as well as value commitment shows the mutual influence of the codes of ethics.) The Code of Ethics and Professional Conduct of the Association for Computing Machinery (ACM) says:

> The honest computing professional will not make deliberately false or deceptive claims about a system or system design but will instead provide full disclosure of all pertinent system limitations and problems

and

> a computer professional has a duty to be honest about her own qualifications and about any circumstances that might lead to conflicts of interest.[3]

The formulation of the provision about honesty in the Code of Ethics of the American Institute of Chemical Engineers (AIChE) is very similar to that of the ASCE, NSPE, and ASME, namely: "Issue statements or present information only in an objective and truthful manner."

These provisions point to two values: the ethical value of being truthful or honest in one's communication to others, and the value of being "objective." Being **objective** or **impartial** is the opposite of being biased and sometimes functions as an ethical value (similar to honesty), and sometimes as a knowledge value (similar to competence).

The discussion above brings out the fact that one is not always ethically blameworthy for being less than fully objective. **Disciplinary bias**, the influence of one's disciplinary training on the concepts one uses, the way in which one sets up problems and tries to solve them, is a well-known phenomenon. It is, however, a

---

[2] Fingarette, Herbert. 1969. *Self-Deception* (original hardcover edition). London: Routledge & Kegan Paul. (In 2000, a paperback edition was issued by the University of California Press.)

[3] http://www.acm.org/about/code-of-ethics. Anderson, R. E., Johnson, D. G., Gotterbam, D., and Perrolle, J. 1993. "Using the New ACM Code of Ethics in Decision Making," *Communications of the ACM* 36(2): 98–107.

form of bias that one cannot remove. Furthermore, one routinely informs others of that potential bias in stating one's degrees and discipline, so deception about one's disciplinary perspective is not an issue. Such bias *may distort* findings, but that bias can be eliminated and does *not* represent a moral vice in a researcher who has the bias. In contrast, a deceptive skewing of results to further one's own financial interest or those of one's personal or business associates would be a bias that can be eliminated. One would be biased in a hidden and ethically blameworthy way if, for example, one were to say that some chemical spill would not affect a certain area in order to allow one's company or one's friend to sell property in that area before the spill reached it.

### Is Complete Objectivity Attainable?

Many scholars in epistemology, philosophy of science, and science and technology studies have questioned whether *complete* objectivity is an illusory, unattainable, and misguided ideal. Perhaps our objectivity is only a matter of degree.[a]

In past ages, many smart people, including Aristotle and Descartes – to name two figures who contributed greatly to mathematics and science as well as philosophy – did not have the concept of mass. For people in industrially developed societies today, the concept of mass is part of common sense. It is difficult for such people to think about the world without using the notion. These concepts are specific to our cultures, however.

---

[a]The most convincing of these arguments is put forward in *Science as Social Knowledge, Values and Objectivity in Scientific Inquiry* by Helen E. Longino (Princeton, NJ: Princeton University Press: 1990), especially chapters 1–4 and 10.

Plans and strategies are common objects of **prudential judgment**. When people speak of a good (prudent or effective) strategy or a bad (foolish, short-sighted) plan, they are making a prudential judgment about the efficacy of the plan or strategy in question, that is, whether it will achieve certain ends or goals. Behind most prudential judgments are value judgments of other sorts that certain ends are *worth* achieving.

Prudential value is somewhat different from the other types of value that we have been discussing in that whatever has prudential value is *instrumental* in achieving the flourishing or well-being of the agent or furthering of the agent's interests. Something has *only* **instrumental value** when its value is *entirely* due to the value of what it brings about, or may be used to bring about. In contrast that which is valuable in itself (rather than merely as a means) is said to have **intrinsic value**. The same thing can be valuable as both a means and an end, of course; a tool may be beautiful (have intrinsic aesthetic value) as well as useful for making other things (have instrumental value). When people speak of aesthetic value, ethical value, epistemic value, or religious value, they are speaking of intrinsic value. Because prudential value is a matter of what the prudentially valuable thing (action, plan, tool, or strategy) can bring about, that which has only prudential value has only instrumental value and not intrinsic value. The practical application of this point is that *one must always consider what the prudentially valuable thing is intended to accomplish to assess the derived value of what is prudentially valuable.*

### Is Disciplinary Bias Blameworthy?

Suppose you are a chemical engineer and given the job of predicting where a certain chemical contaminant will wind up, if it is spilled in a certain area. When you complete and submit your report, the report is criticized as showing the disciplinary bias of a chemical engineer and the critics want a study done by an environmental engineer with a background in civil engineering.

Is this criticism appropriate? Why or why not? If it is, would you have been failing to live up to your professional ethical obligation to give unbiased advice?

Survival, either biological survival or continuance as a member of some group, is generally assumed to be valuable. A plan or idea is generally judged imprudent or stupid (lacking in prudential value) not only if it is ineffective but also to the extent that it neglects the survival, well-being, or interests of the *agent*(s) who perform the act. People do speak colloquially of "survival value." When two people disagree in their prudential judgments, they may be disagreeing about the dangers in some course of action or the importance of that danger. For example, consider the warning, "If you want to survive in this organization, you will not report corruption." If one believed the speaker, one might decide to leave such a corrupt organization, rather than, as the speaker recommends, stay in the organization and keep quiet about wrongdoing. Choosing to leave suggests that one disagrees with the speaker about the value of staying in the organization.

*Each January, the* IEEE Spectrum *chooses the best and the worst technology projects. In 2008, winners were a geophysics project to develop advanced seismic image codes based on the two-way wave equation (thus exploiting the power of supercomputers), and a semiconductor project using air gap technology to insulate microchip wiring. Among the losers were a climate engineering project to fertilize the ocean with iron to stimulate phytoplankton growth, thereby sequestering carbon in the deep ocean, and a project to supply broadcast TV to children riding in the back seat of automobiles. What types of value judgments seem to you to underlie the IEEE's evaluation of these technologies? Give reasons for your answers.*

### Religious Value in Relation to Ethical Value
How, if at all, do you understand religious and ethical values to be related?

The last major type of value to consider before turning to ethical value is **religious value**. The terms of evaluation include "sacred" and "holy" as contrasted with "profane" and "mundane." Purely religious standards are often applied to people, writings, objects, times, places, liturgies, rituals, stories, doctrines, and practices. Religions that emphasize the importance of doctrine (a body of teachings about the divine or the human relationship to it) are called "doctrinal." For some, liturgy (the order of worship) is central. These are called "liturgical." Governance, including the roles of elders, priests, imams, rabbis, or bishops, is a defining factor in some religions. Indeed, the names of some Protestant denominations reflect their form of governance; for example, "Episcopalians," "Presbyterians," and "Congregationalists." Some religions understand life in terms of sacred stories or sacred times and places. Other religions emphasize nonliturgical practices, such as forms of yoga or meditation. Some emphasize care of less fortunate people, or compassion toward all sentient beings (beings that can feel pain). One emphasis may coexist with others. (I note these differences because a surprising number of philosophers write as though doctrine were always the central concern in religion.) Some emphases change over time. For example, in Judaism before the Babylonian exile, a place – the Temple at Jerusalem – had central importance. After the destruction of the Temple, scripture – the Torah – became central.

Most existing religions, and all major world religions, uphold ethical as well as religious standards. These ethical standards apply to moral agents – to their character traits, motives, or actions. Religions vary somewhat in their relative emphasis on such matters as spiritual and moral virtues of individuals, a particular kind of family structure, and the faith or practice of a nation, religion, or congregation as a whole. Confucianism puts great emphasis on the family, for example, and largely defines the virtues of individuals in terms of their place in the family. Family and caste identity have had a defining role within Hinduism. Buddhism, in contrast, emphasizes enlightenment of the *individual*. Judaism emphasizes the relation of the *whole people of Israel* to God, so that praiseworthy individuals are those who strengthen the relation between God and people of Israel. Because Christianity emphasizes individual salvation, it is generally regarded as more individualistic than Judaism, notwithstanding a continuing emphasis on the community of the faithful or "the Church." Islam emphasizes the duty to form an equitable society where the poor and vulnerable are treated decently.[4]

Notwithstanding differences concerning the primary social and spiritual unit, many religions share ethical norms and even some underlying convictions that support the practice of those norms. For example, Hinduism and Buddhism hold that karma, the total effect of a person's actions in successive phases of existence, determines that person's destiny. All major world religions have some version of the "Golden Rule," the admonition to treat others as you would want to be treated.

In addition to generally applicable ethical norms, religions often offer guidance to their members about what they as individuals are particularly called to do. The Native American practice of embarking on a "vision quest" to discern one's life path is an example of a means of seeking spiritual guidance. The Middle English word from which the term "vocation" is derived means a divine calling. Exclusively secular means, such as aptitude tests, also address questions of vocation.

*How, if at all, do you understand religious and ethical values to be related?*

### *Relations among Types of Value*
How, if at all, do you understand truth to be related to other values?

One type of value may be relevant to another. For example, aesthetic criteria, such as beauty and symmetry, are commonly held to enter the assessment of scientific theories. Conversely, many argue that great art gives a profound insight into reality, which brings aesthetic value close to religious, or scientific/knowledge, value. Therefore, although I have distinguished various types of value here, *it is an open question whether there are fundamental connections among them.*

The term "value" has another use with which readers of this book will undoubtedly be familiar. In sentences such as "Solve for the value of x," the word "value"

---

[4]These rough generalizations do not take into account differences among branches of these religions.

## Where Do "Economic Values," and So On, Fit In?

The term "economic value" is sometimes used as a synonym for "market value," but not always. "Economic value" often means the usefulness of the object in question for creating prosperity, and thus is a type of instrumental value. The economic value to a country of having a system of transportation and sanitation is not the price of these systems if sold, but rather the prosperity that the systems help create.

Notions like "nutritional value," "sanitary implications," "security implications," or even "entertainment value" are also types of instrumental value. Nutrition, entertainment, protection of health, and security are the sorts of things that humans have an interest in being able to obtain or retain.

Terms such as "democratic values" or "family values" apply to groups of norms such as those that form part of political ideals or a certain ideal of family life. Those ideals are themselves often taken to have ethical, prudential, or religious value. These terms do not name another kind of value on a par with ethical, prudential, religious, knowledge, or aesthetic value, however.

just means a numerical quantity and has nothing to do with values that we have been considering.

Notice that all of the types of value we have been considering differ from **market value**, which is synonymous with market price. When one assesses market value, one is not making a value judgment of what is good or bad in some respect. Rather, one is simply referring to the price at which the supply of an item equals the demand. The price need not reflect the value of an item. For example, we all need breathable air for health and survival – which are fundamental goods. Because there is no scarcity of air of breathable quality in most areas, no one needs to buy it. Therefore, breathable air has no market value.

Just as air has no market value, high market value may attach to items that are not good by any reasonable standards. Market value depends on the relation of supply to demand. Thus, it depends on the strength of preference of those who have the means to pay for an item and the willingness of those who have it or can make it to sell it. An addictive and physiologically destructive drug with analgesic or euphoric properties might have high market value. Such a drug would not be "good," however, even in the sense of having the properties it would be rational to want in a drug with analgesic or euphoric properties.

*How, if at all, do you understand truth to be related to other values?*

## Section 3. Ethics and Ethical Justification

In deciding to study engineering, what value judgments did you make (or others, such as parents and guidance counselors make for you)? (Such value judgments might vary from ones about the relative importance of the material comforts obtainable with a good starting salary, or of joining a profession that is highly esteemed, or of having relationships with friends and relatives who are engineers already.) Have those value judgments changed as you have learned more about engineering? Have the judgments that engineering actually has these advantages changed?

Having briefly examined other types of value and value judgments, let us consider values that are specifically *ethical values*, and with them the concepts of ethical evaluation and ethical justification. As philosopher Amelie Rorty observes, it is not always a simple matter to classify a value judgment as being aesthetic, moral, prudential, and so on, and not all major philosophers have distinguished among all of these types of value. Using the distinctions among the basic types of

## The Relations among Different Types of Value

Plato argued that the Good, the True, and the Beautiful are ultimately one, thus claiming the ultimate identity of ethical, knowledge, and aesthetic value. Aristotle, in taking the question of what is the good for "man" as the core question for ethics, was interested in clarifying which character traits are the virtues that simultaneously make a good life. He understood the good life as one that achieved εὐδαιμονία, which is variously translated as "happiness," "human flourishing," or "realization of what it is to be human." Thus, Aristotle considered the good life as one that later philosophers would describe as uniting ethical and prudential values.[a]

---

[a] Oksenberg Rorty, Amelie. 1995. "The Many Faces of Morality," *Midwest Studies in Philosophy*, XX: 67–82.

value, we find that considerations of one type often have implications for others. In particular, *prudence can support ethical values where it does not compete with them*, because prudence in fulfilling one set of moral obligations improves the chances that one will be able to fulfill other moral obligations later. Self-sacrifice may be noble if no alternatives are available and the value or cause is worthy, but *unnecessary* self-sacrifice is only foolish. Therefore, we will note prudential considerations as well as ethical considerations in seeking solutions to ethically relevant problems in engineering. For that we will first need a deeper understanding of value judgments that are identifiably *ethical* judgments.

In this book, I follow the common practice of calling a code of behavior an "ethical code" only if its claims, judgments, or rules are *supported with ethical justification* (i.e., only if ethically significant reasons or evidence is given to bolster those claims, judgments, or rules). We will examine the concept of justification later in this section.

> *In deciding to study engineering, what value judgments did you make (or others, such as parents and guidance counselors, make for you)? (Such value judgments might vary from ones about the relative importance of the material comforts obtainable with a good starting salary, or of joining a profession that is highly esteemed, or of having relationships with friends and relatives who are engineers already.) Have those value judgments changed as you have learned more about engineering? Have the judgments that engineering actually has these advantages changed?*

### Ethical Conventionalism and Ethical Relativism(s)
What do you understand by the term "relativism"?

One of the first questions that many people raise about ethical values (and value judgments more generally) is whether those judgments are "relative" so that they are applicable only in certain situations or to certain individuals.

"Ethical relativism" names several quite different views. First, it is used for a view that is better described as "ethical subjectivism." **Ethical subjectivism** holds that whether a certain act is right or wrong in a given situation is determined by whether *the agent* performing that act *believes* the act is right or wrong. This view represents ethics as lacking in objective standards, because all that matters is what the agent believes, without consideration of whether those beliefs are well-founded. When the view is applied to our own value judgments, it seems to have the odd implication that we do not have any ethical beliefs. The reasoning is this: Suppose we hold the view that what makes our acts right (or wrong) is whether we believe them right (or wrong). Then rather than believing that some act A is right

## Loose Uses of the Term "Ethics"

A few authors use the term "ethics" or "morality" for any code of behavior, even one that does not claim to have moral justification. For example, Robert Jackall in *Moral Mazes* describes what he calls a corporation's "ethics" or "morality" and takes it to include such judgments as, "What is right is what the guy above you wants from you."[a] What is "right" then just means what is the most effective way for an individual to survive in such an organization, regardless of what is morally/ethically justified.

---

[a] Jackall, Robert. 1988. *Moral Mazes*. New York: Oxford University Press, 6.

(or wrong), what we believe is *if we believed A to be right (or wrong) it would be right (or wrong) for us to do A*, which is not the same as believing A either right or wrong. So we have no ethical views. In addition to this paradoxical result, this view also undermines any discussion about the *reasons* for thinking some action is right or wrong, because on this view *all* that determines the rightness or wrongness of someone's act are that person's moral beliefs, *regardless of whether those beliefs are supported by reasons*. Therefore, ethical subjectivism would make reasoned discussion of ethical views impossible. The grain of ethical insight in ethical subjectivism is that although believing some act is right does not make it right, believing some act is wrong *is* one good ethical reason *not* to do it, because doing what one believes to be wrong undermines one's moral integrity.

*Consider two different ethical views on some matter familiar in your experience, say about what would be a fair distribution of something. For example, two roommates might disagree about how to divide the space in their shared dorm room. One might argue that the floor space should be divided numerically in half, for example, while the other might argue that either could furnish as much of the space as she liked (consistent with being able to move around easily) so long as the items were available for both to use. Either for this situation or for an example of your own, consider what if any reasons could be given to support each of the two views of what is a fair distribution. Do both positions fulfill the requirement that there are good reasons for the criteria they offer for fairness and thus count as ethical judgments?*

## Is Man the Measure of All Things?

In Plato's *Theatetus*, the figure of Protagorus presents the thesis that "Man is the measure of all things." Many philosophers[a] take Protagorus to give a paradigmatic statement of a relativistic thesis with respect to truth and, derivatively, with respect to ethics.

---

[a] Blackburn, Simon. 2005. *Truth, a Guide*. New York: Oxford University Press.

Some people are attracted to calling ethics "subjective" because they fear that the alternative will be intolerance for others' views. Real respect for others' views involves understanding those views and the criteria that others use in making judgments, not simply ignoring differences between their views and one's own. Understanding the reasons and criteria that underlie others' judgments is to understand what makes them *ethical* judgments, whether or not one agrees with those judgments.

As the philosopher Joseph Raz has pointed out, even if someone wished to argue that some values and value judgments are subjective, the person would have to grant that *instrumental value* at least is objective. Whether some state or action actually works does depend on the way the world is. Thus, although under most conditions healthy people would live longer (so that health would be instrumental in obtaining a long life), under conditions in which the healthiest members of a group were taken as slaves and worked to death, health would not be instrumental to obtaining a long life. Because the way the world is (including social facts as well as distribution of wealth and material resources) varies from one social group to the next, one might say that for that reason instrumental value is "objectively relative" in that it would be relative to one's circumstances *and* objectively testable. The substantive question is whether *intrinsic value*, and not only *instrumental value*, is in some sense *relative*.

Various forms of cultural or social relativism find more advocates than ethical subjectivism. In a pluralistic culture (i.e., a culture in which there are many different subcultures with differing values) such as the United States, various subcultures disagree about what is "a" or "the" good life, that is, what way it is rational to want to live. Subcultures in the United States hold widely differing views on the importance of the arts, of scientific knowledge, of education, of participation in a religious community, of material comforts, of close friends, and living close to nature. Even those who agree on the importance of some of these, say family life and religious participation, may disagree about the ideal form of the family or which religious tradition one should follow.

Disagreements between subcultures are most often about the relative importance of various goods (including benefits and virtues), rather than disagreements about whether something is desirable at all. Pride is one of the relatively few examples of something (in this case a character trait) that some groups view as positive and others view as negative. Roman Catholic ethics lists pride as one of the seven deadly sins, but the U.S. Marines take pride as a defining characteristic of their branch of the armed services. However, it might be argued that the two groups mean something different by the term "pride," and that arrogance is the vice formerly termed "pride," whereas what the Marines seek to uphold when they speak of "pride" are self-esteem and group-esteem.

### The Notion of Justice in John Rawls' Work

The twentieth century philosopher John Rawls in his influential work on social justice sought to define the notion in a way that is independent of notions of the disparate value judgments of what constitutes "the good life." His work, which drew from many previous philosophical contributions to ethics, and is much admired, nonetheless is agreed even by his followers to have failed in this respect. At present philosophy has no concept of justice that is independent of other value judgments. Had Rawls succeeded, there would have been a notion of just procedure that disparate cultural groups might have agreed upon even while disagreeing about what makes a good life. His project is widely seen as (a particularly sophisticated) twentieth century example of Liberal thought.

The contrasting views that some subcultures hold about wealth may be a better example of something that some groups view as bad while others see it as good. Some religious traditions and subcultures regard possessions as a potential distraction from spiritual growth and advise their adherents to at least live simply and perhaps embrace poverty. Other subcultures view wealth as either valuable in itself or a sign of divine favor. These differing views about the good life held by subcultures within democracies frequently coexist with a common agreement on the democratic value of liberty. In that case, people who hold radically different values may agree that (within certain limits) people should be free to do what they choose, even if they make bad choices, as long as they do not harm others.

Let us now consider social or cultural ethical relativism. These views share the belief that ethical evaluations and judgments are relative to one's culture or social group (that is, relative to the social group in which they take place or in which they are judged). (Let us set aside for the moment the real difficulty that many people today are members of more than one distinct culture, subculture, or social group.)

## Liberal Thought

Liberal thought should not be identified with political liberalism. Some forms of Liberal thought, such as Libertarianism, which views maximal freedom for the individual as the highest good, are politically conservative views. "Liberal thought" is the name given to an intellectual movement that began in the seventeenth and eighteenth century Enlightenment and is marked by great faith in human reason and individual freedom and the belief that reason by itself can provide a basis for ethics generally or, as in Rawls' specific project, for social ethics.

On a smaller scale, there exist local standards that apply only to one's local community, but which are based on more general universal principles. For example, a community living where water was scarce might have standards for the fair sharing of water, while another living in an area of plentiful water would not. More tellingly two societies living with a scarcity of water might have different rules for fair sharing of those scarce resources, with each allocating extra water to, say, anyone who used it for designated socially useful purposes. In both instances, both communities would be operating under the more general ethical principle: Be fair in the distribution of scarce resources. They might have different needs, however, and want to encourage different socially useful endeavors and so distribute the water differently.

The view that social groups or cultures may differ in their specific ethical requirements but justify these by appeal *to the same ethical principles* is sometimes called "local relativism." However, because it finds an invariant standard in the shared ethical principles, it is not a relativistic view of *ethical value*. When we examine the difference between the ethical codes of different professions and trace these different ethical standards of different professions to the nature of what it is they must do to be worthy of the trust placed in them and to differences in the problem situations and temptations they commonly encounter, we are committed to something like "local relativism."

That agreement about underlying ethical principles may exist even in the presence of contrary or contradictory specific moral rules or behavioral norms should warn us that the existence of contrary or contradictory specific moral rules might not be evidence in favor of a strong or radical relativism. Lack of options may gravely affect such rules and norms. A social group might seem to condone some practice – for example, suicide by those who are too old and infirm to work – when there are no alternatives that would not put the survival of their society in jeopardy. To see if members of that first society genuinely disagree with members of other groups who abhor such suicide one would want to know whether if that first society had greater resources and possibilities for caring for its elderly members, it would still condone suicide by the elderly.

A more thorough-going relativism is the view that there are *no* shared underlying principles. This view, sometimes called "radical relativism," is problematic for ethical discussion, because, if true, it would mean that ethical discussion or deliberation between cultures would be an impossibility. Perhaps it would be impossible for a culture to criticize even the standards it had held at a previous time.

The view that ethics *does* depend solely on the arbitrary decisions or preferences of a social group is the most extreme of "cultural relativism," sometimes called "naïve cultural relativism." **Naïve cultural relativism** holds that the moral beliefs of one's society by themselves determine what is morally acceptable behavior for a member of that group. This view is similar to ethical subjectivism except that the beliefs that supposedly determine the ethical acceptability of an agent's actions are those of the agent's culture rather than the agent alone. Setting aside the fact that many people are members of several subcultures, each with somewhat different standards of acceptable behavior, naïve cultural relativism, like ethical subjectivism, is incompatible with ethical argumentation. Naïve cultural relativism would make impossible any substantive discussion of ethics, because only societal beliefs, not ethical criteria, would be relevant to deciding whether some action or state of affairs was ethically acceptable.

## Does Dependence of Values on Social Practice Imply Relativism?

A growing number of philosophers who do not embrace relativism, including Annette Baier, have each emphasized that ethics is a cultural product constructed by people in particular historical contexts and can be fully understood only in relation to those contexts. Furthermore, having a value is not primarily a matter of intellectual assent; rather, values are inextricably bound up with social practices.

Their views raise the question of whether a thesis of the dependence of value commitments on social practices must commit one to radical relativism with respect to values and ethical values in particular. Since the first edition of this book, I have read Joseph Raz's argument for the social dependence of value without relativism in his essay "The Practice of Value."[a] Raz's argument might satisfy a philosopher like Baier. It may not satisfy MacIntyre, however. Raz is arguing for value pluralism, and MacIntyre, although seeming to give many examples of value pluralism in recent works, argued vociferously in *After Virtue* that modern pluralism had decimated ethics.

[a] Raz, Joseph. 2001. "The Practice of Value." http://users.ox.ac.uk/%7eraz/Web_publishing/Tanners/Doc7.htm (last modified on December 27).

A second and more plausible view, which can also be described as "cultural relativism," is the view that ethical beliefs, rules, and norms accepted within a culture form part of the societal context in which actions take place. Hence, those beliefs, rules, and norms affect the *actual* options that are available in the cultures from which they are derived. For example, some act, such as showing the soles of one's shoes to another person, might be seen as highly disrespectful in one society but not in another, and thus the ethical significance of showing the soles of one's shoes would be different in the two societies. Those who are cultural relativists in this second sense do not find it impossible that there be ethical standards that would apply universally but *they put the burden of proof* on those who seek to generalize from one social context to another to show that such generalizations do not ignore relevant social and material differences. This form of cultural relativism is compatible with ethical argumentation and ethical reasoning.

Is ethics "conventional" as is often claimed? If conventions chosen by the individual were held to determine what is ethical, that view would be ethical subjectivism. Let us consider group conventions, therefore. "Group conventions" refer to the general agreements or customs recognized by a group and to which there are conceivable alternatives. **Ethical conventionalism** may then be defined as the view that ethical norms are the conventions that groups develop for acceptable behavior, and to which alternatives exist. Unlike ethical subjectivism, ethical conventionalism need not undercut the reasoned discussion of ethics. To know whether some form of ethical conventionalism discounts the role of reason in ethics, we need to understand what is involved in the choice of a convention in that conventionalist view.

Some conventions are *objectively* more convenient or practical than others. The French mathematician and philosopher Henri Poincaré, who held that convention plays a major role in scientific knowledge, argued that the wrong choice of conventions could threaten survival. (He saw the role of convention in science as a reflection of human creativity, not arbitrariness.) The point is nicely illustrated by considering the choice of coordinate systems. On the one hand, polar coordinates can express all of the same locations as do Cartesian coordinates, and vice versa, so the choice is a matter of convention. It is objectively easier to represent certain physical laws in terms of one system of coordinates than another, however, and using the "wrong" system of coordinates might prevent the discovery of those laws.

Similarly, a society that did not have some ethical norms, such as for the nurture of children, would not flourish (although to say that is to give pragmatic basis for its ethically significant conventions). Even conventions that might at one time have been chosen in any of several equally convenient ways – such as whether to drive on the right or the left side of the road – soon become entrenched so it is then difficult to change them. For example, the U.S. Virgin Islands as well as the British Virgin Islands adopted the British convention of driving on the left side of the road. They have retained it, presumably because driving habits are deeply entrenched, even though, because of the proximity of the islands to the United States, virtually all the automobiles in the Virgin Islands are prepared for the U.S. market and, therefore, have the driver's controls on the left, which is suited to driving on the right rather than the left, as Virgin Islanders do. (The advice to drivers in the Virgin Islands is "Keep your shoulder to the shoulder.")

## Multiple Systems of Measurement

It would be simpler and more convenient if the whole world used the same measurement system at least for commonly made measurements such as length, weight, and temperature.

How do you explain the evident fact that there is more than one such system currently in use?

### Getting Started
We have been discussing conventions, convenient and cumbersome. Are multiple systems of measurement explainable by the same factors that explain the persistence of cumbersome conventions?

If someone regarded the fundamental ethical principles as merely *arbitrary* choices, then the form of conventionalism they espouse *would* be close to ethical subjectivism. Plausible alternatives that really might function as ethical principles are not easy or obvious, however.

In contrast to both objectivist conventionalism and subjectivist conventionalism is the view that reason, God, or nature dictate certain basic principles for ethics. The name of that view is "**ethical absolutism**." Ethical absolutism might lead to intolerance, unless one of its basic principles were a principle of tolerance.

Either the objectivist forms of ethical conventionalism or ethical absolutism are compatible with the ethical investigation carried out in this book (and with

## Distinguishing Ethical Norms from Other Types

The philosopher Kurt Baier, in his famous 1954 essay, "The Point of View of Morality," proposed criteria for distinguishing between moral rules and other behavioral norms, such as taboos or rules of etiquette, or fashion, that might be treated the same way as genuine moral rules within a culture. The tests he gives include:

- Is it universally teachable and hence "universalizable"? (In contrast, "Lie whenever it is expedient" is a self-defeating rule, because believing it will result in a general loss of trust, trust which is necessary even for lying to work.)
- Is someone who breaks it considered "bad," "evil," or "irresponsible" as contrasted with, say, rude, corny, weird, uncool, stupid, foolhardy, unthinking, false, trivial, or profane and mundane?
- Is it applied in accordance with principles of exception and modification *so there are established ways of deciding what to do if the rule conflicts with other moral rules?*

Some have argued that the third requirement goes too far, and might suggest that there are no moral rules, because cultures often find that they need to establish such principles of exception and modification, when faced with new circumstances. New medical technologies have presented people with new options, and there is much debate about whether or under what circumstances those options ethically ought to be exercised.

the "local relativism" implicit in comparing the ethics of different professions). We need not choose between them for the purposes of investigating engineering ethics.

In summary, if ethical conventions are supposed to be entirely arbitrary, then ethical conventionalism might seem to be similar to a view that ethics depends solely on the subjective preferences of the members or at least leaders of the social group in question. Societies must have some established expectations about what their members may and may not do to maintain the trust necessary for a stable society. Thus, there are at least some objective constraints on ethical conventions.[5]

Cultural relativism applied to a single culture over time yields the conclusion that it is simplistic to judge an action in some other period *solely* by today's standards without taking account of the differences in conditions. This does not mean that an action can be criticized *only* by the criteria used in the period in which the action occurred, however. For example, informed consent for medical experiments is a standard that has developed in industrialized democracies only since World War II. The implicit prior standard was, "First do it [the experiment] on yourself," a standard that considered the welfare of subjects but not their **right of self-determination** (i.e., their right to decide the practices in which they will participate). Someone who in 1940 used the "first do it on yourself" standard conscientiously rather than the informed consent standard is not subject to the same moral criticism as would someone today who knows or should know about the informed consent standard. Nonetheless, the informed consent standard is arguably a *superior* standard; we would think highly of someone who had sought informed consent from her human subjects for experiments in 1940.

*Identify a standard of responsible behavior for some profession that now exists but that did not exist in a prior age, or one that could not have existed in a prior age because of a lack of knowledge about particular harms or dangers. Alternatively, identify a standard for student behavior that has changed over the last century.*

---

[5] Oksenberg Rorty, Amelie. 1995. "The Many Faces of Morality," *Midwest Studies in Philosophy* XX: 67–82.

### *Ethical Evaluation, Justification, and Excuses for Actions*
What kinds of considerations are relevant to judging an act or course of action *morally justified or unjustified?*

Ethical judgments, judgments about what is right or wrong, ethically good or bad, or what one ought or ought not do, need the support of justifying reasons. *Any* judgment, even a judgment about how fast something is moving, needs the support of reasons/evidence. Availability of explicit reasons or identifiable evidence is what distinguishes judgments (ethical or other) from the operation of intuition. **Intuition** is the ability to immediately recognize what is going on in a situation. There need not be anything mysterious about intuition; it may result from training or experience. The ability to recognize something without being able to articulate the basis for one's recognition is familiar in everyday life. One may recognize an acquaintance at a great distance just from the person's walk, without being able to say what it is that is distinctive about that walk. Many parents can distinguish the cry of their own child from that of other children, although few are able to describe what is distinctive about the cry of *their* child. In contrast to the exercise of intuition, the ability to infer what is going on from other independently identified evidence or premises is called "*reasoning.*"

**Ethical justification**, that is, reasons/evidence or argument to demonstrate that something is ethically acceptable or desirable, is necessary to support any ethical value judgment. As we saw earlier, the presence of justifying reasons is a major difference between a value judgment (a judgment about what is good or bad in some respect) and a statement of one's preference about something. Ethical justification also distinguishes an ethical code from just any set of rules for behavior, such as the rules of a game or rules of etiquette. For that reason ethical justification is a central topic for ethics.

**Ethical evaluation** is a judgment about the extent to which the object of the evaluation is good or bad, ethically speaking. A variety of criteria are relevant to the ethical evaluation of an act or course of action. A reasoned judgment about whether (or the extent to which) some act (or course of action) is morally justified will mention some or all of the following:

- The act produces good or bad consequences
- It respects or violates rights
- It fulfills or shirks obligations
- It honors or ignores agreements and promises
- The act displays or fosters the development of positive (ethical or other) character traits (virtues) or negative ones (vices). (The consequences upon people's character are generally considered separately from consideration of other sorts of consequences.)

Justifications are offered directly for acts or policies, however, and not only for judgments. To **ethically justify** some act or policy is to show that, ethically speaking, it was a good, or at least an *acceptable*, thing to do. Usually we do not bother to offer an ethical justification of an act or policy, unless there is at least *some* reason to think that it might be *bad or wrong*. If an engineer says that she has reviewed some plans that were given to her to review, it would be odd

A variety of criteria are relevant to the ethical evaluation of an act or course of action. A judgment about whether (or the extent to which) some act or course of action is morally justified will mention some or all of the following:

- It produces good or bad consequences
- It respects or violates rights
- It fulfills or shirks obligations
- It honors or ignores agreements and promises
- It displays or fosters the development of virtues or vices

for someone to ask for a (ethical or other) justification for reviewing the plans. However, if she said that she had not reviewed and would not review the plans, then, because by assumption that is part of her job, someone might ask for her justification for not reviewing them. Perhaps she is quitting her job. Perhaps she has reason to believe the plans have been falsified and her refusal is meant to safeguard the public, a reason that would count toward moral justification, too. As this example illustrates, *many ethical and factual assumptions underlie the requesting, giving, and accepting of ethical justifications.*

*What kinds of considerations are relevant to judging an act or course of action morally justified or unjustified?*

### Examples of Justifications and Excuses for Lying

Suppose you are helping to install some equipment. You are to install one component by yourself. Afterward you are criticized for the way you installed it. Which of the following responses are excuses, which are justifications, and which are something else? Give reasons for your answers.

- I was given the assignment late in the day and told to finish by 5. There wasn't time to do it any other way.
- That is what the building/safety code required.
- This was my first time, so I made a few mistakes.
- If you don't like it, do it yourself next time.

Many specific considerations may be relevant to justification of some specific act or course of action. For example, in her book, *Lying, Moral Choice in Public and Private Life*, philosopher Sissela Bok considers under what circumstances lying is justified. Among the factors she evaluates are the extent to which lies undermine general trust, the importance of veracity (being truthful) for personal integrity, and whether any lies are justifiable or excusable. Are "**white lies**," that is, lies about minor matters that cause no immediate harm, justified or at least easier to justify than other lies? What about giving placebos, treatments with no known medical action, when it has been shown that giving such placebos actually does tend to make people feel better? Are exaggerations in letters of recommendation justified? If everyone is exaggerating, does that make a moral difference or does it only increase the temptation to lie without justifying it? Is one ever justified in lying if one could achieve the same results without lying? What about telling a lie, when telling the truth would cause harm? What if the lie would not prevent harm but is expected to produce benefits? Must justifications for lying by public officials pass the test of being a publicly acceptable *kind* of lie? (For example, the public would find acceptable the lies of the *kind* that the government might tell to mislead the enemy in wartime, but which had little effect on the actions of citizens. The public could not be asked to approve *specific* lies, because that would defeat the purpose of telling lies.) Is lying more acceptable than other forms of force and coercion? What lies in a crisis are acceptable (assuming they do prevent harm and because what is done in a crisis is less likely to set a precedent for lying in other situations)? What if the crisis becomes a situation of a prolonged threat to survival? How, if at all, is it morally relevant whether the person deceived is also a liar? Is it morally relevant that the person one lies to is one's enemy? What about lies told by professionals to protect peers and clients? How about the special case of lies to protect clients' confidentiality, when the professional is supposed to preserve the client's confidentiality? This array of considerations shows the range of factors one may consider in judging when if ever doing certain sorts of things is justified.

A particularly important distinction is the one between justification of a lie and excuse for a lie. Think about valid excuses with which you are familiar. Illness is an excuse for missing a test. The illness of the person does not make it a good thing to miss the test, but failure to take the test is not in the person's control, so the person should not be blamed for missing the test. If Chris is given an ultimatum to lie or be killed, then if Chris lies that may be excusable *because Chris has little choice* in the situation. If one's failure to carry out some obligation, say, reviewing plans, were due to being caught up in a life-threatening emergency that would also be an excuse and not a justification. **Excuses** *do not justify acts* or policies, but *may reduce or remove blame from the agent* who performed the act in question. A valid excuse is one that shows the agent to have had diminished opportunity to do something better. Although the act was not a good one, the person was not fully responsible for it and so should not bear full blame for it. It shows that the act does not reflect badly on the agent. We will discuss questions of moral character in Part 2 of this book and examine excuses further in connection with mistakes, when we consider which mistakes are excusable and which are blameworthy.

To say that an act is *justified* is to say that it was the right thing or a good thing to do in the circumstances in question, even if the act would be wrong in most other circumstances. For example, suppose that Leslie suddenly knocks Alex to the ground. Ordinarily it would be wrong to do that, but it would be *justified* if Leslie saw that a piece of machinery was swinging toward Alex and had acted to prevent injury to Alex. In contrast, one would speak of the act being *excused only if it were not justified* (i.e., not a good thing to do in the circumstances). For example, Leslie might be *excused* for knocking Alex down if Leslie's action was unintentional and resulted from slipping.

Figure I.4 summarizes several points about justification or excuses:

- One seeks justification for an act only if there is *some* reason to think that it is wrong. (This might be because it betrays trust, or causes considerable harm, violates rights, etc.)
- If the act is justified, then it is morally acceptable ("okay"). It might actually have been a *good* thing to do, but from the fact that it was justified we only know that it was acceptable.
- Only if an act is not justified (or not fully justified) does the question arise of excusing the agent for performing it.

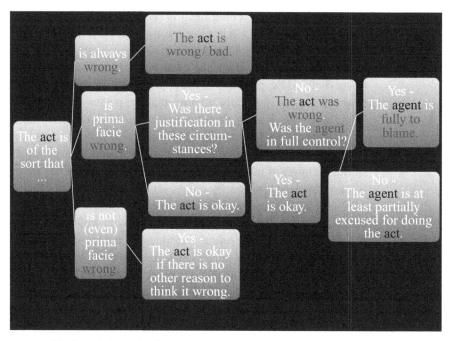

Figure I.4 Justifying Acts Contrasted with Excusing the Agents Who Perform Them

- The agent is blameworthy, if the agent did something wrong *and* the agent was in full control of her actions.

## Evaluation of Justification and the Reverse

Notice that one can inquire into the justification of any moral evaluation. One can always ask "Why is it better to do X than Y in these circumstances?" Similarly, one can ask about the moral evaluation of any justification, "Why are the considerations mentioned in that justification morally relevant?" Therefore, the relationship between evaluation and justification is reciprocal.

The ethical evaluation of an act or course of action can take the form of a judgment about whether (or the extent to which) the action was a good or a bad thing to do, or the act displayed or fostered the development of virtues or vices (such as corruption). Ethical justification, that is, an argument to demonstrate that something is ethically acceptable or desirable, is needed to distinguish an **ethical code or guidelines** from just any set of rules or guidelines about behavior. That is why ethical justification is a central topic for ethics.

Ethical terms, the terms we are examining in this introduction, provide the language for ethical evaluation and justification. For example, "Signing the peace accord was a *good (compassionate, responsible, beneficial)* thing to do" refers to the intended *consequences* of the act or the virtues displayed in doing it. Ethical evaluation may be of the rightness (or wrongness) of the act itself, that is, whether (or the extent to which) it is "the right thing to do."

*Suppose you are helping to install some equipment. You are to install one component by yourself. Afterward you are criticized for the way you installed it. Which of the following responses are excuses, which are justifications, and which are something else? Give reasons for your answers.*

- *I was given the assignment late in the day and told to finish by 5. There wasn't time to do it any other way.*
- *That is what the building/safety code required.*
- *This was my first time, so I made a few mistakes.*
- *If you don't like it, do it yourself next time.*

## Section 4.   Interests and Consequences

The first sort of ethically significant considerations we shall consider are harms and benefits, or more generally, consequences and the concepts related to them such as those of interest, cost, risk, and moral standing.

### Interests and Conflicts of Interest

What is a conflict of interest? Suppose a professional society issues a code of ethics that includes advice to its members on how to handle a situation in which they may have a conflict of interest. What values would underlie such advice? Which, if any, would be ethical values?

The first concepts that we will consider in some detail are those of **interest**, in the sense of an advantage (to someone or something), and the derivative notion of a **conflict of interest** that is central to understanding professional ethics in general and the ethics of the engineering profession in particular. "Conflict of interest" is a **technical term** in ethics, which is to say its meaning cannot be deduced from its component words the way in which the meaning of "conflicting interests" can.

People, other living things, and certain other entities (such as corporations or governments) have interests in securing, obtaining, or enjoying that which is *good* or *good for them*; they have interests in securing, obtaining, or enjoying that which it is rational for them to want or contributes to their thriving. People's interests may or may not coincide with their preferences, and some may not *know* what is in their own interest. Nonetheless, discussion of interests in professional ethics primarily concerns those interests that people or entities commonly recognize and seek to further.

Situations in which some interests (of one party or different parties) conflict, that is, there are conflicts *among interests*, are quite common. For example, an engineering student may have several interests competing for her time, such as an interest in learning engineering and doing well in her courses, an interest in developing her college friendships, and an interest in getting exercise. (The art of simultaneously furthering disparate goals is one we will take up in the second part of this book, in connection with a more complex model of ethical behavior. There we will explore the similarity between the task of meeting moral responsibilities and the task of simultaneously satisfying multiple design criteria in engineering or experimental design.) Such situations of *conflicting interests* are quite different from a conflict of interest situation, however.

The notion of a "conflict of interest" applies to a more specific situation than simply one of conflicting interests. We may say that a person (or perhaps some other party, such as a consulting firm) *has* a **conflict of interest** or *is in* a **conflict of interest position** when that party

- Is in a position of *trust* that requires the *exercise of judgment on behalf of others* (people, institutions, etc.)
- Has interests, obligations/responsibilities, or commitments of the sort that *might* interfere with the exercise of such judgment, and having those interests is neither *obvious* nor *usual* for those in this position of trust.[6]

A party has a **conflict of interest** when that party

- Is in a position of *trust that requires the exercise of judgment on behalf of others* (people, institutions, etc.)
- Has interests, obligations/responsibilities, or commitments of the sort that *might* interfere with the exercise of such judgment, and having those interests is neither *obvious nor usual* for those in this position of trust.

Parties with conflicts of interest are not necessarily guilty of any wrongdoing.

---

**Are Conflicts of Interest Ever "Apparent"**

You will sometimes see references to an "appearance of a conflict of interest" or an "apparent conflict of interest," but as Paul Friedman has argued,[a] this is a mistake. Those who speak of "apparent" conflicts of interest seem to be assuming that "conflict of interest" names a situation in which wrongdoing occurs or at least there is some *actual* influence of the trusted person's other interests, loyalties, and so on, on the decisions entrusted to her. That is not the way in which the term is generally understood nor how it has been defined here.

[a]Friedman, Paul J. 1992. "The Troublesome Semantics of Conflict of Interest," *Ethics & Behavior* 2(4): 245–251.

Parties with conflicts of interest are not necessarily guilty of any wrongdoing. The moral question is how they *handle* the conflict of interest; do they act in a way that makes them deserving of the trust placed in them? The most common ways of handling a conflict of interest are either to recuse oneself from the position of trust, divest oneself of the potentially competing interests, or, if the conflict is one that people can usually be trusted to manage, to openly acknowledge those other interests and obligations. Openly acknowledging conflicts of interest gives others the opportunity to decide if they think the judgment of the person with the conflict can nonetheless be trusted.

The lesser requirement of acknowledgment is adopted when it is too burdensome to require that persons in positions of trust divest themselves of the interest that might influence the decisions entrusted to them. (Holding the patent on an invention that is evaluated in the article would be such financial interest.) Requiring investigators to divest themselves of investments that they may have made (often on the basis of their engineering judgment) would be too burdensome, and might even induce them to forgo publication. Requiring disclosure of financial interests over a certain threshold alerts the journal editor to the presence of a possible bias of the reported research in favor of the author's financial interests, and, if the article is published, allows readers, too, to decide for themselves whether the research seems biased.

Dictionary definitions frequently apply the term "conflict of interest" only to conflicts between a person's *private interests* and those of

---

[6]This definition is a modification of one offered by Michael Davis and is indebted to it. Notice that it is a common situation, too obvious to mention, that people with multiple friends or family members whom they care about may find they have a conflict in seeking to further the well-being of each of several friends or family members simultaneously. Because of the obviousness of the conflict, being in such a situation is not regarded as having a conflict of interest but only conflicting interests, loyalties, or responsibilities.

a public office, and by extension with that person's professional obligations and responsibilities. However, there can also be conflicts of interest in which private interests do not enter. For example, the American Bar Association specifies as part of a general rule on conflict of interest that lawyers should not represent a client if such representation may be materially limited by the lawyer's responsibilities to *another client or to a third party*. (This implies that a lawyer cannot represent one party, call that party "A," in a legal action against another party, B, if the lawyer is doing work for that other party, B. If the lawyer were helping B to write a will, the information the lawyer gains in writing B's will might influence that lawyer's judgment of how to go about suing B on behalf of A.) There is no similar rule saying that engineers or engineering firms ought not to work for two competitors simultaneously, presumably because engineers are not *advocates* for their clients in an *adversarial process* as lawyers are; for an engineer, doing work for one client does not require becoming the adversary of someone else. Engineers might build manufacturing facilities for, or supply parts to, two companies that directly compete in the same market. An engineer or engineering firm in such a position would need to be especially careful to avoid disclosing any **proprietary information** (i.e., information to which one of the parties has an exclusive right) that the engineer learns in building the manufacturing facilities or supplying the parts, however.[7]

The previous example of the difference between the legal and engineering profession illustrates the point that the specific nature of one's obligations and responsibilities determines when conflicting interests become a conflict of interest, and so, when a situation requires a professional to rid herself of or disclose conflicting interests.

In tracing the difference in ethical standards set out for different professions to the nature of what is entrusted to them and the problem situations and temptations they commonly encounter, we are committed to something like the (not very relativistic) view called "local relativism," which we examined in the last chapter. The general principle on which all professions would agree is "Fulfill the trust that is placed in you as a professional," although the specific behavior that would be required to do that might vary with the profession.

Policies requiring financial disclosure, that is, disclosure of *financial* interests that might conflict with one's judgment as a practicing engineer, research investigator, or a public official, are very commonly called "conflict of interest policies," although such financial conflict of interest is only *one* specific type of conflict of interest. One of the Obligations of Reviewers specified in the American Chemical Society's *Ethical Guidelines to Publication of Chemical Research* discusses how reviewers are to handle *another type* of conflict of interest, one that results when a reviewer is a direct competitor of the author in the research under review. It reads:

---

[7]Later we shall consider a judgment by the NSPE Board of Ethical Review on case 80–4 about the information that an engineer should give to two potential collaborators who are competitors. Their judgment shows they expect engineers to show a high degree of consideration for one another, but it does not say that the engineer in question has a conflict of interest in being a potential collaborator of both competitors.

4. A reviewer should be sensitive to the appearance of a conflict of interest when the manuscript under review is closely related to the reviewer's work in progress or published. If in doubt, the reviewer should return the manuscript promptly without review, advising the editor of the conflict of interest or bias. Alternatively, the reviewer may wish to furnish a signed review stating the reviewer's interest in the work, with the understanding that it may, at the editor's discretion, be transmitted to the author.

The alternate course of action might enable a reviewer to show how her own work fits with that of the authors of the article under review and set the stage for a future collaboration. Signing the article would make the reviewer's name known to the authors, something that would not otherwise be done when a journal uses a "blind" reviewing process. (Ethical standards for reviewing grant proposals and manuscripts will be addressed in Chapter 9.)

The topic of conflict of interest is a particularly important one for engineers and computer professionals because their professions are ones whose members are trusted to render **impartial** judgments, that is, judgments that do not favor one party or result over another.

The requirement to avoid conflicts of interest is in some respects comparable to the requirement to conduct an experiment in a **double-blind** manner (so that the investigators as well as the human subjects are ignorant of which subjects are receiving the experimental intervention and which are the controls receiving a placebo.) The requirement is needed, not as a deterrent to intentional falsification of results, but because trying very hard not to be influenced by the knowledge of which patients are receiving the experimental treatment does not work. A sincere attempt not to be influenced could lead to overcompensation, which would also distort the findings.

(The subject of handling various sorts of conflic*ting* interests is a different one and will be taken up in connection with moral responsibility in Part 2.)

*What is a conflict of interest? Suppose a professional society issues a code of ethics that includes advice to its members on how to handle a situation in which they may have a conflict of interest. What values would underlie such advice? Which, if any, would be ethical values?*

### Consequences: Harms, Benefits, and Risks

Suppose your city has money in its traffic equipment budget for two more traffic lights (because it can obtain them at a greatly reduced price). Four intersections have shown, by the incidence of fatal traffic accidents at them each year, that traffic lights would be useful at those locations. Suppose that you have the job of deciding how to distribute the traffic lights. What criteria would be relevant to deciding?

As we noted earlier, at the beginning of Section 2, **the injury** or **benefit** resulting from some action is a morally relevant consideration in evaluating it, *although not necessarily the most important one*. An action may directly or indirectly help or harm others. Someone is harmed directly, for example, by being run over. A person is harmed indirectly if something that she cares about or in which she has an interest is harmed or diminished. Because injuring or benefiting people is morally significant, their interests and values of all sorts – not only moral but also religious, aesthetic, epistemic, and prudential – are often relevant to moral evaluation.

## Kant's Denial That Consequences Are Significant

Immanuel Kant held that consequences make no ethical difference and that what makes an action ethical is its conformity to what he calls the "Categorical Imperative." Kant gives two formulations of that Categorical Imperative (these may not be equivalent): 1. "Act only on those maxims that you could at the same time will to be a universal law." 2. Treat everyone [that is, every person, including oneself] as an end, and not a means only. The motivation for the second formulation may become clearer after we consider the notion of a moral agent in the next section. Kant regards persons as special because they are rational and moral agents (i.e., beings who can act for moral reasons.) Although Kant has contributed many important ethical and philosophical insights, Kant's rejection of consideration of consequences flies in the face of moral reflection in general and moral reflection in engineering ethics or practical and professional ethics in particular.

The rejection of Kant's thesis that consequences of an act are irrelevant to its moral evaluation does _not_ mean that one must go to the other extreme and regard consequences as *the only thing that matters*, as did Classical Utilitarians. The late philosopher Bernard Williams has nicely made the point in the following way:

> [W]hy should [the truth about the subject matter of ethics] be simple, using only one or two ethical concepts, such as *duty* or *good state of affairs*, rather than many? Perhaps we need as many concepts to describe it as we find we need, and no fewer.[a]

---

[a] Williams, Bernard. 1985. *Ethics and the Limits of Philosophy*. Cambridge, MA: Harvard University Press.

The formal technique of cost-benefit analysis is applicable to a special class of problems for which the consequences under consideration are ones that can be assigned arithmetic quantities.

For such problems, this technique may clarify the tradeoffs involved in following alternative courses of action. (The same action may produce both harms and benefits, of course. For example, some measures currently used to control bacteria in the water supply introduce minute quantities of carcinogens into the water.) In **cost-benefit analysis** one compares different courses of action by multiplying the *probability* that a given course of action will produce some outcome, by the *magnitude of the harm* (or benefit) of that outcome, and comparing this quantity to the quantity resulting from alternative actions.

Although the word "cost" in the name "cost-benefit" analysis means harm generally, and not only monetary costs, we shall see that the technique does favor consideration of monetary costs, and other *readily quantifiable* harms rather than those that are not readily quantifiable. In the past, some product design decisions were made by assigning a dollar amount to a human death to generate an amount that could be compared to the cost of making a product safer. An infamous case of such a calculation was in Ford Motor Company's decision about the explosion hazard posed by the location of the gas tank on the Ford Pinto.[8] After management realized the design of the Pinto left the car unusually vulnerable to explosion of the gas tank if hit from the rear, they decided against adding an inexpensive safety feature that would have lessened the risk. Management sought to justify its decision by a dubious cost-benefit calculation that assigned $200,000 as the monetary compensation for the pain and suffering of a burn death.[9] (The Pinto gas tank case was loosely paralleled by the safety problems of automotive design in the film, *Class Action*, where those suggesting cost-benefit techniques were derisively called "bean counters.") Ford's decisions about the

---

[8] DeGeorge, Richard T. 1991. "Ethical Responsibilities of Engineers in Large Organizations: The Pinto Case." In *Ethical Issues in Engineering*, edited by Deborah Johnson (Englewood Cliffs, NJ: Prentice-Hall, Inc.), 174–186.

[9] Dowie, Mark. 1977. "Pinto Madness," *Mother Jones* (September/October): 19–32.

In **cost-benefit analysis** one compares different courses of action by multiplying the *probability* that a given course of action will produce some outcome, by the *magnitude of the harm* (or benefit) of that outcome, and comparing this quantity to the quantity resulting from alternative actions.

## Other Names for Harms or "Costs"

When an action has unintended bad consequences, these "side effects" are technically termed "externalities" in economics.

## The Reliability of Probability Estimates

Estimates of the probability that a course of action will achieve a given result may be quite unreliable, however. The field of **risk assessment** has developed many sophisticated means for estimating these probabilities more accurately. Such assessments may also shift the focus of attention from consequences that cannot be assigned arithmetic quantities, to those that can, however.

Risk (in the technical sense) is the *probability* that a given course of action will produce some harm multiplied by the *degree* of that harm.

Pinto gas tank proved to be shortsighted (a prudential concern) as well as ethically suspect, because they took account of neither the monetary expense of liability judgments against Ford nor the damage to the company's reputation. The assignment of any dollar amount to human life is ethically suspect, as we shall discuss later when we consider rights and human rights. Without such assignment it is difficult to use cost-benefit analysis in such a case, however.

If the type of harm or benefit is held constant, the task is somewhat simpler. For example, one might compare the effect that each of several business plans will likely have on the share of the market for some product that one's company will have. When the harm or benefit is held constant, the technique is called "risk-benefit" analysis rather than "cost-benefit analysis."

The *probability* that a given course of action will produce some harm multiplied by the *degree* of that harm *defines* **risk, in the technical sense**. This technical notion of risk is a bit different from several other senses in which the term is used. "Risk" is *commonly* used to mean a danger or hazard that arises unpredictably, such as being struck by a car or capsizing in a boat. The "unpredictable" element in this colloquial sense of risk links it to the notion of an accident. The term "risk" is also used colloquially for the *likelihood* of a particular hazard or accident, as when someone says, "You can reduce your risk of capsizing by sailing only in light or moderate winds."

Risk analysis, risk assessment, and risk management use the technical sense of "risk" defined previously. In technical risk analysis one focuses on the *resulting harm* and not just the harmful event. One would consider such risks as the *risk of death by drowning or exposure* resulting from capsizing, rather than simply the *risk of capsizing*. The harms (or benefits) that are commonly considered are those that can be readily quantified, such as increased (or decreased) probability of death ("mortality risk") or monetary loss (or gain). Using the technical notion of risk one can compare, say, the relative chance of dying when traveling between two points by automobile and by commercial airline. One can also compare the risks associated with harms of different magnitudes. For example, consider two monetary risks: the rather common event of losing money in a broken vending machine, and the rarer event of having one's money stolen in a holdup. In many locales, there is a greater risk of monetary loss from malfunctioning vending machines than from being held up and robbed.

The difficulty in finding a meaningful way of quantifying some harms is important to bear in mind when using either cost-benefit or risk-benefit analysis. For example, in the comparison of monetary loss to vending machines with loss to robbers, we did not consider the greater emotional trauma associated with being held up. Assigning an arithmetic quantity to such trauma is very difficult. As a result, consideration of trauma is easily neglected in favor of monetary loss. (Recall the saying that if your only tool is a hammer, you will see every problem as a nail.) It would be a mistake to conclude that someone was behaving irrationally in taking greater precautions against being held up than against using malfunctioning vending machines, even if the risk of monetary loss from machines were greater. It is important to understand the limitation of a tool such as cost-benefit analysis to use it responsibly.

### Classical Utilitarianism and "Utility"

In 1823, Jeremy Bentham proposed the utilitarian calculus in an attempt to provide a moral justification for legislative reform. This calculus, which was later refined by John Stuart Mill, requires the measurement of "utility." By "utility" Bentham meant the property that tends to produce the benefit of pleasure or happiness. However, Bentham did not formulate a program for satisfactorily measuring utility.

In 1906, Vilfredo Pareto, in *Manuale d'economia politica*, showed that for classical economics all that was needed to measure utility was a statement of preferences by the individual or group whose interests were in question. This measure of preferences, however, provides no generalized measure of happiness for *humanity*, but only the preferences of the *groups and individuals who are sampled*.

Comparison of different sorts of harms or benefits is difficult, in part because in a **pluralistic society** – a society with diverse ethnic, religious, cultural, or social groups – there are *different notions of the good life*. Therefore, people disagree about what promotes or frustrates the achievement of the good life, and, therefore, about what values are most important. In addition, specific harms have different implications for people in different circumstances. Consider whether it is worth increasing one's chance of death by 25 percent, or to be painfully disabled for ten years. When such choices arise in health care, we say that the individual patient has the right to make them.

Some decisions about harms and benefits affect the general population, such as ones about the acceptable side effects of purifying the public water supply. Such decisions often made use of a measure or estimate of the degree that each consequence would be *preferred* by most people in the affected group. Those preferences are often quantified as the dollar amount that people would be willing to pay to achieve the benefit (in our example, clean water) or avoid the harm (in our example, avoid the side effect). However, as we have already seen, preferences are subjective, because they depend on characteristics (such as personal history) of the subject who has the preferences, rather than characteristics of the thing preferred. They are not a measure of magnitude of harm or benefit. By considering average preferences within a population those who use this technique seek to avoid that drawback of preference measure. Measurement of harms and benefits in terms of willingness to pay also ignores the fact that people vary both in the importance that money has for them and their ability to pay. So two people might both want expensive medical treatment, but one could readily pay for it, while for the other, paying for the treatment would mean sacrificing all the family assets and so he would not be willing to pay for it. Considering what the *same people* would be willing to pay for clean water and what they would be willing to pay to avoid the side effect helps correct for this drawback, however.

Risk-benefit and cost-benefit calculations may obscure another morally significant consideration: whether some measure harms one group while benefiting another, which is a question of fairness. If the population that stands to benefit is not the population that incurs the risk, "**risk shifting**" has occurred. Even if the net risk is lessened by some action, there is an ethically significant question of the fairness of any shift of the risk from one group to another. The invisibility of risk shifting when using cost-benefit and risk-benefit analysis is another limitation of those tools.

*Your city has money in its traffic equipment budget for two more traffic lights (because it can obtain them at a greatly reduced price). Four intersections have shown, by the incidence of fatal traffic accidents at them each year, that traffic lights would be useful at those locations. Suppose that you have the job of deciding how to distribute the traffic lights. What criteria will you use to decide?*

### Consequences for Whom? Moral Standing

Some scientists burn and maim animals to devise treatments for burned and maimed people. Furthermore, because anesthesia and analgesics would interfere with some of these experiments, the animals in those experiments receive nothing to relieve their pain. Just because these acts are experiments does not decide the question of whether they are also acts of cruelty. Should we view these acts as cruel? If they are, do moral obligations toward animals forbid such cruelty? If they do, could such treatment nonetheless be justifiable in some experiment that promised great advances in human medicine?

### Contractarians and the "Social Contract"

The school of philosophical ethics called "contractarian" (dating from the work of Thomas Hobbes in the seventeenth century) regards agreement to a social contract as the bedrock of ethics and legitimate legal authority. They hold that if some beings are incapable of agreeing to act according to certain rules, those rules do not have to be followed in treatment of them. They have no moral standing and are said to be "outside of the social contract."

The social contract is not an actual agreement. Much of the original argument about social contracts applies better to laws than to ethical rules, because the agreement actually is to live in a society where certain rules are *enforced*.

To say that some animals are not capable of acting morally or immorally is not to deny that there are moral constraints on the way moral agents should treat *them*. Moral constraints on the way some animal is treated is a matter of the animal's **moral standing**, that is, its intrinsic **moral worth**, *rather than its moral agency, its ability to act morally or immorally*. Another way of expressing the view that a being has moral standing is to say that its well-being (or at least some aspects of it) is of value in itself, and not merely a *means* to other desirable ends. Some being might have moral standing without having *the same* moral standing as people and thus having "human" rights, of course. (The well-being of something may be sought as a means to some other goal or end. For example, a person might want her vegetables to be healthy in order to eat them or to show them off, without believing that vegetables have moral worth or moral standing.)

Any moral agent has moral standing, but the prevalent view is that some beings that are not moral agents *do* have moral standing. For example, it is generally agreed that it is wrong to be cruel to nonhuman animals – even though they, or at least most species of them, are incapable of moral action.

The moral prohibition on cruelty to animals, mentioned in the last section, is widely recognized and has some legal backing. The question of the moral limits on experimentation with animals is of importance for the ethics of research.

*Some scientists burn and maim animals to devise treatments for burned and maimed people. Furthermore, because anesthesia and analgesics would interfere with some of these experiments, in those experiments, the animals receive nothing to relieve their pain. Just because these acts are experiments does not decide the question of whether they are also acts of cruelty. Should we view these acts as cruel? If they are cruel, do moral obligations toward animals forbid such cruelty? If they do, could such treatment nonetheless be justifiable in some experiment that promised great advances in medicine?*

### Treating Some Animal Like a Person

Suppose it is the case that some other beings should be treated as persons. What does it mean to treat a nonhuman the way one would treat a person? For example, if you come across an injured wild rabbit, ought you leave the animal to its natural devices or get veterinary help for it? There are those who claim that at least some other animals have the same moral standing as people and claim people should not interfere with animals at all. Others who defend the moral equality of animals hold that one should show the same concern for relieving animals' pain and suffering as one would for a human.

What determines whether experimentation on some animal is morally justifiable? What the animal undergoes in the experiment – whether it is disabled, killed, or caused pain is one relevant factor. Another is the condition of life of the experimental animal: what sort of life it has while awaiting an experiment. Although wanton cruelty to any creature may be objectionable, the rule against causing severe pain to animals when those animals will not benefit from the experiment may apply only to animals with certain moral standing. Some experiments that are currently conducted using crawfish would be shocking if carried out using vertebrates. (See http://www.onlineethics.org/CMS/research/rescases/gradres/gradresv2/subject.aspx for an example of such an experiment.[10])

Rules about the use of experimental animals have been formulated only for certain species. Is this because the animals that come under regulation have higher moral standing than those that are not? By what criteria would it be reasonable to judge that the members of some species have higher moral standing than others?

The history of ethical thought shows that those in power have often recognized the moral claims only of those who are similar to themselves. The human rights of many people have been ignored because of their race, class, or gender. That behavior is now described as racism, classism, or sexism, understood as

[10]This case originally appeared in Volume 2 (1998) of *Graduate Research Ethics: Cases and Commentaries*, edited by Brian Schrag. These volumes resulted from an NSF-funded project conducted by the Association for Practical and Professional Ethics (APPE) and by the Office of Research, University Graduate School, and the Poynter Center for the Study of Ethics and American Institutions, all at Indiana University.

### How Moral Standing Constrains Action

Consideration of moral standing is relevant in determining what is permissible to do in seeking to achieve desirable ends, that is, what ends are justifiable in themselves. For example, suppose that the moral standing of some animals makes it wrong to needlessly cause them to suffer. Then anyone who claimed that it would be morally acceptable to perform painful experimental procedures on them without giving them anesthesia would need to show that the value of the results to be obtained by withholding anesthesia outweigh the pain to the animal. Such an argument is required in U.S. research institutions for approval of experiments that cause pain to certain types of animals, where the institutional animal care and use committee (IACUC) reviews proposals to use animals in experiments to determine whether the treatment of the animals in those experiments is justified.

unwarranted preferential treatment of the race, class, or gender in power. The philosopher Peter Singer has called the claim that humans are the only group with moral standing "speciesism" to suggest that the view is unfairly biased in favor of humans. To show that the claim of the presence or absence of moral standing of members of some species is more than an exercise in prejudice or preference, one would need to show that differences in moral standing are based on *morally relevant* differences among the beings in question.

As Robert Proctor points out, the Nazis held that it was wrong to victimize healthy specimens of other species by using them for scientific experiments. They thought it better to use supposedly "defective" humans as experimental subjects. This example illustrates the point that cruelty to one group can easily coexist and even seek justification in compassion or respect for another.

*1. You are siting a new facility and wherever it is placed, it will require clearing 100 acres of trees. All the possible sites are habitat for some local fauna and none are adjacent to areas that provide similar habitat, so clearing the land is likely to mean the death of many of the creatures that live at that site. Assuming the other ethical considerations about possible harms associated with clearing the candidate sites are equal, how do you weight the harm of directly killing various creatures or of depriving them of habitat necessary for their survival? Which ones count for their own sake (i.e., have intrinsic value)? For example, do any insects? Do all insects? Explain the criteria you used in deciding which creatures are most worthy of protection.*

*2. You are assigned to test a vehicle that is designed to travel at high speed with frequent acceleration (changes of direction). Before you test it with a test pilot, you are to run it with an automatic pilot and experimental animals inside. Assume that the greater the physiological similarity between the animal and humans, the better this first test will detect any unanticipated hazards to the test pilot. By what criteria will you choose the type of animal? Will its physiological similarity to humans be any indicator of its moral worth or moral standing? If you choose a type of animal that has moral standing, what justification can you offer for potentially harming the animal, which cannot volunteer, to protect a human test pilot, who is a volunteer?*

## Section 5. Moral Obligations and Moral Rules in Engineering

### *Moral Obligations and Moral Rules*

It is widely agreed that in presenting engineering and scientific results, the central obligation is to present an accurate account of the research performed and an objective discussion of its significance. What moral rule for research investigators expresses this moral obligation?

| Agent A has a **moral obligation** $\leftrightarrows$ | There is a **moral rule** that applies to A |
|---|---|
| A has an obligation to do X | Do X |
| A has an obligation to refrain from doing Y | Do not do Y |
| Unless Q, A has an obligation to do Z | Do Z, unless Q |
| In circumstances C, A has an obligation to refrain from doing W. | Whenever C, do not do W |

**Figure I.5**    Moral Rules and Moral Obligations

## The Effect of Cruel Actions on Character

A *different sort* of argument against harming animals is that doing so corrupts the people who do the harm. That is an argument about *character* and goes beyond the scope of our present consideration of the morality of *acts*. It is worth mentioning here only to show that denying that some beings have moral standing does not necessarily commit one to the view that, ethically speaking, "anything goes" with regard to their treatment.

An ethical duty or **obligation** is a moral requirement to follow a certain course of action, that is, to do or refrain from doing certain things. It may arise from making a promise or an agreement or from entering a profession. For example, according to many engineering codes of ethics, engineers not only have a moral *right* to raise issues of wrongdoing outside their organizations, but, additionally, they have an *obligation* to do so when public health and safety are at stake. (The National Society of Professional Engineers [NSPE], in its 2006 code of ethics, lists nine main entries under "Professional Obligations." Each has two to five more specific obligations, making thirty-eight general and specific obligations.)

**Moral obligations** and **moral rules** are interdefinable, that is, if you have a moral rule, there exists a corresponding statement of obligation and vice versa. This point is represented in Figure I.5. (As we shall see in Section 6 of this introduction, rights share some of the same logic of obligations and moral rules.)

> **Moral obligations** and **moral rules** are interdefinable, that is, if you have a moral rule, there exists a corresponding statement of obligation and vice versa.

Obligations and rules may be **institutional** or **legal** rather than moral. For example, at many colleges there is an institutional rule obliging all students to see their advisors on or before Registration Day. Some workplaces have rules about where various categories of employee may park or how employees earn the right to park in certain desirable locations. Certain institutional rules, such as the designation of the parking spaces reserved for emergency vehicles, may have an ethical as well as institutional basis, but institutional rules *need not* have ethical significance. Legal and institutional rules share the logic of moral rules, so legal obligations and legal rules are interdefinable in the same way as moral obligations and moral rules. Moral obligations and most moral rules specify what acts one is morally forbidden, or morally required, to perform (without consideration of the consequences of the action – except in so far as these consequences are part of the characterization of an act itself; killing, for example, is an act that results in death).

The *Ethical Guidelines to Publication of Chemical Research** gives examples of ethical obligations to serve *knowledge values* such as truth and accuracy as well as ethical values such as fairness (in assigning credit) and protection of others from harm. These guidelines were first issued in 1985 by the American Chemical Society (ACS) to guide those involved in publishing in any of its numerous journals and have gone through numerous revisions, which were primarily augmentations. Some of the ACS publications, such as *Chemical and Engineering News*, have a large engineering readership. (These *Ethical Guidelines to Publication of Chemical Research* have served as a model for many other societies, some of which have issued guidelines that are virtually identical to those of the ACS.) The ACS latest revision (2010) includes the following moral obligations and rules among the fourteen obligations it lists for authors:

1. An author's central obligation is to present an accurate account of the research performed as well as an objective discussion of its significance.

2. An author should recognize that *journal space* is a precious resource created at considerable cost. An author therefore has an obligation to use it wisely and economically.

5. Any *unusual hazards* inherent in the chemicals, equipment or procedures used in an investigation should be clearly identified in a manuscript reporting the work.

9. An author should *identify the source of all information quoted or offered, except what is common knowledge.* Information obtained privately, as in conversation, correspondence, or in discussion with third parties, *should not be used without explicit permission* from the investigator with whom the information originated. Information obtained in the course of confidential services, such as refereeing manuscripts or grant applications, should be treated similarly.

12. The authors should reveal to the editor and to the readers of the journal any potential and/or relevant competing financial or other interest that might be affected by publication of the results contained in the authors' manuscript. Sources of funding of the research reported should be clearly stated. In addition, all authors should declare (1) the existence of any significant financial interest ($>\$10,000$ or $>5\%$ equity interest) in corporate or commercial entities dealing with the subject of the manuscript; (2) any employment or other relationship (within the past three years) with entities that have a financial or other interest in the results of the manuscript (to include paid consulting, expert testimony, honoraria, and membership of advisory boards or committees of the entity). The authors should advise the editor in writing either that there is no conflict of interest to declare, or should disclose potential conflict of interests that will be acknowledged in the published article, whether by insertion of a footnote, or incorporation of a sentence or paragraph in the

*The current version (2010) of the entire set of guidelines (nine for editors, fourteen for authors, eleven for manuscript reviewers, and an additional three for those seeking to publish in the popular literature) are available as a pdf file from the ACS Web site at http://pubs.acs.org/page/policy/ethics/index.html.

"acknowledgments" section, or by other format of disclosure to the reader as specified by the journal.[11] (Italics added.)

Italics have been added to indicate *key values and concerns* in the quoted passages. Item 12 deals with conflict of interest, a topic that we discussed earlier in Section 4.[12]

> *Consider the five obligations of authors quoted previously from the ACS's Ethical Guidelines. For each obligation, state the moral rule for research investigators that expresses this moral obligation.*

### Prima Facie and Absolute Obligations and Rules: The Burden of Proof

Do the codes of ethics of the ACM, ASCE, ASME, and the NSPE regard the obligation to protect the public safety as one that must be honored no matter what?

Moral obligations and moral rules (and moral rights, too, as we shall discuss) are subject to further classification. In particular, an obligation or rule is classified as either absolute or prima facie. It is absolute if its direction for action always overrides other considerations, but prima facie if other moral considerations might be weightier and so justly override those directions. For example, it is often argued that lying to a potential murderer would be justified to save an innocent person.

**"Prima facie"** is a Latin term meaning "on first appearance." One might wonder whether it can be very important to specify only what, *on first appearance*, one is morally required to do. Knowing that someone has a prima facie obligation does not settle the ethical question of what that person is required to do, as does knowing they have an absolute obligation. However, a prima facie obligation or moral rule establishes the burden of proof. Where the burden of proof lies is a major consideration in ethics and elsewhere (e.g., the law). When it is said that in U.S. law a person is innocent until proven guilty, it means that the burden of proof is on the prosecution. The *burden of proof* determines what the expectation or judgment will be, if no arguments or evidence is given. It establishes the **default expectation**, that is, what is assumed in the absence of other information. To say something is prima facie true is to say that the burden of proof is on those who say it is not true. To say that some act (such as lying) is prima facie wrong is to say that it is wrong to do it unless an adequate justification can be given for thinking it was right to do in specific circumstances. If causing severe pain to animals when they will not benefit from one's actions is prima facie morally wrong, performing painful experiments on animals is wrong unless a sound ethical argument is given to justify causing them pain.

---

[11] ACS, 2010, "Ethical Obligations of Authors" in their *Ethical Guidelines to Publication of Chemical Research*, which is available as a pdf download at http://pubs.acs.org/page/policy/ethics/index.html.

[12] The latest version (2010) of the Guidelines, from which these provisions are quoted, may be accessed at http://pubs.acs.org/page/policy/ethics/index.html.

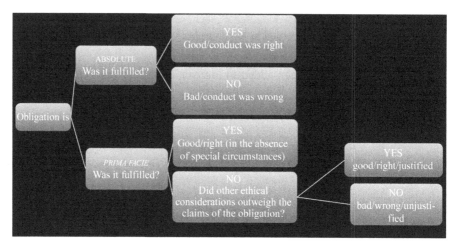

**Figure I.6**    How the Prima Facie versus Absolute Distinction Influences Justification

---

What a prima facie obligation or moral rule does is establish the burden of proof. Where the burden of proof lies is a very major consideration in ethics.

The question of **moral justification** arises when a prima facie obligation (or a prima facie right or moral rule) is infringed. The prima facie obligation/right/moral rule places the burden of proof on those who say that the infringement was *justified*. In the case of an absolute obligation (or absolute moral rule or right) there would be no possibility of a justified exception. This point is summarized in Figure I.6.

The codes of ethics of many engineering societies, including the ACM, ASCE, ASME, and the NSPE, direct engineers to keep sensitive information confidential, including information from clients or employers. However, those same codes also say that engineers, in fulfillment of their professional duties, shall "hold paramount" the safety of the public. That means that the safety, health, and welfare of the public are of greatest importance. Therefore, if in some circumstances, the only way for engineers to protect the public safety is to disclose some confidential information of their clients or employers, engineers should disclose it. Thus the obligation to preserve confidentiality is shown to be one that those codes take to be prima facie, rather than absolute.

> *Do the codes of ethics of the ACM, ASCE, ASME, and the NSPE regard the obligation to protect the public safety as one that must be honored no matter what? Give reasons for your answer.*

### Negative and Positive, and Universal and Special, Obligations and Rules

Is the obligation to keep a client or employer's privileged information confidential an obligation to *refrain from doing* certain things or is it an obligation *to do* something?

In addition to the classification of an obligation as either prima facie or absolute, obligations are also classified as "negative" or "positive" (also called "affirmative"). Roughly, an obligation is **positive/affirmative** if one has to do something to fulfill it. If the agent under the obligation has only to refrain from doing certain things, the obligation is **negative**.

> An obligation is **positive/affirmative** if one has to do something to fulfill it. If the agent under the obligation has only to refrain from doing certain things, the obligation is **negative**.

The obligation to keep some information confidential might sound like a negative obligation, because often it would require only that one refrain from acts of disclosure. However, under some circumstances one might have to take special precautions to avoid disclosing a client's or employer's confidential information. One might need to shield a new model from public view for example. In this case, the obligation would require positive action and so would have the characteristics of a positive obligation. This example illustrates some of the *judgment* that must be exercised in applying ethical concepts.

As we saw, obligations have counterpart moral rules. An engineer's obligation to keep a client's privileged information confidential corresponds to the moral rule to keep confidential the information of one's client or employer. This rule appears in the codes of ethics of many engineering societies.

**Moral rules** or **rules of ethical conduct** specify the acts or course of action that are ethically required, forbidden, or permitted. Following the usage of engineering societies, this book uses "rule of practice" or "rule of conduct" to mean a moral rule that precisely delineates the acts or courses of action in question. General admonitions such as "Be honest" or "Treat every person as an end and not as a means" are also moral rules but they are so general that they tend to be called "basic ethical considerations" or "**ethical principles**," or (in some engineering codes of ethics) "fundamental canons." The NSPE in its Code of Ethics classifies as "fundamental canons" such general imperatives as that engineers shall perform services only in areas of their competence or that they shall issue public statements only in an objective and truthful manner. In contrast to the fundamental canons, the NSPE calls specific moral rules (grouped under each of the Fundamental Canons), "Rules of Practice." For example, under the canon enjoining truthfulness and objectivity are three rules:

### Ethical Principles in Engineering, from the Societies

Engineering societies, such as ASME, ASCE, and the NSPE, list both "fundamental principles" and "fundamental canons." The ASME Code, like the NSPE Code, places specific rules under each of the canons, but, unlike the NSPE and ASCE, puts these rules of practice in a separate document called "Guidelines to Practice." The principles are general statements, some of which express a concern with the well-being of the profession. For example, the first of the ASME fundamental principles is "Engineers uphold and advance the integrity, honor, and dignity of the engineering profession by using their knowledge and skill for the enhancement of human welfare." This statement presents enhancement of human welfare as a *means to* the end of enhancing the profession.

Engineers shall be objective and truthful in professional reports, statements, or testimony. They shall include all relevant and pertinent information in such reports, statements, or testimony, which should bear the date indicating when it was current.

Engineers may express publicly technical opinions that are founded upon knowledge of the facts and competence in the subject matter.

Engineers shall issue no statements, criticisms or arguments on technical matters which are inspired or paid for by interested parties, unless they have prefaced their comments by explicitly identifying the interested parties on whose behalf they are speaking, and by revealing the existence of any interest the engineers may have in the matters.

The first rule of practice speaks to the truthfulness and thoroughness expected in the reports issued by engineers. The second supports an engineer publicly voicing ethical concerns. The third rule requires engineers to disclose any conflict of interest.[13]

### Autonomy

Some philosophers, including Immanuel Kant and John Rawls, have regarded the capacity to govern one's own behavior as the essence of being a person and a moral agent and what makes a person an "end" (in the sense of Kant's imperative to treat everyone as an end and not a means only), that is, valuable in themselves. They call this capacity, "autonomy." Others use the term "autonomy" more narrowly as synonymous with possessing a right of self-determination.

Other codes of ethics for engineers are organized differently and use a different language. The Code of Ethics of the Institution of Engineers, Australia (IEA) states three general "cardinal principles," such as "to respect the inherent dignity of the individual" (which they say should apply in personal as well as professional life), and states nine "tenets" that are general rules for the ethical conduct of professional practice. For example, "members shall apply their skill and knowledge in the interest of their employer or client for whom they shall act as faithful agents or advisers, without compromising the welfare, health and safety of the community." The IEA code also offers clarification of some conceptual points, such as:

Members should understand the distinction between working in an area of competence and working competently. Working in an area of competence requires members to operate within their qualifications and experience. Working competently requires sound judgment.

– Code of Ethics of the Institution of Engineers, Australia (IEA)

> Because the point of this book is to help you think about the ethics of engineering, I recommend that you rephrase any ethical principles that are phrased in technical jargon in terms that *connect the subject matter under discussion to the moral categories you already recognize.*

Many writers state principles in the form of obligations or general rules of behavior, but you will also see references to, for example, the "principle of honesty" or "the principle of respect for persons." It is easy enough to put such principles in the form of rules, however. The principle of honesty clearly translates as the basic rule, "Be honest." "The principle of respect for persons" is a somewhat jargoned way of expressing the rule to respect other's right of self-determination (i.e., their right to make decisions for themselves). Cultural and religious heritages often influence the name or formulation of a principle. Because the point of this book is to help you think about the ethics of engineering, and not to change your culture or religion, I recommend that you rephrase any ethical principles that are phrased in technical jargon (i.e., words that you will not find in any and all good collegiate dictionaries) in

---

[13] The NSPE code of ethics is available in the ethical codes section of the Online Ethics Center, which is available at www.onlineethics.org.

## How Moral Rules Are Learned and Known

As philosopher Alasdair MacIntyre argues, moral rules and principles are learned and formulated in the context of situations to which they apply, rather than being known abstractly, like principles of logic. He says a "moral principle or rule is one which remains rationally undefeated through time, surviving a wide range of challenges and objections, perhaps undergoing limited reformations or changes in how it is understood, but retaining its basic identity through the history of its applications. In so surviving and enduring it meets the highest rational standard."[a]

---

[a] MacIntyre, Alasdair. 1984. "Does Applied Ethics Rest on a Mistake?" *The Monist* 67(4): 499–512; pp. 508–509.

## The Empirical Quality of Ethical Reasoning

The eighteenth century philosopher Thomas Reid argued that systematic ethical reasoning resembles systematic knowledge in the natural sciences, but is unlike a deductive chain of reasoning system, such as a system of geometry.[a]

---

[a] Reid, Thomas. 1995. "On the Active Powers of the Mind." In *Philosophical Works*, Vol. II (Hildesheim: Gekorg Olms Verlagsbuchanlung), 642.

## Pluralism's Question: "Who Should Decide?"

Recent discussions of informed consent in the United States and *other pluralistic cultures* have tended to replace the question of "What should be done?" with the question of "Who decides?" and to accept different answers to the question "What should be done?" given by those with value commitments of the party who "owns" the decision.

terms that *connect the subject matter under discussion to the moral categories you already recognize*. (Disciplinary jargon in scholars' formulations of ethical principles generally reflects their positions in scholarly disputes, which are rarely relevant to deepening your understanding of ethical problems in engineering.)

Specific rules of practice may evolve into ethical principles in much the way that mathematician Henri Poincaré argued *empirical* scientific laws (i.e., laws based on observed regularities, rather than derivation from other laws) may evolve into principles. In both cases, we give the name "principle" to those relationships that we regard as fundamental and the truth of which we assume in making other observations and inferences.

Poincaré described this process for some fundamental physical laws and principles. For example, Newton's second law, $f = ma$, was an empirical law when Newton first proposed it, but it quickly evolved into a principle, so that now it functions virtually as a *definition* of force.

On the ethical side, people often speak of "the principle of informed consent." The rule to obtain informed consent, although formulated only in the mid-twentieth century, quickly became an ethical requirement for enrolling human subjects in experiments (and more loosely applied as a precondition for medical treatment). It soon became a basic element in reasoning about many ethical matters other than health care and human experimentation.

The character of the situations for which a rule is offered influences the formulation of that rule. (Chapter 1, Professional Practice in Engineering, contains illustrations of the influence of problem situations commonly encountered in engineering on the rules of practice for the engineering as reflected in the codes of ethics of engineering societies.)

The classification of moral and legal obligations and rules is shown in Figure I.7. In summary: A **moral rule** is an ethical standard stated as a rule of behavior. An ethical principle or "fundamental canon" is a general moral consideration that provides the framework for more specific rules of practice. Such principles are sometimes stated as rules,

1. Who has this obligation or is subject to this rule?

All moral agents → The obligation is a universal obligation

Only some moral agents (with special characteristics) → It is a special obligation

The relevant special characteristics include

- particular agreements—e.g., Obligation to fulfill a promise
- special relationships—e.g., Parental obligations
- specific expertise/knowledge—e.g., Obligation of researchers to disclose any hazards in conducting the experiments on which they are reporting
- involvement in an established process—e.g., Obligation to answer a charge

2. Can the claims of the obligation *ever* justly be overridden (whether often and easily or very rarely and only in highly unusual circumstances) or can the rule ever be justifiably ignored?

| | | |
|---|---|---|
| Yes | → | prima facie obligation |
| No | → | absolute obligation |

3. Does it require that the moral agent under the obligation or subject to the rule do something or only refrain from doing certain things?

Do something          → positive or affirmative obligation

Refrain from doing certain things    → negative obligation

**Figure I.7**    Categories of Moral Obligations

## Bernard Williams on the Modern Notion of Obligation

The distinguished late philosopher, Bernard Williams in his 1985 book, *Ethics and the Limits of Philosophy*, raised questions about the proliferation of supposed sources of moral obligation in modern philosophical ethics. He recognized obligations based on one's promises, but argued that much modern philosophical ethics has exaggerated the conditions under which we actually have obligations and distorted the understanding of ethics. He may have a good point, but the initial task in Part 1 of this book is to introduce the subject of engineering ethics, including its discussion by engineers and engineering societies, and those societies make frequent use of the concept of an obligation. Therefore, that concept of obligation is discussed here.

but also may be stated as ideals of behavior. Thus, the "principle of veracity," which mentions the virtue of veracity (honesty, speaking the truth), implies the moral rule against lying. Thus, the term "moral rule" may apply either to a general rule, in which case it may be called an "ethical principle," or to a more specific rule of ethical conduct.

Although any obligation has a corresponding moral rule, and, as we shall see in the next section any right imposes obligations, moral rules need not have corresponding rights. This will be the case if the obligation is toward a being that does not have rights. There are moral rules that apply to the behavior of moral agents toward beings who, although they have moral standing (and therefore their welfare must be considered), are not the sort of beings that have rights. For example, if insects do not have rights, but it is wrong to be cruel to insects (say, by tearing off their wings), then the moral rule against being

## Types of Legal and Institutional Obligations

The distinctions between types of moral obligations such as special and universal can be applied to legal and institutional obligations and rules as well as moral ones, but in practice the prima facie-absolute and positive-negative distinctions are the only ones you are likely to encounter in legal and institutional contexts.

The law has many special distinctions that apply to legal rights and obligations.

cruel to insects is a moral rule that has no corresponding right.

Several years ago, the question of moral constraint on the treatment of human corpses was discussed with practical application to product development when it was decided to resume using human cadavers in auto safety test crashes to test the crash worthiness of car designs and safety devices.

The ethical questions about the treatment of corpses are also of practical importance in setting practices of teaching hospitals. Some medical school faculty members encourage student physicians to practice medical procedures on corpses before *rigor mortis* sets in. This practice gives student doctors the opportunity to become more proficient in clinical skills before they use those skills on living patients. Laws requiring the consent of the family for any procedures done to the corpse are common and reflect the repugnance with which many people view such instrumental use of corpses. (This legal restraint is commonly circumvented, however, by delay in pronouncing the patient dead, and telling the medical students to perform certain tasks on the not-yet-officially-dead corpse.)

> *Consider the two cases of the instrumental use of human corpses. Do they differ in any ethically significant ways? Choose one and evaluate the adequacy of the justification for using human corpses in that way.*

## Section 6. Categories of Moral (and Legal and Institutional) Rights

### Moral Rights

*Suppose you are an engineer in a large company that makes technologically sophisticated manufacturing equipment, and you discover that one of the company's marketing representatives is knowingly misleading a client company about how soon your company can supply certain manufacturing equipment. Do you have a right to protest this lying? If so, to whom? Does your company have a right to dictate that you keep silent? If not, what (if anything) does your company deserve from you and how does that affect your actions?*

One of the most familiar concepts in contemporary ethical discussion, along with the concepts of benefits and harms, is the concept of a moral right. A **right** is a justified claim or assertion of what a rights-holder is due. We are particularly concerned with moral rights here, but there are also legal rights and official or institutional rights. As we shall see in Section 7, legal considerations have some claim to being moral considerations as well. The bases or justifications for moral/ethical rights are ethical, the bases or justifications for **legal rights** are legal, and those for **official or institutional rights** derive from the definitions of the offices or institutions in question. The same right might be simultaneously a moral/ethical right, a legal right, and an institutional right, but one sort of right *need not* be another sort, because only one sort of justification might exist for it.

Many engineering societies recognize an engineer's moral right to protest matters other than safety risks. (As we saw, there is wide agreement that engineers have an *obligation* to protest safety risks.) The matters that engineers have a right to protest and bring to light are generally serious defects and wrongdoing, such as poor quality or financial fraud. Saying that they have a moral right to do this means that if they bring such matters to light, they are at least prima facie morally justified in doing so. (The right would be only prima facie, because there might be special circumstances in which publicly disclosing some defect would obviously make it possible for a terrorist to cause massive destruction by exploiting that defect, in which special circumstances disclosing the defect would be both unwise and unjustified.) Suppose the engineer has some affiliation with the organization that would suffer from disclosure of the defect or wrongdoing (such as being an employee of that organization) and the engineer had good reason to think that the company did not already know of the defect. In that case, loyalty to the organization would prima facie oblige the engineer to bring the information or protest to responsible persons in the organization to give the organization a chance to remedy the situation, before "going public" with the information or protest. What does the right of engineers to protest a matter imply about the obligations of others? It means that others ought not to try to prevent the engineer from doing so. It also means that, as a moral matter, no one should retaliate against an engineer who does so. For protests or disclosure of certain sorts, such as disclosure of evidence that someone has committed research misconduct, protections against retaliation against those who make such protests **in good faith** (i.e., on the basis of evidence of misconduct, rather than motivated by malice) are necessary.

### Does Every Obligation Have a Counterpart Right?

Some philosophers, such as Peter Singer, argue that nonhuman animals do have rights, although they do not have the ability to choose to exercise a right as requisite for having it. Those who take such a position might argue that for every obligation there is a moral right.

Rights share some of the same logic that applies to obligations and rules. As we saw in the last section rights are not interdefinable with obligations and rules, however. For every moral/ethical, legal, or institutional right there is a corresponding moral/ethical, legal, or institutional obligation and a moral/ethical, legal, or institutional rule – see Figure I.8. *An obligation or rule may not have a corresponding right*, however. This is most obvious in the case of moral rules and obligations toward beings with moral standing that are not moral agents. Part of the special characteristic of a right is that it can be either exercised or **waived** (i.e., intentionally not exercised). If one were required to exercise the right to do something, by definition it would be an obligation and not simply a right. Because it is characteristic of rights that they can be waived, beings without the ability to decide whether to exercise a right are for that reason held to be incapable of having moral rights (although such a being might have moral standing). In that case, moral agents would have some

---

**Moral Right (possessed by A)** ➡ **Moral Rule or Obligation (on others)**

| | |
|---|---|
| A has a right to do X. ➡ | Do not interfere with A doing X. |
| A has a right to receive S from B. ➡ | B is obligated to supply S to A. |
| If Q, then A has a right to do Z. ➡ | If Q, then everyone is obligated to refrain from interfering with A doing Z. |
| . . . . | . . . . |

---

**Figure I.8**    Moral Rights and Their Counterpart Moral Rules or Obligations

obligations toward such beings, at least the obligation to refrain from wantonly destroying them.

> For every moral/ethical, legal, or institutional right there is a corresponding moral, legal, or institutional rule, but *an obligation or rule may not have a corresponding right.*

Some have claimed that moral rights always override, or at least always have more ethical weight, than other moral considerations. Philosopher Judith Jarvis Thomson closely examines this view in her book *The Realm of Rights*[14] and convincingly argues that it is mistaken. One of the counterexamples that she develops in detail is that of a person who is ill and needs medical attention and who takes a shortcut over another's property, thus violating another's property right by trespassing. This example certainly seems to be a case in which a moral (and legal) right is ethically outweighed by other considerations.

*Suppose you are an engineer in a large company that makes technologically sophisticated manufacturing equipment. You discover that one of the company's marketing representatives is knowingly misleading a client company about how soon your company can supply certain manufacturing equipment. Do you have a right to protest this lying? If so, to whom? Does your company have a right to dictate that you keep silent? If not, what does your company deserve from you and how does that affect your actions?*

### Human and Special Rights

*When competent human beings are recruited for an experiment, such as the testing of a new biomedical device, the accepted ethical standard is to seek their informed consent, even if the risks to the research "subjects" or "participants" are minimal. This means that to enlist subjects/participants in one's research study, one must first give them full information\* about the study and their own ability to drop out of the study at any time; the subjects must freely consent to participate. Respecting the ability of people to refuse to participate in an experiment is understood as necessary to respecting their human rights. Do you agree that the right to refuse to participate in an experiment is a human right? Why or why not?*

---

[14] Jarvis Thomson, Judith. 1990. *The Realm of Rights*. Cambridge, MA: Harvard University Press.
  \*In some experiments in which it is important that participants be unaware of exactly what is being tested, such information may be withheld, but not information about risks or the general nature of what they will experience.

## "Human Rights" and the Concept of a Person

The term "human rights" is one that Eleanor Roosevelt brought into widespread use. Previously these rights were called the "rights of man" or "**natural rights**." She chose "human" as a more inclusive modifier. There is now international and cross-cultural agreement that all people have some rights simply because they are people.

Although the term "human rights" is widely accepted, it is being a person rather than simply being human that is generally regarded as necessary for having "human" rights. A human tissue culture, although human and alive, would not have human rights, for example.

What counts as a person? The answer is controversial. There is general agreement that *any moral agent* would count as a person and that normal reasonably mature human beings are persons. What is most controversial is the question of whether immature humans (those too immature to act for moral reasons) or humans who are too profoundly mentally disabled to act for moral reasons should count as people. For legal purposes in the United States, newborns and those with profound mental disabilities *are* considered people (although people who need guardians to look out for their interests and reason on their behalf).

---

A **human right is** a right that all people have simply because they are people. In contrast, a **special right** is a right possessed only by some.

---

To better understand how rights function in ethical deliberations, we will examine four distinctions among rights. Any moral right may be classified in terms of all four distinctions. For a start, consider how the sort of justification for a right will vary depending on whether the right in question is a **human right**, that is, a right that all people have simply by virtue of being people, or whether it is a **special right**, that is, a right possessed only by some people.

The consideration of human rights provides part of the background for discussion of professional ethics. The international recognition of human rights provides endorsement of an ethical standard that transcends cultural differences. Consideration of human rights is implicit in the formulation of responses to a broad range of problems. Among those problems are the ethical problems of engineers and computer professionals that are the focus of this book.

Engineers and scientists encounter issues of human rights explicitly in the requirement to obtain the informed consent of any person who is to be an experimental subject in their research.

Special rights may have any of several different bases. The agreements or the so-called contracts that a person makes are one basis that some philosophers especially emphasize. The right to occupy an apartment is a special right that usually derives from an agreement. Other people's initiatives may also be the basis of some special rights, for example, inheritance rights; so may the existence of (chosen or unchosen) relationships – for example, "parental rights." Membership in a certain group might be the basis of a special right. The group in question might be a group that one was *born into* or *drafted* into, as well as a group that one has *agreed* to join, so the basis in group membership does not reduce to the agreements or "contracts" basis. The rights to wear certain uniforms, insignia, or plaids are examples of rights deriving from group membership.

In the last section, we examined an analogous distinction between moral obligations (and moral rules) that apply to everyone and those that apply only to some. An obligation that everyone has (or a moral rule that applies to everyone) is usually called "universal," rather than "human," however.

## What Is the Basis for Human Rights?

The Declaration of Independence says simply that certain human rights are god-given. Although some writers on ethics begin with the assumption of the existence of rights, especially human rights, philosophers typically want to know the ethical basis for rights. An account of the basis for rights is said to provide a "vindication" or "warrant for" those rights. John Ladd gives one such argument. He argues that human rights are the claims that need to be honored if people are to be able to meet their moral responsibilities and maintain their moral integrity.[a]

The late philosopher Gregory Vlastos argued that all people possess human worth that is quite independent of whether they possess valuable qualities, and that human worth is the basis of human rights.[b] His argument has been criticized for not supplying an argument to support the view that people have worth or equal worth.

---

[a] Ladd, John. 1979. "Legalism and Medical Ethics." In *Contemporary Issues in Biomedical Ethics*, edited by J.W. Davis, Barry Hoffmaster, and Sarah Shorten (Clifton, NJ: Human Press).
[b] Vlastos, Gregory. 1962. "Justice and Equality." In *Social Justice*, edited by Richard Brandt (Englewood Cliffs, NJ: Prentice-Hall).

## Recognition of Human Rights in the United States

The United States has not always recognized every person's human rights. In particular, the rights of enslaved people were denied. The Fourteenth Amendment to the U.S. Constitution recognized the citizenship of people who had been enslaved and affirmed that all citizens possess the right to life, liberty, and property (although only men could vote). It also provided that naturalized citizens have the same legal rights as native-born Americans.

**Civil rights** are the legal rights of citizens or more generally the members of some civil society. Many societies give legal recognition to human rights by making them civil rights. (Not all civil rights are human rights, however, and the content of civil rights that are not human rights, such as the right to vote, varies widely with the enacting government.)

The term "right" itself is a relatively modern notion (although it had some parallels in the Roman concept, "*ius*" or "*jus*"). The concept of a right is one that you will not find in premodern books of moral teachings from any cultural or religious tradition, books such as the Bible, or Koran (Quran), or in moral philosophy before the modern period.

The notion of moral rules (along with those of a virtue or vice, which will be discussed in Part 2) has been explicitly used in a larger range of cultures than has the notion of a right. Virtually every major ethical or religious tradition employs some notion of a moral rule as well as that of virtue. The notion of a moral rule that specifies what act must or must not be performed, and hence specifies an obligation, is most relevant for the moral evaluation of *acts* in engineering practice and research.

The discussion of rights has a particular prominence in comparatively individualistic societies, such as the United States, which is sometimes described as a "culture of rights" as contrasted with Japan, which is described as a "culture of duties." In a homogeneous society, there is often agreement on what each person owes others. In a pluralistic society there are many subcultures with differing views of the good life – the subcultures may disagree about who should supply or safeguard some aspect of a specific person's welfare, and agree only on what that person is due. Members of different cultures may agree that elderly people deserve certain care, for instance, but differ on the moral duties of family members, churches, communities, and the state in providing such care.

Nonetheless, today when people with diverse political and religious views, such as the representatives to the United Nations, formulate basic moral requirements,

## The Idea of Human Rights Arises in the Eighteenth Century

The view that there are human rights first gained wide acceptance in the eighteenth century Enlightenment. It strongly influenced the U.S. Declaration of Independence, the framing of the Constitution, and the Bill of Rights.

"Modern thought" and "modern philosophy," unlike "modern dress" or "modern architecture," date from the sixteenth century.

they frequently formulate these moral requirements in terms of human rights. This fact shows that the notions of rights and human rights are in broad use today even though the notion of moral rights came into wide use only in the modern period in relatively individualistic societies.

As we have seen, rights are either human rights or special rights (by definition, rights that belong to some people and not all). For example, if Cory and Hilary have a carpool and Cory drives in January with the understanding that Hilary will drove in February, Cory has a special right to be driven by Hilary in February.

Here we have considered moral rights, the distinction between legal and moral rights, and a contrasting pair of categories, "human right" and "special right," such that one or the other will apply to any moral right. In the next section, we will consider three other such pairs that apply to moral rights.

*When competent human beings are recruited for an experiment, such as the testing of a new biomedical device, the accepted ethical standard is to seek their informed consent, even if the risks to the research "subjects" or "participants" are minimal. This means that to enlist subjects/participants in one's research study, one must first give them full information\* about the study and their ability to drop out of the study at any time; the subjects must freely consent to participate. This is understood as necessary to respect their human rights. Do you agree that the right to refuse to participate in an experiment is a human right? Why or why not?*

### Alienable/Inalienable and Absolute/Prima Facie Rights

*Are you aware of any current controversy in which each side claims that the burden of proof is on the other side? This may be the ethical burden of proof or it can be simply the burden of proof as to what is the case.*

*Suppose that the Acme Company asks independent contractor Leslie to write a software program for its back office operations (or to write plans to expand the septic system for their office). After the plans are written and Leslie is paid, Leslie has second thoughts about whether the company should be allowed to do what it plans to do with the software (or septic system plans). What, if any, rights does Acme have vis-à-vis Leslie and what sort are they? What, if any, rights does Leslie have vis-à-vis Acme and what sort are they?*

The contrast between human and special rights is the first of four major contrasts in the categorization of moral rights. In addition to being either human or special, any right must also be either an alienable or inalienable right, a positive or negative right ("a liberty"), and an absolute or prima facie right. Knowing how a right is classified in terms of these four contrasts helps to clarify what the presence of such a right does or does not morally *justify*.

---

\*In some experiments in which it is important that participants be unaware of exactly what is being tested, such information may be withheld, but not information about risks or the general nature of what they will experience.

## Are Rights Overemphasized in the United States?

Many argue that, in recent decades especially in the United States moral rights have been emphasized to the neglect of other ethical considerations. Philosophers such as Annette Baier, John Ladd, Alasdair MacIntyre, and Martha Nussbaum; legal theorist Clare Dalton; and psychologist Carol Gilligan have argued that rights are overemphasized. This overemphasis is held to stem from:

- The assumption that the moral evaluation of an act can be made without reference to the context of social practices in which the act is performed.
- A view of human relationships as being tenuous or adversarial.
- Disregard of the fragility of human existence or of the good.

The objection to an excessive reliance on the concept of a moral right does not imply that the concept has no proper application, however.

Annette Baier agrees with philosophers such as Stanley Benn and Richard Peters that the language of rights is the language of users who are becoming conscious of themselves as individuals.

An **alienable right** is one that a person can trade away. For example, tenants who sublet their apartments to others alienate their right to occupy the apartment for the period of the sublet. You may give up your ownership of a car by selling it, showing that this property right, like property rights generally, is an alienable right. If a right is **inalienable**, the possessor of the right cannot divest herself of the right or trade it away.

The U.S. Declaration of Independence assumes the existence of human rights and sees at least some of them as inalienable. It asserts that all persons are created equal, for all are endowed with certain inalienable (or "unalienable") rights, which include life, liberty, and the pursuit of happiness. The inalienable right to liberty is generally agreed to mean that people cannot make a morally valid agreement to sell themselves into slavery. Although the writers of the Declaration of Independence counted the rights to life and liberty among those rights, the fact that they countenanced not only imprisonment but also execution as punishment for some crimes raises the question of whether they meant that inalienable rights might be forfeited. If they thought that life and liberty are inalienable rights, how could they think it just to imprison or to execute people?

There are two possible explanations. The first is that by saying those rights are inalienable they meant only that the possessor could not trade it away, but meant to allow that the right could be forfeited as a result of the possessor's actions. Another explanation employs the third basic classification of rights, the contrast between absolute and prima facie rights. Just as any moral right is either a human right or a special right, and either an alienable right or an inalienable right, so any right is either absolute or prima facie. An **absolute right** is one the claims of which can never be morally outweighed by other factors. As we saw in Section 5, prima facie (literally, "on first appearance") places the burden of proof on the side of whatever is prima facie. If some statement is prima facie true, it is regarded as true unless and until the contrary is shown. Similarly, **a prima facie right** is a right the claims of which are prima facie, that is, it is possible that they may be morally outweighed by other sufficiently important moral considerations, but the burden of proof would be on those who would argue that the claims of the right are outweighed by other considerations in given circumstances. The signers of the Declaration of Independence may have thought that the inalienable rights to life and liberty were prima facie rather than absolute. If they had thought that the rights were absolute (and believed that others could not remove those inalienable rights for any reason), then they could not have countenanced imprisonment or

## A Minimal Conception of Rights

Philosopher Annette Baier points out that people make claims and give moral justifications for them in every human group with any social organization.[a] Therefore, there are claims with moral justification in every society. Because rights are justified claims, then in every society there is an equivalent to the notion of a right, even in those societies that do not have a ready term for justified claims. Cultures that formulate basic moral considerations in terms of moral rules, obligations, and duties have a concept that functions much as the concept of a moral right, in that the moral requirements they do recognize provide moral justification for certain claims of individuals. For example, "I have a right not to be murdered, because murder is morally prohibited."

Theologian Stanley Hauerwas argues that the language of rights does not do justice to the moral convictions of believers. For example, some religions prohibit murder because of the conviction that only God may legitimately take a human life, rather than because they believe that the right to life is inherent in persons. Nonetheless, Hauerwas accepts the formulation of human rights in Baier's culturally generalized sense.[b]

---

[a] Baier, Annette. 1994. "Claims, Rights, Responsibilities," reprinted in *Moral Prejudices*, 95–129. Cambridge, MA: Harvard University Press.
[b] Personal communication, January 1996.

---

> If a right is **inalienable**, the possessor of the right cannot divest herself of the right or trade it away.

execution, because the claims of an absolute right cannot be outweighed by other considerations.

The right to travel freely illustrates that an inalienable right need not be an absolute right. To say that a person has a right that is inalienable only means that there is always moral justification for that person's claim, but there might be an even greater moral justification for overriding that claim in some particular situation. Consider the right to travel freely; we regard this as a basic liberty and an inalienable right, but it is only a prima facie right. If there is reason to believe that people are carrying a dangerous and highly contagious disease, our society accepts that temporarily overriding the right of the disease-carrier to travel freely by putting them under quarantine is justified. Overriding the claims of some right is not disregarding that right, because *overriding those claims requires showing* that countervailing ethical considerations are more important than the considerations that justify the claim that is to be overridden.

The same point is illustrated by the fact that we regard it as just for people to be fined or imprisoned in some cases, notwithstanding their inalienable right to liberty and to property. No court could justly take away their right to own property, however, or deny them all liberty by enslaving them, even if they were imprisoned for life. Justice also requires that the amount of the fine and the extent of imprisonment or probation must be in proportion to their offense, and not be cruel or unusual.

In cases of quarantine or in a case of the police taking over a private automobile in an emergency, the person whose liberty or property is taken has done nothing to *deserve* forfeiture. For such actions to be just, they must be morally warranted by the importance of the competing considerations – in these cases, protecting the health of the public and coping with the emergency.

Although a prima facie right, such as the right to travel freely, does not trump other considerations, it does establish a burden of proof. Recall from our discussion of burden of proof in connection with obligations that if one bears the burden of proof, the opposing conclusion is accepted unless one can give acceptable reasons or evidence in support of one's conclusion. (As an illustration of the burden of proof concept, consider how U.S. law sets the burden of proof in a criminal

> An **absolute right** is one the claims of which can never be morally outweighed by other factors. **A prima facie right** is a right the claims of which are prima facie, that is, it is possible that they may be morally outweighed by other sufficiently important moral considerations, but the burden of proof would be on those who would argue that the claims of the right are outweighed by other considerations in given circumstances.

case on the prosecution and assumes the accused is innocent until proven guilty.) In the case of the burden of proof arising from a prima facie right, one would need to give reasons, argument, and perhaps evidence that an infringement of that right would be justified in certain circumstances. (The burden could be shifted back by showing that the reasons given were inadequate or the evidence was flawed.) The question of where the burden of proof lies is a very important one in ethics. Often, when two parties who stand on opposing sides of an issue are reasonable, they will grant that the considerations that the other party cites are of some moral significance. They come to different conclusions, however, because each sees the issue as one in which the burden of proof is on the other side and hold that the considerations brought forward by the other side are not weighty enough to discharge that burden of proof.

*Look in the newspaper and other news reports for a current controversy where each side claims that the burden of proof is on the other side. This may be the ethical burden of proof or it can be simply the burden of proof as to what is the case. What does each side claim the other party still needs to show?*

> An inalienable right need not be an absolute right.

Because an **absolute right**, in contrast to a prima facie right, is one whose claims, ethically speaking, must be honored in all circumstances, no circumstances would ethically justify overriding the claims of an absolute right. Earlier in this section, we considered the view that moral rights always override other moral considerations. If that were so, all moral rights would be absolute.

Instances of rights that are widely agreed to be absolute rights are few. The right to be free of torture, rape, and other sexual violation, and the right of people to refuse to participate as subjects in research studies are commonly offered as examples of absolute rights. Some also regard the right to life as an absolute right and so regard all homicide as unjustifiable. (Note that *if people were to violate some right for any reason other than an ethical reason, their statement would not count as an attempted ethical justification for overriding the claims of that right*. They would be saying that they placed other considerations over ethical considerations.)

Most of the commonly considered rights are regarded as prima facie rights. For example, the right to travel freely, the right to control property that one owns, the right to drive, and the right to service according to one's place in line are (moral or legal) rights that can be justly overridden under certain circumstances. Nonetheless, the circumstances that would justify overriding a right are common for some prima facie rights, but rare for others. A right is prima facie rather than absolute if there *might be* other considerations that would outweigh the claim of that right, but it may not be easy to provide such weighty considerations.

Often multiple criteria are relevant to judging whether the claims of some prima facie right can be justly overridden. For example, a **copyright** is a special, alienable, prima facie legal property right of the holder; specifically it is the right to prevent others from making copies of the material to which one holds the copyright. "Fair use" is a *justified* exception, that is, copying that meets the criteria to be a "fair use" is morally (and legally) acceptable. As we shall see when we examine intellectual property in Chapter 6, four different sorts of considerations bear on deciding whether some copying has discharged the burden of proof placed by the *copyright* to be justified as a fair use. Some of them, such as whether the use significantly reduces the sales, licenses, etc. from which the copyright holder would normally profit, have more weight than others. Nonetheless, all must be considered in deciding the question of whether some copying is justified, and there is no simple algorithm or recipe for making that decision. Thus, the application of the concept of fair use to copyrighted material requires *judgment.*

> *Suppose that the Acme Company asks independent contractor Leslie to write a software program for its back office operations (or to write plans to expand the septic system for the office). After the plans are written and Leslie is paid, Leslie has second thoughts about whether the company should be allowed to do what it plans to do with the software (or septic system plans). What, if any, rights does Acme have vis-à-vis Leslie and what sort are they (i.e., how should they be categorized)? What, if any, rights does Leslie have vis-à-vis Acme and what sort are they (i.e., how should they be categorized)?*

### Thinking about Digital Rights Management (DRM)

"Digital rights" is a widely used term for certain property rights. Today in computer fields, "digital rights management" (DRM) receives much attention. DRM refers to the measures that owners may take to protect their (special alienable) property rights in a digital recording or piece of software and prevent users from making unauthorized copies of them. Controversy surrounds the question of what form DRM should take. Some measures that have been taken to protect property rights to digital information have interfered with the licensee's equipment. The use of the term "rights" in the term "digital rights" stakes an ethical or legal claim. The first question is about the justification for such a claim. Presumably, it rests on the claim to legal ownership of the intellectual property embodied in the digital information that others might seek to copy. Understanding the justification for such claims is a first step to understanding the scope and limits of what protection of them might be justified.

One point that may cause confusion is that alienable special rights are sometimes called "privileges" and contrasted with rights! For example, one may hear that "driving is a privilege and not a right." By this is meant that one's license to drive (i.e., one's *special right* to drive) is alienable and one can lose it if one abuses the right. Sometimes people contrast a privilege with a right when they wish to emphasize the power of some granting agency to grant or remove the right as they see fit, perhaps with no possibility of appeal of their decision. For present ethical purposes, **"privilege"** means "an alienable special right."

As we have seen, rights are *justified claims*, and those claims may or may not be pressed. Rights can be either exercised or waived. One may fail to exercise a right for many reasons, including not getting around to it. For example, if you obtain a driver's license you may decide that you do not want to do any driving and not exercise that right. If one *acts* voluntarily to give up the claim on some occasion, one **waives** the right. The question of whether one waives a right usually arises when the exercise of that right comes into conflict with something else – for example, the need of some person standing in line to be served out of turn.

From an ethical point of view, it is crucial for professionals to distinguish between those clients, patients, and students who wish to waive or choose not to exercise some right and those who do not realize that they have the right in question, or who do not know how to go about exercising the right. To waive a right, a person must be aware of the right and choose not to exercise it. In some cases, others, usually practicing professionals, have an ethical obligation to inform people of their rights. Situations that are unfamiliar to most people, or situations that they do not enter willingly, are situations in which people are likely to be ignorant of their rights. Being a patient, being arrested, and being accused of some wrongdoing are examples of such situations, and so there are often provisions to ensure that people know their rights, so they can make a decision about whether to exercise those rights.

Prominent among recently recognized employee rights is the right to know the nature of the occupational hazards to which one is exposed. This right is comparable to a patient's right to be informed of the risks associated with health care and to refuse that care if they wish. Right-to-know legislation requires that workers be informed of the hazards associated with their jobs. The intent of these requirements is to enable workers to make informed choices both about which jobs to accept and how to reduce their risks from hazards.

Recall what makes some claim a *moral (or ethical)* right: When there is *moral* justification for some claim, then that person has a *moral* right. From this definition, we see that in order for a person to have some moral right, the person's claim must be morally justified.

A person's claim (usually) continues to be morally justified even if that person chooses to waive the right in some circumstances. The decision to waive a right in some circumstances does not mean that one waives it in others. For example, in the United States students have a legal right to see records concerning their performance. A student may waive the right to see a particular letter of reference, but the general right remains in force and may be exercised with respect to other material. However, certain rights, such as the legal right to keep others off one's land, are forfeited if the right is not exercised for a given period. The justification for this forfeiture is that if people make a habit of using a path through one's property, they grow to depend on it and make other plans and commitments on the expectation that they can use it.

Consider whether an inalienable right would have to be exercised. That a right is inalienable means that the person's claim is always justified, but not that the claim must always be pressed. That is, the right does not have to be exercised by the person who has the right, even if it is inalienable.

## Infringement of a Right Needs Justification

Note that if a right is infringed, this puts a burden of proof on the party doing the infringing to show that the infringement was justified and so not a violation of the right. Although an infringement may be justifiable, infringement calls for justification.

## The Term "Positive" in U.S. Law

Although the distinction between a negative right or liberty and a right to obtain something is recognized in U.S. law, legal scholars regularly use the adjective "positive" applied to law, in a rather different sense. "Positive law" means law that has been enacted, that is, statutory law or legislation.[a] This is in contrast to case law or common law. Perhaps for this reason philosopher Philippa Foot has coined another term for positive or affirmative rights. She calls them "claim rights."[b]

---

[a]Nolan, Joseph R. and Jacqueline M. Nolan-Haley. 1990. *Black's Law Dictionary: Definitions of Terms and Phrases of American and English Jurisprudence, Ancient and Modern*, sixth edition St. Paul, MN: West Publishing Co.
[b]Foot, Philippa. 1977. "Euthanasia," *Philosophy and Public Affairs* 6(2): 85–112.

Consider rights of **confidentiality**. These are rights to have one's confidential information shared only within a certain restricted group. For example, it is common for engineering firms designing a new manufacturing facility or computerized billing system for a client to learn some of the client's business plan. The client's right to confidentiality requires that members of the engineering firm share the client's information among themselves only as necessary to complete their work. A right of confidentiality is a special right and imposes a special obligation on others not to disclose the information outside the defined group. Some disclosure of confidential information is always wrong. For example, it is always wrong for medical personnel to discuss a patient's case in public areas, such as hospital elevators where others may overhear, even if those overhearing the discussion did not understand it. Nonetheless, the right of confidentiality is not generally understood to be an absolute right. As we say, engineering codes of ethics commonly say that the engineer's obligation to ensure the public safety generally overrides the client's right of confidentiality and so imply that if the *only* way to ensure the public safety is to disclose the client's information, doing so is justified.

When the claim of some right is not met, it is common to say that the claim (and the right) is **infringed**. For example, if Alex refuses to return Bea's car keys to her because Alex thinks Bea is too drunk to drive, Bea's right to her car has been infringed. If a moral wrong is done in infringing a right (i.e., if there are no adequate moral reasons for infringing the right), it is said to be **violated as well as infringed** (although usage does not consistently distinguish infringement from violation). For example, if A refuses to turn over B's car keys simply because A is in a bad mood, A has violated B's right to the use of her property. If an absolute right is infringed, it is necessarily violated, because the claims of an absolute right by definition *cannot* be justly overridden. Notice, however, that even if a prima facie right *were* infringed, *the burden of proof would be on the infringers* to justify that infringement.

### Negative/Positive Rights

*Does a person's right to travel freely require the existence of a free public transit system? Why or why not?*

A fourth and final distinction among rights is that between **negative rights** or **liberties** and **positive or "affirmative" rights. Negative rights** or **liberties** require of others only that they not interfere with or restrict the rights-holder. **Positive or** "affirmative" rights are claims to receive something. To respect others' negative rights requires only that one not interfere with the person's exercise of the right in question. One need not provide the rights-holder with particular opportunities to practice the right in question. Examples of these negative rights include a person's rights to free speech and to religious expression. In the case of positive rights, it is not enough to leave the rights-holders alone; something must be done for them. Usually some goods or services must be supplied. If you pay for the future delivery of an automobile, you have a positive right to the automobile and the seller has a positive obligation to provide it to you. Your right to life, on the other hand, is a negative right, that is, everyone else must refrain from killing you – a universal negative obligation. Your right to life does not impose on others any positive obligation to *save* your life, although saving your life would usually be morally praiseworthy.

**Negative legal rights**, which are the legal counterpart of moral liberties, specifically forbid acts of interference (including threats of harm for exercising one's right) – for example, the right to remain silent. **Positive legal rights** are legally warranted claims to *receive* something – for example, a due process proceeding, or compensation for past injury.

The distinction between positive and negative legal rights is an important one in U.S. law. Although many legal rights are also moral rights, the two notions are not coextensive. For example, people have a moral obligation to keep their promises, although not all promises are covered by legal statute. Furthermore, as the former laws upholding slavery illustrate, legal and moral standards may conflict.

It is often held that all people have a right to certain necessities, and thus, a society that is able to provide them is obliged to do so. Such rights are called **economic rights**, as contrasted with **political rights**. Economic rights are positive human rights. Political rights include physical liberty, or the right to travel freely; freedom of association; and freedom of speech. These examples are all negative rights/freedoms/liberties, and may suggest that political rights are negative rights – that is, they require only that others not interfere with the rights-holder's activities. However, some political rights, such as the right to vote, require the provision of services, in this case services that ensure the secrecy of one's ballot choices and security of the ballots to ensure that they are counted accurately.

Positive rights to health care and to education are often held to be basic human rights. The nature and extent of a right to health care are now

## Political and Economic Rights

The relative importance of political rights, the power to participate in the establishment or administration of government, as compared with economic rights, is widely debated. For example, is a country ever justified in curtailing political rights to make major improvements in its citizens' economic well-being? Democracies, by their very nature, emphasize political rights. Some democracies, like Sweden, also emphasize economic rights.

## What Is the Basis for a Right to Education?

One argument for public education is a rather different argument from one based on a human right to education; the survival of a democracy requires the education of citizens so that they can participate in the democratic practices.[a] This argument says that a democracy, if it is to survive, must educate its citizens, so any individual's right to an education at public expense stems from what a democracy does in its own self-interest.

[a] A recent formulation of this argument is found in Amy Gutmann's *Democratic Education* (Princeton, NJ: Princeton University Press, 1987).

## A Right to Life Contrasted with an Obligation to Live

The distinction between positive and negative rights helps to clarify some common confusions, for example, one about the right to life and suicide. The view that suicide, or assisting suicide, is wrong requires more than a belief in the right to life. The right to life, even if an absolute right, could nonetheless be waived. Hence, having a right to life does not by itself imply that one also has an obligation to live.

Suicide, and requesting or assisting in euthanasia, would be compatible with a right to life, as long as the euthanasia was performed at the uncoerced request of the person whose life was at stake. This point is often obscured because the belief that it is always wrong to take any person's life, including one's own, is often wrongly described as a "right to life" position. The view is better described as a belief in the sanctity of (human) life so that one is obligated not to destroy any human life, including one's own.

widely debated in the United States. Many people also claim that everyone has the right to a basic education, and indeed public education in the United States is legally mandated for all. Furthermore, people with disabilities have a legal right to education in the least restrictive environment possible. This innovation illustrates how views of the scope of positive rights continue to evolve.

The rights that were taken to be principal human rights at the end of the eighteenth century – the rights of life, liberty, and the right to own property – were taken to be liberties, not positive rights. What was prohibited was the interference with the *continuance* of another's life, freedom to travel, pursuit of happiness, or retention of property. The Declaration was not taken to imply a general moral obligation to save other people's lives, ensure their liberty, promote their happiness, or provide them with property.

In summary, consider what is at issue in the contrast between:

1. Alienable and inalienable rights;
2. Human rights and special rights;
3. Negative rights or liberties, and positive rights;
4. Absolute rights and prima facie rights.

The first contrast turns on whether or by what means (e.g., only by forfeiture) the right may be removed from the person, the second on whether the right belongs to all people, the third on whether the claim of the right is to receive something or just to be left alone, and the fourth with whether it can ever be just (morally acceptable) to override the claims of that right. The relationship between additional moral rights and their counterpart moral rule or obligation is summarized in Figure I.9.

*Does a person's right to travel freely require the existence of a free public transit system? Why or why not?*

One can ask all four of the questions shown in Figure I.10 about any right. The answers will classify that right along the four dimensions shown. For example, a student's right to the privacy of her college transcript is a special inalienable negative right. I think it is prima facie rather than absolute, although no plausible instance in which it might be justly overridden comes to mind.

Moral rights, along with moral obligations, moral rules, moral responsibilities, and moral standing, constrain how far it is ethically permissible to go in seeking to bring about a good outcome. For example, suppose you find yourself in some sort

**Moral Right (possessed by A)**  ➡  **Moral Rule or Obligation (on others)**

A has an absolute (negative) right to do X.  ➡  Do not interfere with A doing X, no matter what.

A has a prima facie (negative) right to do X.  ➡  Do not interfere with A doing X, unless other considerations justify it. In that case, minimize the interference.

A has a (positive) right to receive S from B.  ➡  B is obligated to supply S to A.

If Q, then A has a (negative) right to do Z.  ➡  If Q, then everyone is obligated to refrain from interfering with A doing Z.

. . . .                                                                  . . . .

**Figure I.9**  More Moral Rights and Their Counterpart Moral Rules or Obligations

1. <u>Who has this right?</u>
   All moral agents (are people the only moral agents?)  →  human right
   To some people (because of their special characteristics)  →  special right
   Such special characteristics, as
   • particular agreements—e.g., A right to what was promised
   • membership in special groups—e.g., The right to wear the uniform
   • special relationships—e.g., Parental rights
   • specific work/knowledge—e.g., Right of researchers who are reading a research article to know about any special hazards in conducting the reported experiments, so they will be forewarned should they want to replicate them
   • due to the actions of others—e.g., Right to confront one's accusers
   • possession of privileges—e.g., Right to receive a government-issued picture I.D.

2. <u>Can the holder trade away the right?</u>
   Yes  →  alienable right
   No  →  inalienable right

3. <u>Can the claims of the right *ever* justly be overridden (perhaps often and easily or perhaps very rarely and only in highly unusual circumstances)?</u>
   Yes  →  prima facie right
   No  →  absolute right

4. <u>Does it require that the holder of the right receive something or is it simply free of interference by others?</u>
   Receive something  →  positive or affirmative right
   Be free from interference  →  negative right

**Figure I.10**  Summary of Categories of Ethical/Moral Rights

## Categories of Rights, Obligations, and Moral Rules

These distinctions are more frequently applied to rights than to moral rules and obligations, and, as we saw, the counterpart to the human-special distinction regarding rights is a universal-special distinction with regard to obligations and moral rules. Only for moral rules and obligations does the question of whether they are prima facie or absolute arise, and the negative-positive distinction is sometimes relevant, but the alienable-inalienable distinction is not applied to obligations and moral rules.

of emergency where you can act to save one person's life or act to save the lives of four other people, but you do not have time to do both. (Other things being equal) you ought to save the four people, rather than one. However, the greater value of four lives as compared with one would not allow you to violate one person's right to life by harvesting that person's organs and transplanting them into four people who each need the organs to survive.[15] In contrast, although both human life and great art have value, art does not have moral rights or moral standing. Therefore, it might be justified to destroy one great painting to save four others – for example, by using the first to wrap the other four.

*Classify the following rights according to the four distinctions summarized in Figure I.10:*
*The right to be served according to one's place in line*
*The right to bring fraud and waste of public funds to public attention*
*The right to travel freely within the country*
*The (legal) right not to have to testify against oneself*

## Section 7.   Rights of Privacy/Confidentiality and Intellectual Property

### Rights of Privacy and Confidentiality

*The Family Educational Rights and Privacy Act of 1974 As Amended (FERPA) protects student grade information. To ensure compliance with this law, universities instruct their faculty members to refrain from publicly posting grades or leaving graded material outside of office doors for students to claim. Do you agree that grade information is private information? Why or why not? Because college students are (mostly) over 18 years of age, and so (mostly) legal adults, their parents also do not have a (legal) right to their grade information. Some parents believe they have a moral right to this information, at least if they are paying some of the son or daughter's tuition, and make it a condition of paying that tuition that the son or daughter give the university permission to send the grades to the parent. Do you think that parents making grade revelation a condition of tuition payment is morally justified? Why or why not?*

As we saw in the last section, human rights reflect moral claims people need to have honored if they are to act as moral agents and be accountable for their actions. Often included among these basic rights are the right or rights of privacy.

The **right of privacy** is, roughly, the right to freedom from intrusion, especially in matters that affect people's ability to act as moral agents, and be accountable for their behavior. (As such, it is a negative right.) Privacy is a notion that receives more attention in individualistic cultures, however. As mentioned earlier, some cultures do not have a word for this notion although they usually would find objectionable those acts that people in Western democracies would describe as

---

[15] I am indebted to Judith Jarvis Thomson for this example.

## The Origins of a Right of Privacy

The question of when the right of privacy was first conceptualized as a legal right in the United States at least and whether there is a right of privacy guaranteed in the U.S. Constitution is controversial. The word "privacy" does not appear in the Constitution or early amendments to it. Some argue that Article IV (and perhaps even Article III) of the "Bill of Rights" (the first ten amendments to the Constitution) designates certain spheres (one's person, one's papers, and one's home) as deserving special protection from (government) intrusion or seizure, however. Article IV states: "The right of the people to be secure in their persons, houses, papers, and effects, against unreasonable searches and seizures, shall not be violated, and no Warrants shall issue, but upon probable cause, supported by Oath or affirmation, and particularly describing the place to be searched, and the persons or things to be seized." [*Capitalization as in the original.*]

invasions of privacy. They would describe the offensive behavior in other terms, such as "rudeness."

The claim to privacy finds moral justification in the recognition that in order to function as moral agents people need to have control over matters that intimately relate to them. Therefore, privacy is closely connected to the right of self-determination. How one shows respect for another's privacy varies from culture to culture, as does any tradeoff between respecting privacy and pursuing other ethical values. In some cultures, parents oversee their children's affairs more than in others. Thus, in a traditional Chinese culture, parents expect to do such things as read mail addressed to their adolescent children as part of their responsible oversight of the children. In contemporary Anglo-American culture, such acts would be viewed as intrusions of the adolescent's privacy.

Privacy has become a larger issue with the rise of technology and the specific threats to privacy posed by particular technologies. Advances in surveillance technologies and technologies that can be used for intentional invasions of privacy are only one type of threat. As we shall see in Chapter 5, computers and other information and communications technology make it possible to collect, correlate, and transmit quantities of personal information in ways that previously were impossible. Such comprehensive correlations create dossiers of information on people that raise new issues of privacy. Furthermore, scientific discoveries and technological innovations have made possible the acquisition of new *types* of "private" information, such as genetic information. (Some of the special ethical problems concerning genetic information arise because it is information not only about the person sampled but usually also information about that person's biological relatives.) Many questions of employee privacy arise in the twenty-first century workplace involving communications technologies, including privacy of email, telephone, and fax communications, and drug testing of employees.

It is common to distinguish among physical, informational, and decisional privacy.[16] **Physical privacy** is a restriction on the ability of others to experience a person through one or more of the five senses. **Informational privacy** is a restriction on facts about the person that are unknown or unknowable. **Decisional privacy** is the exclusion of others from decisions, such as health care decisions or marital decisions, made by the person and her group of intimates. Philosopher and legal theorist Anita Allen argues that **dispositional privacy**, which excludes

[16] Allen, Anita L. 1995. "Privacy in Health Care." In *Encyclopedia of Bioethics*, 2nd ed. New York: Macmillan: 2064–2073, 2065f.

## What Is a Privacy Violation?

The preceding discussion of privacy adopts the usual conception of privacy violation as an invasion directed toward an individual and harming that individual in some way.

In Chapter 5, we will take up questions of privacy again in the specific context of digitalized information and the aggregation of data that it makes possible. Some organizations, such as the Electronic Frontier Foundation, consider the acquisition of data itself a privacy violation. Others, such as NAE past-president Bill Wulf,[a] argue that it is what one *does* with such information that may be a privacy violation.

---

[a] Bill Wulf, personal communication, June 16, 2010.

others from knowing one's dispositions or states of mind, should be thought of as distinct from the first three.[17] As an example of how dispositional privacy differs from informational privacy, consider the Lotus Corporation's proposal in the 1980s to offer copies of its MarketPlace Data Base for sale. That database contained extensive information about the consumption patterns of large numbers of people. This proposal was widely criticized as an intrusion of the privacy of those profiled, *even though the individual items of information aggregated were not the sort of information that is considered private.*[18] (As we shall see in Chapter 5, Computers, Software, and Digital Information, currently existing databases go even further than Lotus had proposed, but those companies have less concern about the ethical consequences of such databases than did Lotus, and its head, Mitch Kapor, who also founded the Electronic Frontier Foundation.)

Although taken individually the component items of information in the Lotus aggregation did not give dispositional access, the aggregated data offered by Lotus did give information about what people were disposed to do. Specifically, it gave information about their consumer preferences. The Lotus example shows the greater adequacy of Allen's scheme: If knowing a person's dispositions is a species of privacy invasion, then the aggregation of nonprivate facts can result in a privacy invasion – taken collectively, those nonprivate facts do reveal a person's dispositions.

Once the desired levels of privacy protection have been determined, the question of how to ensure that level of privacy is a matter of security. The **security** of a system is the extent of protection afforded against some unwanted occurrence such as the invasion of privacy, theft, the corruption of information, or physical damage. Conversely, keeping some information private may enhance security. For example, the goal of measures to safeguard the informational privacy of children online is to safeguard children from predators. As the threat of terrorism has

---

[17] Allen, Anita L. 1987. *Uneasy Access: Privacy for Women in a Free Society*. Totowa, NJ: Rowman and Allanheld, 15–17.

[18] "Lotus – New Program Spurs Fears Privacy Could be Undermined," *The Wall Street Journal*, November 13, 1990, p. B1 and "Lotus is Likely to Abandon Consumer-Data Project," *The Wall Street Journal*, January 24, 1991, p. B1. It is important to distinguish information that is private (rather than public) from information that is personal in the sense that it is information that would be intrusive for others to demand, obtain, or discuss. For example, in the United States in the early part of this century some people (especially women) rarely disclosed their age. They considered this information highly personal even though their birth dates were matters of public record. The judgment of what matters are personal is highly cultural. For example, the Dutch consider it intrusive to look over the books in a person's bookshelf without first asking permission. In some cultures, it is considered impolite to speak of a woman's pregnant condition even when her pregnancy is evident.

increased, a different relationship between privacy and security has emerged in which citizens are asked to give up some physical or informational privacy, for example, have inspectors scan or search their person at airports in order to be more secure against terrorist attack.[19]

Complete secrecy is one strategy to preserve informational privacy. Another is to keep sensitive information *confidential.* Information kept **confidential** may be shared only within a restricted group, usually those who have a need to know the information to be able to perform their part of some joint task. For example, information in a patient's medical record is confidential and properly shared by health care workers caring for that patient. Those health workers are bound not to disclose the information outside of the health care team except for such purposes as insurance reimbursement. Should someone, say a biomedical engineer, wish to report on a clinical case at a conference, that engineer must first remove identifying information about the patient unless the patient explicitly agrees to reveal some of it. (For example, a patient might allow the use of a photograph of some part of the patient's body that showed signs of some medical condition.)

*The Family Educational Rights and Privacy Act of 1974 As Amended (FERPA) protects student grade information. To ensure compliance with this law, universities instruct their faculty members to refrain from publicly posting grades or leaving graded material outside of office doors for students to claim. Do you agree that grade information is private information? Why or why not? Because college students are (mostly) over 18 years of age, and so (mostly) legal adults, their parents also do not have a (legal) right to their grade information. Some parents believe they have a moral right to this information, at least if they are paying some of their son or daughter's tuition, and make it a condition of paying that tuition that the son or daughter give the university permission to send the grades to the parent. Do you think that parents are morally justified in making grade revelation a condition of tuition payment? Why or why not?*

*Does FERPA legislate secrecy or confidentiality of student grade information?*

### Intellectual Property Rights
*Are there circumstances in which it would be reasonable for an inventor to keep the invention (or the formula or design for it) a trade secret rather than seek to patent it?*

*What is it about (some) software that makes it, and nothing else, both patentable and copy-rightable?*

A clause of the U.S. Constitution provides for encouraging the development of science and the useful arts by granting to authors and inventors a time-limited exclusive right to their writings, discoveries, and inventions, and federal statutes specify those rights, which are legal rights, and for that reason, moral rights as well.

In engineering practice, a common reason for holding information confidential is preservation of rights to intellectual property, roughly intellectual creations that are subject to ownership. (We will examine Locke's theories of property and various views for and against extending the concept of property to intellectual

---

[19] At least one prominent writer has argued that terrorism is now a source of design constraints for engineers and architects. See Henry Petroski, 2004, "Technology and Architecture in an Age of Terrorism," *Technology in Society* 26(2–3): 161–167.

creations in Chapter 6.) An example of how confidentiality might protect property rights is the use of confidentiality to protect a company's trade secrets and its business plan.

A **trade secret** is a device, method, or formula used in one's business that gives one an advantage over the competition and that must be kept secret to preserve that advantage. The formula for Coca-Cola is a trade secret. The courts of the various states protect trade secrets as a form of intellectual property if the holder of a trade secret takes sufficient precautions to keep the secret secure. (If they reveal it to others, they have only themselves to blame.) Others are legally restricted from wrongfully taking, using, or disclosing the trade secret. Disclosure of confidential information by a previous employee or contractor is an example of a wrongful disclosure of a trade secret. Acceptable means of learning a trade secret include reinventing it or learning it through reverse engineering of a purchased product. If the secret can be readily learned by acceptable means, secrecy is a poor way of restricting use of an intellectual creation.

Copyrights, patents, and trademarks give legal protections in the form of property rights to those creations that, unlike trade secrets, are fully disclosed. The disadvantage of copyrights and patents is that they are time-limited, although recent legislation has made copyright renewable to such an extent that many current works may never come into the public domain. (Items are in **the public domain** when they are not or are no longer under copyright and so may be copied and republished without permission. The works of Shakespeare are examples of works in the public domain that are commonly republished.)

A **patent** is a special, alienable legal right granted by the government to make, use, and sell, or at least (when the patent holder's use of the new patent would infringe other patents) to bar others from making, using, or selling a device, design, or type of plant that one has created. In the United States, this is 20 years for useful devices, and 14 years for designs. (Here a "design" means a "new, original, and *ornamental design* for an article of manufacture."[20]) In applying for a patent, one makes the nature and details of one's invention explicit. In the United States, to be eligible for a patent an application must be initiated within one calendar year of "public disclosure" of the idea. The European and Japanese patent laws require application *before* public disclosure. Most countries honor the patents taken out in other countries, however. Although the name(s) of inventor(s) always appear on patents and they can always claim *credit* for their inventions, the *property rights* (ownership) may be assigned to others. Ownership rights to a created work are distinct from the *credit* that belongs to an author, composer, artist, or inventor. The ownership rights, like other property rights, are alienable, but the credit due an author, composer, artist, or inventor is always due that person. Both ownership rights and the credit are ethically significant. Codes of ethics of professional engineering societies require both general crediting of intellectual work, and honoring property rights or "proprietary interests." For example, the 2006 NSPE Code of Ethics specifies the following as the ninth of engineers' professional obligations:

[20] U.S. Patent Office Web site, http://www.uspto.gov/ (last modified November 8, 2010).

Engineers shall give credit for engineering work to those to whom credit is due, and will recognize the proprietary interests of others.

> Ownership rights to a created work are distinct from the *credit* that belongs to an author, composer, artist, or inventor. The ownership rights, like other property rights, are alienable, but the credit due an author, composer, artist, or inventor is always due that person.

To patent a device one must prove that it is useful, novel, and not obvious. Therefore, not all inventions are patentable. If some invention is not patentable, that is a reason to try to hold it as a trade secret instead.

To establish that a patent is valid, it must often survive a court challenge. Obtaining a patent is a significant expense, and fighting or defending one in court is extremely costly, often costing a million dollars or more in legal fees. This is another reason that many inventors either try to protect their intellectual property in other ways or hold the patent jointly with some organization that has a legal staff.

A **trademark** is an officially registered name, symbol, or representation the use of which in commerce is legally restricted to its owner.

A **copyright** is a legal right to exclusive publication, production, sale, or distribution of some work. As we saw, it is a special, alienable, prima facie legal property right of the holder; specifically it is the right to prevent others from making copies of the material to which one holds the copyright. Copyright protection extends to "nonliteral" copying, which is copying that is paraphrased rather than word for word. A copyright is most commonly held by the author, composer, or publisher of a work, but may be assigned to others or inherited. Most of the law of copyright is developed in the context of literary works such as novels, plays, and films. (Today the term "**literary works**" includes computer databases and computer programs, however.) Notice that the copyright holder, like the owner of a patent, need not be a party who deserves credit for creating the work.

The intellectual property protected by the copyright is the "expression," not the idea. For example, one can copyright the architectural *drawings* for a house, not the *house layout*. Ideas cannot be copyrighted. Taking credit for another's idea *is plagiarism* – the appropriation of another's ideas or writings and representation of them as one's own – so copyright violation fails to be the legal counterpart to plagiarism.

Under the U.S. 1976 Copyright Act, copyright protection of a work begins as soon as a work exists in a concrete form and remains in effect until 50 years after the death of the author. In 1998, the Digital Millennium Copyright Act (DMCA) made changes to copyright protection, including the extension of copyright protection by 20 years and made provisions for further renewal of the right. This act and other issues of intellectual property will be discussed at length in Chapter 6, but its provisions are incorporated into Table I.1, which applies to works created at the time of this writing (previously written works may be covered by laws in effect when they were created).

*Are there circumstances in which it would be reasonable for an inventor to keep the invention (or the formula or design for it) a trade secret rather than seek to patent it?*

*What is it about (some) software that makes it, and nothing else, both patentable and copyrightable?*

Table I.1  Summary of categories of intellectual property recognized in U.S. law

|  | Patent | Copyright | Trade secret | Trademark |
|---|---|---|---|---|
| Applied to | Designs or useful devices/processes | Literary/artistic works and some software | Processes, plans, or devices | Name/symbol |
| Term length | 20 years (designs-14 years) | 28, renewable up to 95 years | No time limit | Indefinite |
| Application required | Yes & qualification (e.g., sufficiently original) | Registration is automatic | Does not apply | Yes |
| Disadvantages | Must qualify & the right is time-limited | Applies only to certain sorts of works Time-limited, but now renewable | Must protect against disclosure | Must determine that it is novel |
| Advantages | A period of legal protection | A period of legal protection | No qualification or time limit | Legal protection |

## Ethics, Conscience, and the Law

*What is the difference between breaking a law as an act of civil disobedience and breaking a law to test its constitutionality? What is the difference between breaking a law to test its constitutionality and simply violating that law? What three conditions are necessary to make some violation of law an act of civil disobedience?*

In the earlier discussion of moral rights, we distinguished **moral rights** from **legal rights**. A person may have a legal right to do something, but not a moral right to do it. There are many historical examples of unjust laws, and some laws today are said to favor special interests. Does law have *no* moral force?

The view that seeing the influence of particular interests on the formulation of legislation can lead one to doubt the moral authority of the law is expressed in the sentiment that if one respects the law and likes to eat sausage, one should never watch either one being made, a view widely attributed to Otto von Bismarck. Most societies attempt to have laws that are just and morally sound, notwithstanding the multiple influences on specific pieces of legislation. Many specific provisions of laws are neither just nor unjust when considered by themselves. For example, requiring everyone to drive on the right side of the road is not morally superior to requiring everyone to drive on the left side of the road, although it is certainly generally beneficial and therefore justified to make laws that require everyone to drive on the *same* side of the road. Societal standards embodied in law help people know what they can expect from others and so foster the trust necessary for people to confidently engage others in public and private endeavors. The entire system of legislation (making law) and judging cases (applying laws) is supposed to be just, but it is fallible. At least in functioning democracies, law has moral authority even if particular laws may have no moral significance, may be poorly written (so they do not accomplish what they were intended to), or even favor some person or group without justification. The *burden of proof* is on any person who claims that some law is unjust or that one ought not obey it. Ethically speaking, people are expected to be law-abiding except where they can show very good reasons for thinking some law unjust or immoral.

Openly breaking a law to test its constitutionality respects the legal system, but uses part of the legal system, the court system, to test whether the law is consistent with other law, including the constitution of the country (or, in the United States, of the state). Breaking a law to raise some moral objection to either the law one has broken or some other law or policy is called "civil disobedience." For an act to count as **civil disobedience**, it must break the law publicly and nonviolently in an attempt to draw public attention to an injustice. Civil disobedience aims to bring about a change, often a change in the law that was broken. It requires a willingness to undergo whatever punishments the law provides for those who break that law. The term "civil disobedience" is applied even when the alleged injustice that a person protests is not the injustice of the law that the person violates, but another law or legally sanctioned activity that one believes to be unjust. Recent examples include trespass on the grounds of nuclear weapons facilities and abortion clinics. Henry David Thoreau's 1849 essay on civil disobedience is a classic statement on the subject.[21]

It is also possible to change a law by a public protest that breaks no laws but attempts to use nonviolent means to draw attention to a perceived injustice, usually injustice in some law. **Nonviolent protest**, such as the Alabama bus boycotts that protested segregated busing, uses many of the same methods as civil disobedience but may not break any laws.

**Conscientious refusal** is a second related notion. Examples of conscientious refusal include refusal to carry out work or a military order that one believes to be immoral, and the refusal by those concerned with animal welfare to eat or wear the skins of higher animals or use products that have been tested on them. It can occur in work or nonwork situations and may or may not involve breaking any law. It may be done either simply from a motive of not participating in what one sees as a moral wrong or it may be done with the hope of making a public protest that will draw attention to the situation one believes is wrong.

Finally, there is **outright breaking or evasion of the law on grounds of conscience**. Those individuals who refused to identify and turn over Jews, homosexuals, or gypsies to the Nazis for extermination broke the law on grounds of conscience. Such law-breaking is covert (as contrasted with the publicity of civil disobedience, nonviolent protest, and conscientious refusal). It is morally justified only under conditions in which public protest would certainly be futile and a grave wrong is done if one complies with the law.

Sometimes all of the actions discussed here are loosely referred to as "conscientious objection" or "civil disobedience," but if this is done, it is still important to distinguish between what are here called nonviolent protest, conscientious refusal, and a conscientious attempt to evade the law, because there are relevant ethical differences in the conditions that justify their performance.

*What is the difference between breaking a law as an act of civil disobedience and breaking a law to test its constitutionality? What is the difference between breaking a law to test its constitutionality and simply violating that law? What three conditions are necessary to make some violation of the law an act of civil disobedience?*

[21] Thoreau, Henry David. 1894. "Civil Disobedience," reprinted in *Civil Disobedience in Focus*, edited by Hugo A. Bedau. New York and London: Routledge, 1991, 28–48.

The purpose of this introduction has been to present basic ethical terms and distinctions that are used in the discussion of many ethical questions. These terms and distinctions are intelligible from a variety of cultural and religious perspectives, although some assume the context of a democracy. Concepts with application to specific contexts of engineering practice and research will be introduced as needed in later chapters.

# 1 Professional Practice in Engineering

## Professions and Norms of Professional Conduct

You chose engineering with the hope of being able to address the need for energy sources that do not pollute the environment or contribute to climate change. Your interests have brought you to a project that addresses the fundamental drawback to solar energy: the lack of a cheap and efficient way to store that energy. Your R&D group has been looking to the photosynthesis of plants for a model of how this is accomplished. The group is making good progress on developing a process to use the sun's energy to split water into hydrogen and oxygen. These gases could later be recombined in a fuel cell to create electrical energy for a variety of uses including powering an automobile.[1]

You have the technical work well in hand and you are confident that you are doing work that is likely to benefit society. However, you are wondering what it means that you are a professional and what the implications of being a professional are for the way you and other team members handle the rewards for making this breakthrough. (For example, what you owe to the company for which you previously worked and at which you first worked on a similar problem; what you should expect in the way of credit to you personally for the contribution you have made to this project.) Where do you begin finding out what you need to know about your rights and responsibilities as a professional?

> Professions are those occupations that both require advanced study and mastery of a specialized body of knowledge, and undertake to promote, ensure, or safeguard some aspect of others' well-being.

Professions are those occupations that both require advanced study and mastery of a specialized body of knowledge, and undertake to promote, ensure, or safeguard some aspect of others' well-being. This chapter examines the norms and standards of good conduct in professional practice. Ethical (and sometimes legal) requirements also exist for nonprofessionals when their work immediately affects the public good. For example, food handlers are bound by sanitary rules. Arguably,[*] many moral rules apply equally in all work contexts. All should be honest, for example. What is distinctive about the ethical demands professions make on their practitioners is the combination of the responsibility for some aspect of others' well-being and the complexity of the knowledge and information that they must integrate in acting to promote that well-being.

---

[1] Trafton, Anne. 2008. "'Major Discovery' from MIT Primed to Unleash Solar Revolution," *MIT News*, July 31. Accessed at http://web.mit.edu/newsoffice/2008/oxygen-0731.html.
[*] "Arguably" means "there are some good arguments for thinking that."

## Engineering Societies and Professional Societies

Engineering societies in the United States may be professional societies, disciplinary societies, or some combination of the two. A society that focuses on technical or scholarly advances in its discipline is a disciplinary society.

A professional society is one that focuses on professional conduct and professional issues (e.g., conditions of employment). Only professional societies have codes of ethics.

The NSPE is a professional society. The IEEE is both a professional and a technical society, although in recent years the IEEE has stepped back from its involvement with professional ethics. The IEEE has not only given up its Ethics Hotline and stopped giving advice to *individual* engineers on ethical problems,[a] but has recently omitted from its Web site valuable advice to engineers in general on handling such ethical problems as dissenting from their supervisors on ethical grounds. One of the most valuable of these will be summarized and discussed later in Chapter 7. The IEEE does still maintain a code of ethics, however.

---

[a] Notice that despite the many statements on ethical problems previously issued by the IEEE Ethics Committee through the mid-1990s, the successor committee, the IEEE Ethics and Member Conduct Committee, now says on its Web page, "Neither the Ethics and Member Conduct Committee nor any of its members shall solicit or otherwise invite complaints, nor shall they provide advice to individuals." http://www.ieee.org/about/ethics/ethics_mission.html.

*Professional* practice requires acquisition of the *special knowledge and skill* peculiar to one's profession and application of that knowledge to achieve certain ends. The further requirement for an occupation to be a profession, namely, that the ends it seeks are to preserve or promote some aspect of human well-being, *distinguishes professions from disciplines, such as mathematics or philosophy.* Societies that are only "disciplinary," "scientific," "scholarly," or "learned," such as the American Philosophical Association, have no ethical codes and guidelines, because they are organized to advance learning on some subject only, rather than to guide professional practice. Some scientific societies, such as the American Physical Society (APS) in 1991, in the wake of high-profile scandals, became aware of the effect on society of their research and so issued their first ethical guidelines, thus becoming professional societies as well as disciplinary societies. (Since the 1980s, research investigation has been recognized to be a profession, one that advances knowledge [often with the support of public funds]. Research investigators bear a public trust and have a corresponding moral responsibility.)

The first professions to be identified (medicine, law, teaching, and the ministry) were seen as clearly addressing aspects of human well-being. The more recently emerged professions, such as engineering, nursing, and financial advising, also marshal knowledge to promote, safeguard, or achieve the well-being of others. Because being a "professional" carries social status, many other occupations – florists, for example – have recently claimed to be professions and support their claim by having a code of ethics. (The codes of ethics of occupations that do not fulfill the requirement of being professions because they do not directly influence a major aspect of human well-being may, nonetheless, have ethical content. That content commonly articulates norms of fair *business* practice. Business and the specific sorts of business, for example, banking, are not professions.)

Professional societies such as the National Society of Professional Engineers (NSPE), the American Medical Association (AMA), and the American Bar Association (ABA) issue explicit codes of conduct. Although these statements have names like "code of ethics," they can vary in the extent to which the matters they address are matters of ethics.

## Can an Employee Be a Professional?

Some writers on professionalism designate some occupations as "professions" based only on the high social status of those occupations. For example, sociologist Bernard Barber, who has written extensively about trust in the professions,[a] counts scientists as professionals but classifies engineering and nursing as "quasi-professions," because most of their members are employees and have lower social status. (Later in this chapter, we will consider the question of whether employee status prevents one from being a professional.)

As we have just seen, scientific societies like the APS formulated a code of ethics only in the 1990s, but nursing ethics has a long and rich history. Therefore, from the point of view of *professional ethics*, the social status of an occupation is irrelevant to whether it is a profession.

---

[a] Barber, Bernard. 1983. *The Logic and Limits of Trust*. New Brunswick, NJ: Rutgers University Press.

You chose engineering with the hope of being able to address the need for energy sources that do not pollute the environment or contribute to climate change. Your interests have brought you to a project that addresses the fundamental drawback to solar energy: the lack of a cheap and efficient way to store that energy. Your R&D group has been looking to the photosynthesis of plants for a model of how this is accomplished. The group is making good progress on developing a process to use the sun's energy to split water into hydrogen and oxygen. These gases could later be recombined in a fuel cell to create electrical energy for a variety of uses including powering an automobile.[2]

You have the technical work well in hand and you are confident that you are doing work that is likely to benefit society. However, you are wondering what it means that you are a professional and what the implications of being a professional are for the way you and other team members handle the rewards for making this breakthrough. (For example, what you owe to the company for which you previously worked and at which you first worked on a similar problem; what you should expect in the way of credit to you personally for the contribution you have made to this project.) Where do you begin finding out what you need to know about your rights and responsibilities as a professional?

## How Norms of Ethical Conduct Vary with Profession

*Identify two professions other than engineering. What characteristics do they share with engineering such that those occupations are all considered professions? What similarities or differences are there in the rules or obligations stated for those professions in the ethical guidelines or codes issued by the corresponding societies?*

In professional codes of ethics, rules of practice specify the acts that must or must not be performed. Below are such items from codes of professional behavior of several engineering societies requiring engineers to respect confidentiality and prohibiting bribery (the offering of a payment or inducement to obtain something to which one has no legitimate claim).

> Engineers shall not disclose, without consent, confidential information concerning the business affairs or technical processes of any present or former client or employer, or public body on which they serve.

– National Society of Professional Engineers (NSPE), *Code of Ethics for Engineers*[3]

---

[2] Trafton, *op. cit.*

[3] References to the NSPE code are to the latest revision (2006). References to other codes are to the versions current as of January 2007. Full text of the engineering society codes of ethics discussed here is available in the Online Ethics Center.

An engineer shall not accept a mandate which entails or may entail the disclosure or use of confidential information or documents obtained from another client without the latter's consent.

– Ordre des ingenieurs du Quebec (OIQ), *Code of Ethics of Engineers*[4]

1.8 Honor confidentiality.

The principle of honesty extends to issues of confidentiality of information whenever one has made an explicit promise to honor confidentiality or, implicitly, when private information not directly related to the performance of one's duties becomes available. The ethical concern is to respect all obligations of confidentiality to employers, clients, and users unless discharged from such obligations by requirements of the law or other principles of this Code.

– ACM, *Code of Ethics and Professional Conduct*

We, the members of the IEEE ... do hereby commit ourselves to the highest ethical and professional conduct and agree ... [10 items, including] to reject bribery in all its forms.

– Institute of Electrical and Electronics Engineers (IEEE), *Code of Ethics*

Engineers shall not solicit or accept financial or other valuable consideration, directly or indirectly, from contractors, their agents, or other parties in connection with work for employers or clients for which they are responsible.

– NSPE, *Code of Ethics for Engineers*

[Members] shall neither pay nor offer directly or indirectly inducements to secure work.

– Code of Ethics of the Institution of Engineers, Australia (IEA)

Some rules of practice in one profession have counterparts in others, but not all do. For example, rules about maintaining client confidentiality do exist in law and medicine as well as engineering.

A lawyer shall not reveal information relating to the representation of a client unless the client consents after consultation, except for disclosures that are impliedly [sic] authorized in order to carry out representation . . . .

– American Bar Association (ABA), *Model Rules of Professional Conduct*

The information disclosed to a physician during the course of the relationship between physician and patient is confidential to the utmost degree . . . .The physician should not reveal confidential communications or information without the express consent of the patient, unless required to do so by law.

– American Medical Association (AMA), *Principles of Medical Ethics*

Unlike engineering codes, however, the codes for medicine have no rule against paying or accepting bribes, although the codes do identify certain other payments as improper. Of course, it is no more ethically acceptable for physicians than for engineers to accept bribes. However, being offered or being tempted to offer others an outright bribe *is not a common ethical pitfall for physicians*. As this example illustrates, rather than being intended as an exhaustive list of what one

---

[4]The provisions of the codes of ethics of Canadian provincial engineering societies have the force of legal regulation. The Quebec society, the OIQ, has four mounted police who carry out the police functions of their work.

should or should not do, codes serve as guidance on *handling common temptations* and *avoiding common pitfalls* in addressing problems that arise in the profession but that may be unfamiliar in other contexts.[5] They do delineate areas in which someone in that profession must be especially scrupulous in order to fulfill the public trust. For example, because any lawyer is also an "officer of the court," lying in court or to the court is an especially grave offense.

> Ethical codes and guidelines from professional societies give moral guidance on handling situations that commonly arise in the practice of their professions but that may be unfamiliar in other contexts.

Other sorts of payments could create a conflict of interest for physicians. In particular physician organizations are concerned about **kickbacks** for referrals (i.e., payments for referrals of patients after the referrals have been made). Among physicians, the practice is called "**fee-splitting**." In fee-splitting a physician to whom another physician refers a patient pays part of the fee received from the referred patient back to the physician who made the referral. The American Medical Association Code of Ethics prohibits fee-splitting. Fee-splitting might tempt physicians to refer patients to those who will split the fee received from the patient, rather than to the person who will best care for the patient. Fee-splitting is not ethically equivalent to bribery, however, because a physician who receives a referral and pays part of the resulting fee back to the referring physician *may* have actually merited the referral.

A more remote analog to bribery is the lavish entertainment of physicians by drug companies. For example, until the 1990s, many drug companies commonly invited physicians to all-expense-paid "educational" seminars in resort locations. The practice among physicians has been widely criticized and therefore is becoming rarer, because at a minimum it creates the appearance that the drug companies are buying the favor of physicians, and the drug companies may actually be doing so.

Such lavish entertainment has long been unacceptable by the standards of the engineering profession. The NSPE has strict limits on acceptable levels of hospitality offered by vendors, to prevent not only bribery but conflict of interest or the potential for such a conflict.

Recall the definition of a conflict of interest in Section 4 of the introduction: Some party has a **conflict of interest** when that party

- is in a position of *trust* that requires the *exercise of judgment on behalf of others* (people, institutions, etc.)
- has interests, obligations, or responsibilities of the sort that *might* interfere with the exercise of such judgment, and having those interests is neither *obvious* nor *usual* for others in the same position of trust.

---

[5] As Heinz Luegenbiehl has observed, the justification for areas of professional ethics, such as engineering ethics, is that moral problems arise in professional practice that are unfamiliar in ordinary life. See Heinz C. Luegenbiehl, 1983, "Codes of Ethics and the Moral Education of Engineers," *Business and Professional Ethics Journal*, 2(4): 41–61; reprinted in *Ethical Issues in Engineering*, edited by Deborah Johnson (Englewood Cliffs, NJ: Prentice-Hall, Inc., 1991), 137–154.

## A Commission Payment to an Engineer for Marketing

Inaka, a licensed professional engineer, has been extensively engaged in engineering activities in the international market. Inaka intends to draw on this experience and on personal contacts to represent U.S. firms that wish to practice in Inaka's country, but that lack sufficient background knowledge of the culture, law, and business practices required for that purpose.

Inaka drafts a marketing agreement to address the need of many U.S. firms to develop their overseas potential without a large expenditure. For a basic fee plus a retainer fee Inaka proposes to develop contacts within stated geographical areas, evaluate potential projects, coordinate project development, and arrange contract terms between the client and the represented firm. The two fees are to be negotiated on an individual firm basis. Inaka will also receive a marketing fee, which is to be a negotiated percentage of the fees actually collected by the firm for projects that Inaka successfully markets.

Suppose you are in a decision-making position for an engineering firm that wishes to do business in Inaka's home country. Ought you to enter into a marketing agreement that agrees to pay Inaka a percentage of fees collected for projects Inaka marketed? Does the commission situation create a conflict of interest for Inaka? If so, how might you recommend rewriting the marketing agreement?

*Source:* Adapted from NSPE BER Case No. 78–7[a]

### Getting Started

The NSPE Board of Ethical Review has often expressed concerns about engineers receiving commissions that might create a conflict of interest position for them. Is that concern well founded in this instance?

---

[a]This case is based on NSPE BER Case 78–7. The NSPE Board of Ethical Review (BER) cases for 1976–2007 with judgments offered by the BER based on application of the then current NSPE code of ethics are available in hard copy in volumes 5–9 of *Opinions of the Board of Ethical Review*, from the National Society of Professional Engineers. See the reference guide with an index of cases through 2009 at http://www.nspe.org/resources/pdfs/Ethics/EthicsReferenceGuide.pdf. Cases and opinions from 1976 through 2001 (only) are available online at http://www.niee.org/pdd.cfm?pt=NIEE&Doc=EthicsCases.

In making a judgment on the case (78–7) that was the basis for the previous case, "Commission Payment to an Engineer for Marketing," the NSPE's Board of Ethical Review judged it to be a violation of professional ethics for an engineer to be paid a contingency fee based on the projects the engineer successfully markets. It had made a judgment in a somewhat similar case (62–4) that it was ethically acceptable for an engineering firm to pay a combined salary-commission to a marketing employee who was not an engineer. It saw it as crucial that the person doing the marketing on commission not seem to be offering a professional engineering opinion that could be influenced by the incentive of a commission. The board stated the concern that " . . . this method of compensation is undesirable because it could lead to loss of confidence by the public in the professional nature of engineering services." Its discussion and previous judgments suggest that it might be acceptable for such marketing services to be provided by a nonengineer (in which case there would be no possibility of corrupting engineering judgment with the temptation of a higher commission). These examples illustrate the variation in the problems that commonly arise in different professions and the moral rules that professional organizations formulate to address the problems that the members of their profession commonly encounter.

The practice of giving or receiving a commission or contingency payment for engineering services receives great attention in many provisions of past and present versions of the NSPE code of ethics. One concern is the prevention of kickbacks. For example, the present NSPE code of ethics says, under the fifth "rule of practice" [against deceptive practice]:

> For an engineer to receive a commission payment is viewed with great suspicion by engineering professional societies such as the NSPE because they fear that a commission payment might bias that engineer's professional judgment.

b. Engineers shall not offer, give, solicit, or receive, either directly or indirectly, any contribution to influence the award of a contract by public authority, or which may be reasonably construed by the public as having the effect or intent of influencing the awarding of a contract. They shall not offer any gift or other valuable consideration in order to secure work. They shall not pay a commission, percentage, or brokerage fee in order to secure work, except to a bona fide employee or bona fide established commercial or marketing agencies retained by them.

Under category 5 of Professional Obligations (on conflict of interest) the code reads:

b. Engineers shall not accept commissions or allowances, directly or indirectly, from contractors or other parties dealing with clients or employers of the engineer in connection with work for which the engineer is responsible.

Under category 6 of Professional Obligations (forbidding improper means of obtaining employment or advancement) it reads:

a. Engineers shall not request, propose, or accept a commission on a contingent basis under circumstances in which their judgment may be compromised.

As we saw earlier, professional societies issue codes of ethics to provide guidance for addressing moral problems that arise in the practice of the profession in question. Useful ethical codes or guidelines do not provide a random collection of ethical advice, but seek to address the actual temptations and moral pitfalls that commonly arise in that profession.

**Ethical Variation among Professions**

The view that the moral rules that are relevant and the temptations one must guard against vary from one profession to another contrasts with the view that being ethical just means acting in accord with a set of moral rules that specify what *everyone* should do or refrain from doing. Furthermore, because the discharge of responsibility is more than rule-following, fulfilling professional responsibility requires attention to the ends entrusted to a specific profession.

To understand the ethical significance of some action, such as how an offer of money or goods might corrupt professional practice, or how carelessness might lead to failure in some matter entrusted to the profession in question, one needs to understand both *the ethical standards of a profession and the conditions of its practice.*

To understand why certain obligations and moral rules have special importance in some profession and not others or why certain moral lapses in members of that profession are particularly injurious, one must understand the practice of that profession. The difference that the professional context makes is most vividly illustrated by rules in the ethical code of one profession that have *no* counterpart in those of others. We saw one such rule in the prohibition against lawyers representing two parties who are legal adversaries. As another example, physicians are forbidden to terminate their relationship with a patient under their care without first referring that patient to another provider.

> Once having undertaken a case, the physician should not neglect the patient, nor withdraw from the case without giving notice to the patient, the relatives, or responsible friends sufficiently long in advance of withdrawal to permit another medical attendant to be secured.
> – American Medical Association, *Principles of Medical Ethics*

## Bribes and Extortion

The "Foreign Corrupt Practices Act" of 1977 (15 U.S.C. 78dd-2) makes it a crime for U.S. corporations to accept or offer payments to *foreign governments/officials* in order to obtain or retain business. (It does not forbid making minor payments to low-level officials, although the latter might, depending on the situation, also count as a bribe or as extortion.)

The Foreign Corrupt Practices Act makes illegal the payment of extortion as well as bribes, notwithstanding the ethical difference that we considered in the introduction. Presumably, this act is intended to discourage foreign officials from *attempting* to induce U.S. companies to pay either bribes or extortion.

As C. E. Harris has pointed out, there is an important moral difference between bribes and capitulating to extortion in that bribes are paid to obtain something to which one does not have a right, such as a special advantage in awarding a contract.[a] In contrast, extortion is paid to secure something to which one does have a right (or at least a legitimate expectation), such as freedom from arson or the return of equipment one has legally brought into a country but which a corrupt customs official alleges to have been "lost."

Although one can imagine emergency circumstances under which an engineer ought, other things equal, to ensure that the necessary engineering services would continue to be available, there is not a generally recognized obligation of engineers to continue rendering service until a replacement engineer takes over.[6] Here is an example of a specific moral rule that applies to medicine but not to the engineering profession. Patients, as compared with the clients of engineers, are likely to be more physically and emotionally vulnerable and likely to fear abandonment in their vulnerable state. Part of what the public entrusts to physicians is the care of people in this vulnerable state. Engineers are not expected to provide a good "bedside manner" (i.e., tender regard for a client's or employer's special vulnerability) and are not required to ensure a continuity of engineering services in the general case. The provisions in ethical codes and guidelines of professional societies are justified insofar as they provide guidance on what is necessary to fulfill the trust placed in members of a given profession.

Not only are some moral rules more germane to certain professions than to others, and some ethical provisions applicable only to certain professions, but also the criteria for the application of some category may vary with the profession or with the customs of the larger culture in which the profession is practiced. For example, the determination of the sort of gift that exceeds appropriate

[a] Harris, Charles E., Michael S. Pritchard, and Michael J. Rabins. 1995. *Engineering Ethics*. Belmont, CA: Wadsworth, p. 108.

[6] I have found one engineering code of ethics that does mention withdrawal from service to a client (although even it does not require ensuring that the client has the services of another engineer). The 1983 Code de deontologies des ingenieurs du Quebec states:

hospitality varies with profession and with culture. A twenty-five dollar cutoff was commonly considered a limit for appropriate gifts and hospitality in U.S. engineering circles in the 1980s. (Some limits on acceptable payments to foreign officials are discussed by the NSPE Case 76–6, "Gifts to Foreign Officials." That case deals with an apparent request that an engineer *pay* a bribe or extortion, rather than an offer of a gift, bribe, or extortion payment *to* an engineer.[7]) As further examples, the United States recognizes freedom of speech as a right and interprets this right more broadly than do other technologically developed democracies. In some countries, it is considerably easier to prove defamation. This is true in Australia, for example, where the truth of one's allegation is not a defense against the charge of defamation. The IEA code of ethics reflects the Australian view in its stipulation that its members shall "neither maliciously nor carelessly do anything to injure, directly or indirectly, the reputation, prospects or business of others."

> *Identify two professions other than engineering. What characteristics do they share with engineering such that all three occupations are considered professions? What similarities or differences are there in the rules or obligations stated for those professions in the ethical guidelines or codes issued by the corresponding societies? Explain the differences between professions that you have identified.*

## Responsibilities, Obligations, and Moral Rules in Professional Ethics

> *What sort of value judgments and practical judgments must engineers make in carrying out their work?*

We have considered rules of conduct and statements of professional obligation from engineering and nonengineering professional societies. Such rules and obligations specify what *acts* a professional is ethically required or forbidden to do, such as: "Engineers should not sign off on work unless they have checked and approved it" or "Surgeons should not operate on patients without obtaining their consent." These express some aspects of the ethics of a profession. Checklists of rights, obligations, and rules of conduct set minimal standards for ethical practice and provide a start for ethical standards of professional practice. One need not make any complex judgments to recognize that signing off on unchecked work or operating without consent departs from responsible practice. If rights, obligations, or rules about what acts to perform or refrain from performing were all

3.03.04 An engineer may not cease to act for the account of a client unless he has just and reasonable grounds for so doing. The following shall, in particular, constitute just and reasonable grounds:
  (a) the fact that the engineer is placed in a situation of conflict of interest or in a circumstance whereby his professional independence could be called in question;
  (b) inducement by the client to illegal, unfair, or fraudulent acts;
  (c) the fact that the client ignores the engineer's advice.
3.03.05 Before ceasing to exercise his functions for the account of client, the engineer must give advance notice of withdrawal within a reasonable time.

[7] The case was originally published in *Opinions of the Board of Ethical Review* Volume V. Alexandria, VA: National Society of Professional Engineers. 1981. 11.

## Standard of Care and State of the Art

Some professions, and medicine in particular, make regular use of the concept of standard of care as a measure of competent practice and regard failure to practice in accord with the then current standard of care for one's profession a matter of negligence, which leaves such a professional open to liability for injuries or damages resulting from that person's conduct.[a]

As Stephen Nichols has pointed out the notion of **standard of care** is distinct from and contrasts with **state of the art** (i.e., what is technically feasible).

As an example, Motorola includes the following language in its design process [11, p. SG-5–1]:

> Identify the physical and functional requirements of the end product which are necessary to satisfy the requirements of: Customer's intended use of the product; Foreseeable misuse of the product; Environment in which the product is used.[b]

---

[a] Nolan, Joseph R. and Jacqueline M. Nolan-Haley. 1990. "Standard of Care." *Black's Law Dictionary: Definitions of Terms and Phrases of American and English Jurisprudence, Ancient and Modern*, sixth edition (St. Paul, MN: West Publishing Co.), 1404–1405.

[b] Motorola. 1992. *Concurrent Engineering: A Designer's Perspective*, Motorola University Press, p. SG-5–1, quoted in Steven P. Nichols, "A Design Engineer's View of Liability in Engineering Practice: Negligence and Other Potential Liabilities," Online Ethics Center, October 4, 2006, National Academy of Engineering, available at http://www.onlineethics.org/Topics/ProfPractice/PPEssays/designnichols.aspx.

there were to professional ethics, it would be a simple matter and hardly worthy of attention in college courses. Exercising professional responsibility is demanding, however, because it requires professionals to formulate a course of action that will achieve the desired *outcome*; they have to figure out what to do and not do. The exercise of responsibility, which we will consider in Part 2, typically requires the exercise of discretion and consideration of both technical matters and matters of value – such as how safe is safe enough.

The statement of ethical obligations or rules of professional conduct provide help in distinguishing malpractice from acceptable practice. To go further and differentiate good/responsible practice from minimally acceptable practice we will need to consider the concept of professional responsibility, which will be introduced in Part 2. Here we will just consider two stories of engineers who sought to marshal their knowledge for the benefit of others.

The case of Peter Palchinskii illustrates how the extent to which professions have control over the projects they work on affects the ability of their members to exercise judgment and discretion and carry out professional responsibility.

## The Case of Peter Palchinskii

The case of Peter Palchinskii appears in Loren Graham's book, *The Ghost of the Executed Engineer: Technology and the Fall of the Soviet Union*.[a] Peter Palchinskii was a multifaceted and extremely talented engineer in the U.S.S.R. during the Stalinist era. Palchinskii frequently criticized government policy for such things as inattention to the health and safety of workers, as well as for shortsighted planning. Although he was a committed Marxist, he was charged with treason and executed for pressing what were simply the concerns for matters like worker safety.

Subsequently, engineering education in the U.S.S.R. narrowed significantly, perhaps to lessen the chance that other engineers would recognize the broader implications of their work and raise criticisms that the government did not want to hear. Palchinskii's story and the subsequent changes in engineering education in the Soviet Union provide an example of how social and political context affect the character of professional education and the scope of professional competence, and hence the capacity of professionals to recognize problems and act in the public interest.

---

[a] Graham, Loren. 1993. *The Ghost of the Executed Engineer: Technology and the Fall of the Soviet Union.* Cambridge, MA: Harvard University Press.

---

Another example of an engineer who died for his ethical concerns is Benjamin E. Linder.

## The Case of Benjamin E. Linder

As an undergraduate studying mechanical engineering at the University of Washington, Benjamin Linder became intensely interested in the human consequences of engineering and the introduction of technology in undeveloped areas to meet human needs. After graduation in 1983, he went to Nicaragua to work as a volunteer under the sponsorship of the Nicaraguan Appropriate Technology Project. (The name "appropriate technology" is the term widely used for technology suited to the needs of small producers, rural and urban, especially in the developing world.) In the spring of 1984, Linder joined a project to provide power to a rural area in the mountains of northern Nicaragua that had no reliable source of electric power. Refrigeration for medical supplies and electric lights to hold evening classes both required electricity.

A small-scale hydroelectric plant was feasible, but because electricity had not been available, there were neither machine shops nor skilled mechanics. Plans were made to accomplish the construction by teaching local people how to build, operate, and maintain the plant themselves. Linder taught local people how to work with concrete and use hand tools. By May of 1986 when the plant was operational, many peasants had new skills and several were fully competent to run and maintain the plant.

The plant was used to power a small machine shop and support a medical center with a refrigerator. Plans included a future sawmill, carpentry shop, and facilities to make cement blocks, bricks, and roof tiles for the local area.

During the 1980s, the *contras* were working to overthrow the Nicaraguan Sandinista government. Their strategy was to attack farmers, teachers, and medical workers in outlying areas to weaken the government. The *contras* had been especially active in the area where Linder was working. When an organization of American citizens living in Nicaragua sued in U.S. court to stop the U.S. government from funding the *contras,* Linder joined the suit. In his affidavit, he said he believed that his life was endangered. The suit was unsuccessful, but Linder continued to be committed to his work. Two years later, he was killed by the *contras* while making rainfall and flow rate measurements.

In 1988, the IEEE SSIT Award for Outstanding Service in the Public Interest was awarded to Benjamin Linder for his "courageous and altruistic efforts to create human good by applying his technical abilities."[a]

---

[a] This account is based on that by Stephen H. Unger, in his book *Controlling Technology: Ethics and the Responsible Engineer*, second edition (New York: Holt, Rinehart and Winston, 1994), 43–48. In that work, Unger also recounts stories of other engineers facing extreme situations.

---

The stories of Palchinskii and Linder together vividly illustrate how the larger society may fail to support the responsible actions of engineers. There are no good alternatives to having professionals behave responsibly. Therefore, the general population has a strong interest in fostering legal and other supports for responsible behavior by professionals.

Although reported cases in which engineers were killed for their aspirations are few, such cases dramatically illustrate the importance of general societal support for ethical behavior. The American Association for the Advancement of Science's Human Rights Program monitors human rights abuses against science professionals around the world. According to their records, engineers significantly outnumber physicians as victims of human rights abuses. Some of these human rights violations are for political actions of the engineers rather than for their practice of engineering.

*What sort of value judgments and practical judgments must engineers make in carrying out their work?*

## Which Mistakes Are Culpable?

*Which mistakes that an engineer might make would be culpable (i.e., morally blameworthy) and why would they be worthy of blame?*

Everyone makes mistakes. Some are trivial, many are regrettable, but the gravity of the consequences does not determine the extent to which a mistake is culpable (i.e., morally blameworthy). The term **"honest mistake"** is used for a mistake to which little or no blame or guilt is attached. Consider the mistakes that led to the injury and death of patients treated with the Therac-25.

### Lethal Treatment: The Therac-25 X-Ray Machine

The Therac-25, a radiation therapy machine, killed or injured patients at several North American health care facilities between June 1985 and January 1987.

When the technician operating the Therac-25 made a typographical error in entering instructions and tried to correct this mistake by using the delete key, the filter on the machine dropped out of position. The result was that the patient undergoing radiation treatment received a massive dose of X-ray. Several patients were injured or killed as result before it was realized that the machine was dangerously defective.

The Therac-25 had been poorly designed and inadequately tested. The story is a complicated one that highlights many subtle as well as gross mistakes. In particular, the design and testing of the linking of the hardware and software were totally inadequate. Competitive machines had a shield that would engage if the power were at a high level. Furthermore, management decisions in the face of evidence of safety problems varied from shortsighted to negligent.

The manufacturer, Atomic Energy of Canada, Ltd., had many problems and has since gone bankrupt.[a] (A fuller account of this case is available at http://computingcases.org/case_materials/therac/supporting_docs/therac_case_narr/therac_toc.html.)

---

[a] Leveson, Nancy G. and Clark S. Turner. 1993. "An Investigation of the Therac-25 Accidents," *Computer* (published by IEEE) (July): 18–41, and Helen Nissenbaum. 1996. "Accountability in a Computerized Society," *Science and Engineering Ethics*, 2(1). An abstract of the study is available in the Online Ethics Center.

Although there were instances in which the operating technicians used poor judgment, those mistakes that were merely operators' typographical errors were honest errors[8] even though some of them caused the patients to receive massive doses of radiation; typographical errors are the sorts of mistakes that humans make even when they are appropriately attentive. Furthermore, the way in which the technicians attempted to correct their error was entirely reasonable, and they could have had no way of knowing the disastrous consequences that would result. Negligence occurred in the design and testing of the Therac-25 and in the mishandling of reports of patient injury.

Not surprisingly, many of the technicians who operated the Therac-25 machines and whose errors caused the deaths or injuries were emotionally devastated by the realization that their error had caused death or serious injury to their patient. Philosopher Bernard Williams has called psychological reactions of this type **"agent regret."**[9] From an ethical point of view, guilt for doing something that is morally blameworthy is importantly different from the normal psychological reaction of regret when great harm results from one's innocent actions.

### Competence and Ethics

It may be negligent or otherwise blameworthy to undertake work that is beyond one's competence. The engineering profession generally regards it to be an ethical failure to take on work that is beyond the engineer's competence, and the NSPE states it as one of their fundamental canons or principles that one should perform services only in the areas of one's competence. Professions such as medicine, in which learning by doing is common, are less explicit on this point.

Notice that "negligence" is a term of negative moral judgment. It is applied to mistakes or errors that are morally blameworthy. Carelessness shows inadequate attention to something. Only if one is morally obliged to do something is one negligent for failing to do it. For example, if I forget to water my house plant and it dies, that is careless of me – I did not give it the care it needed to survive – but my act is negligent only if I had a moral responsibility to give that care. Moral responsibility is not ordinarily a part of a person's relationship to her houseplants. Whereas a **negligence** or a **negligent** act shows insufficient care in a matter in which one has a moral responsibility, **a stupid mistake** shows a lack of judgment, and **an ignorant or incompetent mistake or error** shows a lack of knowledge or experience. A careless *person* is one whose care and attention cannot be trusted. A stupid *person* is one whose judgment cannot be trusted; an ignorant one lacks some knowledge, skills, or experience necessary to be trustworthy in given circumstances – a person may be competent in one area and incompetent in another, of course. Incompetence itself is not a moral failing unless one is morally required to be competent in that area or that area is one in which one *claims* to be competent. Because of their special education and training, professionals are expected to achieve a higher standard than the average citizen in their area of professional knowledge and practice. They have moral responsibilities for matters of great human consequence, such

---

[8]The term "error" is reserved for mistakes in relatively simple tasks where the standards of correctness are fairly clear, require no expert knowledge, and leave little room for "judgment calls," that is, discretion – for example, spelling errors and dialing errors. However, the distinction is often blurred, so you will hear people speak of "errors in judgment."

[9]Williams, Bernard. 1993. *Shame and Necessity.* Berkeley: University of California Press, 70–71, 93.

as public safety. Furthermore, they are morally obligated not to take work that lies beyond their competence. Therefore, an engineer or computer professional can be considered negligent for making certain mistakes that might simply show ignorance if performed by a person outside the engineering and computer science profession.

People, not only careless or stupid ones, may do careless or stupid things under circumstances that put unusual demands on their judgment or attention. These lapses may be more or less excusable. As we saw in Section 3 of the introduction, to say that an act is excusable is to say that there were special features of the situation – what are called "**extenuating**" or "**mitigating" circumstances** – that reduce the blame on the agent performing the act. To excuse an agent is to say that the act was more the product of special circumstances than the agent's choice or character. So, a person who has been drugged or distracted by a major personal tragedy might be excused for even an act of gross negligence. If, however, the agent is partially responsible for creating the special circumstance, the excuse has less force. Thus, if people do something stupid or negligent because they are drunk, drunkenness does not excuse their behavior. Being overly tired or in a rush to meet a deadline is a frequent cause of carelessness in engineering practice and research. How far tiredness and being rushed may excuse a mistake depends on the extent to which the person who is tired or rushed had a hand in creating these circumstances.

> Whether, or to what extent a person is blameworthy for having made some mistake depends on any extenuating circumstances and on how egregious the mistake is, that is, the extent to which it shows a failure to take the care that is morally required.

Thus far, we have established that although people recognize some mistakes as innocent or "honest" mistakes, others make the agent (the one who made the mistake) morally blameworthy. The extent of blame and the possibility that the mistake is excusable (and so it does not reflect on the agent's character) depend in part on the extenuating circumstances. They also depend on how egregious the mistake is – that is, whether it shows extraordinary disregard of others' welfare, which in turn depends on the care that is morally required. Philosopher John Austin illustrated this point by contrasting the plea that one had trod on a snail by mistake with the plea that one had trod on a baby by mistake. More care is required in dealings with babies.

There is a legal counterpart to morally blameworthy negligence in the notion of criminal negligence. Professionals can lose their licenses for negligence, or even be prosecuted. In rare cases, engineers have even been criminally prosecuted for negligence in complying with a law, as the following case illustrates.

---

## Prosecution of Three Engineers for Negligent Violation of the RCRA

In 1988 Carl Gepp, William Dee, and Robert Lentz, three chemical engineers at the U.S. Army's Aberdeen Proving Ground in Maryland, were criminally indicted for violating the Resource Conservation and Recovery Act (RCRA), which the U.S. Congress had passed in 1976. All three were civilians and specialists in chemical weapons work. At issue were the storage, treatment, and disposal of hazardous wastes at the chemical weapons plant, the Pilot Plant where all three worked. Although they were not the ones who were actually performing the illegal acts, they were the highest-level managers who knew of and allowed the improper handling of the chemicals.

In their defense, the three engineers said that they did not believe the plant's storage practices were illegal, and that their job description did not include responsibility for specific environmental rules. They were just doing things the way they had always been done at the Pilot Plant.[a]

Each defendant was charged with four counts of illegally storing and disposing of waste. They were tried and convicted in 1989. William Dee was found guilty on one count of violating the RCRA. Robert Lentz and Carl Gepp, who reported to Dee, were found guilty on three counts each. Among the violations observed were:

" . . . flammable and cancer-causing substances left in the open; chemicals that become lethal if mixed were kept in the same room; drums of toxic substances were leaking. There were chemicals everywhere – misplaced, unlabeled, or poorly contained. When part of the roof collapsed, smashing several chemical drums stored below, no one cleaned up or moved the spilled substance and broken containers for weeks."[b]

---

[a]Harris, C. E., Pritchard, M. S., and Rabins, M. J., *op. cit.*

[b]Weisskoph, Steven. 1989. "The Aberdeen Mess," *Washington Post Magazine*, January 15, p. 55, quoted in Harris, C. E., Pritchard, M. S., and Rabins, M. J., *Engineering Ethics.* "Aberdeen Three" in *Introducing Ethics Case Studies into Required Undergraduate Engineering Courses*, C. E. Harris, Department of Philosophy and M. J. Rabins, Department of Mechanical Engineering, Texas A&M University, NSF Grant Number DIR-9012252.

---

Another example of gross negligence is A. H. Robbins' design and marketing of its "IUD" (intrauterine device), the Dalkon Shield. An IUD implanted in the uterus prevents conception or implantation. All IUDs have a string attached to them to facilitate removal.

### The Wrong Stuff – The Dalkon Shield

A. H. Robbins, the makers of the Dalkon Shield, had first used multifilament polypropylene strings on this IUD. This was a reasonable choice, because this material was used for some surgical stitching. The polypropylene was somewhat stiff, however, and sometimes caused penile trauma to the women's partner during sexual intercourse. Robbins then substituted a string with a sheath made of nylon 6 and fibers made of nylon 66. This material is similar to fishing line. It is a poor choice of material for the human body because it decomposes in such an environment. Worse yet, the area within the sheath provides an optimal environment for the culture of anaerobic bacteria. These can multiply within the sheath, burst the sheath, and cause massive pelvic inflammation, sterility, and even death.

Robbins eventually went bankrupt under the pressure of the liability judgments against the company.[a]

---

[a]I thank Robert M. Rose, Professor of Materials Science and Engineering at MIT, for information on this case. Professor Rose served as an expert witness in the case against A. H. Robbins.

---

Another factor differentiating innocent from blameworthy mistakes is the degree to which the agent should have known that a mistake that results in serious harm could occur.

Another factor differentiating innocent from blameworthy mistakes is the degree to which the agent should have known that the mistake could occur. Thus, it matters if the party making the mistake has made the same mistake previously; people of normal intelligence are expected to learn from their mistakes. This is a point that Charles Bosk makes powerfully in his book, *Forgive and Remember,* with examples from the training of new surgeons. Sometimes, as in most of the cases Bosk discusses, a mistake once made is never repeated. Sometimes, however, what we learn is that we are prone to certain kinds of mistakes, so we add safeguards to prevent making them.

> *Which mistakes that an engineer might make would be culpable (i.e., morally blameworthy) and why would they be worthy of blame?*

## The Autonomy of Professions and Professional Codes of Ethics

> *What is meant by professional autonomy? How does one know whether the provisions in some "code of ethics" actually have ethical significance? How does one know whether a provision in such a code is relevant to engineers' ethical practice?*

Mastery of a specialized body of knowledge – which often includes practical experience in some sort of internship – is the basis for professional judgment. Because those outside of the profession do not have the same practical and theoretical knowledge, they cannot adequately evaluate the professional judgment of someone in that profession. For example, although a layperson might be able to recognize obvious flaws in a bridge design, the thorough evaluation of the plans for the construction of a bridge requires engineering knowledge. Although those outside a profession may possess some aspects of expert knowledge – for example, nurses frequently know enough about some medical procedures to recognize some subtle mistakes made by physicians[10] – it is members of a given profession who are in the best position to evaluate one another's performance. This is the rationale for so-called "**autonomy of professions**": the control of professions over the norms of practice of their members. Ethical codes and guidelines from professional societies are issued to inform and remind members of ethically significant norms of their professional practice. When a profession proves incapable of instilling and maintaining high standards in its members as occurred when the public lost confidence in the accounting profession after the Enron and other accounting scandals, the public generally seeks greater regulation of members of that profession.

> The **autonomy of professions** is the control of professions over the practice of their members.

Silly as well as sound reasons are given for believing that professions should be autonomous. For example, it is sometimes falsely alleged that professionals are inherently more moral than others. This view was more popular when morality and social status were often assumed to vary together. The term "professional," like the terms "gentleman" and "lady," carries prestige and connotations of relatively high socioeconomic status. Not everyone who has such status lives up to a high moral standard, however.

---

[10] Recognition of this fact has led nursing educators to include much more attention to the issue of raising ethical concerns in the professional nursing curriculum than is in the curriculum of most other professions.

---

**Maintaining Professional Standards and Writing Letters of Recommendation**

Meyer is an engineer working for a medium-sized manufacturing company, and is being considered for a promotion. Meyer's employer contacts other engineers who had worked previously with Meyer for their comments. One of these, Singh, is currently employed by another company and does not have any current direct professional relationship with Meyer. Singh replies that he will not submit a comment on Meyer's qualifications or engineering competence because Meyer has dropped her membership in the state professional engineering society. Singh says that all engineers have an obligation to support their profession through membership in the professional organization. Meyer alleges that Singh acted unethically in submitting that reply to Meyer's employer.

What is the basis for the obligation to review or comment upon a colleague's work? How stringent is the obligation, that is, do many other considerations outweigh the demands of that obligation?

Are engineers who fail to participate in their engineering societies undeserving of such effort from their colleagues?

What is the extent of an engineer's responsibilities for maintaining the profession, professional organizations, or professional standards? What, if any, sanctions are appropriate to use against an engineer who fails to live up to this standard?

*Source:* Adapted from NSPE BER Case No. 77–7[a]

---

[a] This case is adapted from NSPE BER Case 77–7.

Just as professionals vary in their trustworthiness, so professions and professional organizations may do a better or worse job of overseeing the ethics and competence of members of that profession. Sociologists have written extensively about the self-protective behavior of professional organizations, and the self-serving character of some provisions within some codes of ethics.

Until about the middle of the twentieth century, some engineering and medical codes of ethics contained provisions that undercut the self-regulation of those professions, for example, prohibitions against criticizing the work of another member of the profession. I know of no current code that still contains a prohibition on the criticism of peers, although some have replaced such provisions with those that prohibit *biased or unfair* criticism. For example, in the current (2006) code of the NSPE we find:

> Engineers shall issue no statements, criticisms, or arguments on technical matters that are inspired or paid for by interested parties, unless they have prefaced their comments by explicitly identifying the interested parties on whose behalf they are speaking, and by revealing the existence of any interest the engineers may have in the matters.
>
> – item 3 c under Rules of Practice

> Engineers shall not attempt to obtain employment or advancement or professional engagements by untruthfully criticizing other engineers, or by other improper or questionable methods.
>
> – item 6 under Professional Obligations

Engineers shall not attempt to injure, maliciously or falsely, directly or indirectly, the professional reputation, prospects, practice, or employment of other engineers. Engineers who believe others are guilty of unethical or illegal practice shall present such information to the proper authority for action.

– item 7 under Professional Obligations

The closest to a general prohibition on the criticism of other engineers is a prohibition on "indiscriminate criticism" found in the ASME code.

Engineers shall not maliciously or falsely, directly or indirectly, injure the professional reputation, prospects, practice or employment of another engineer or indiscriminately criticize another's work.

– item g under canon 5, which forbids engineers from competing unfairly with other engineers

Current engineering codes of ethics have generally dropped self-serving provisions like those preventing criticism of others' work. Some recent engineering codes of ethics have contained provisions that lack ethical significance, however. An example of a provision without ethical significance from a recent version of the NSPE Code of Ethics is a prohibition against engineers advertising their professional practice with "slogans, jingles, or sensational language or format."[11] This clause replaced one in the 1974 revision that had prohibited engineers from *any* advertising.[12]

The NSPE code underwent major revision after two Supreme Court decisions in the late 1970s. In 1977 in *Bates v. State Bar of Arizona,* the Court ruled against the Arizona Bar Association for attempting to prohibit advertising by two lawyers. The Court said that such prohibition violated the Sherman Anti-Trust Act. In April 1978 the Supreme Court struck down Section 11c of the NSPE Code of Ethics that had prohibited engineers from engaging in competitive bidding, saying that this, too, was restrictive of trade. The NSPE's prohibition of competitive bidding may have been motivated by a concern that competitive bidding or other price competition among engineers could undermine the quality of the engineering services they offered. An erosion of quality would occur if engineers tried only to do a job more cheaply rather than doing it well. However understandable was the NSPE's interest in maintaining a high quality of engineering work, its effort to limit competition among engineers conflicted with the U.S. laws promoting free trade. This example illustrates how value commitments of engineering societies may come into conflict with other value commitments, including those that have been given legal force. Notice that the Supreme Court decision against the NSPE

---

[11] Professional Obligations 3a. For a copy of two versions of the Code of Ethics that both contained this clause see the *Opinions of the Board of Ethical Review of the NSPE* Volume V (1981) and Volume VI (1989).

[12] This was item a under section 3 of the NSPE Code as revised January 1974. The prohibition on advertising was followed by clauses permitting such means of identification as professional cards, signs on offices or equipment, brochures stating qualifications, and brief telephone directory listings. A copy of the 1974 revision of the code is printed on the inside cover of the *Opinions of the Board of Ethical Review of the NSPE* Volume IV (1976).

did not indicate corruption on the part of anyone at the NSPE, but there are other cases in which the courts have found corruption in engineering societies.

Probably the most famous case in which an engineering society was found to have been corrupt was the precedent-setting 1980 Supreme Court decision that the ASME had acted from a conflict of interest when it gave an interpretation of ASME Boiler and Pressure Vessel Code, which the firm of McDonnell and Miller used to drive Hydrolevel Corporation out of business.

---

### The Supreme Court Decision on the Hydrolevel Corporation Suit against the ASME[a]

For most of the twentieth century the engineering firm of McDonnell and Miller Inc. had been the leading manufacturer of "float" cutoff valves.

In 1971, Eugene Mitchell, the then vice president for sales at McDonnell and Miller, asked the ASME Boiler and Pressure Vessel Codes Committee for an interpretation of the Boiler and Pressure Vessel Code. Mitchell had hoped the opinion would show the boiler control device of a competitor, Hydrolevel Corporation, did not meet the ASME standard. John James, the vice president for research at McDonnell and Miller, served on the ASME Boiler and Pressure Vessel Codes Committee at the time and had agreed with Mitchell to ask that committee for a determination. Furthermore, the chairman of the ASME BPV Code Heating Boiler Subcommittee discussed Mitchell and James' plan to seek an opinion with them and even advised them on the wording of their letter. Hardin also wrote *the original response* to McDonnell and Miller's request for a determination. The ASME's written opinion implied that Hydrolevel's devices did not meet the ASME Boiler and Pressure Vessel Code. Salesmen at McDonnell and Miller used the opinion to argue to potential customers that the Hydrolevel cutoff valve was a potential hazard. After the opinion had been made public, Hydrolevel lost customers and went bankrupt.

Hydrolevel sued the ASME arguing that it was motivated by a conflict of interest and violated the Sherman Anti-Trust Act. The litigation against ASME went all the way to the Supreme Court where the case was settled for $4.75 million in favor of Hydrolevel.[b]

---

[a] A full account of the case is available at http://ethics.tamu.edu/ethics/asme/asme1.htm.
[b] The legal case, *American Soc. of Mechanical Engineers, Inc. v. Hydrolevel Corp.*, 456 U.S. 556 (1982) is described in detail at http://supreme.vlex.com/vid/soc-mechanical-engineers-hydrolevel-19979461.

---

When professions do actually abuse their power by failing to serve the public good, the public eventually loses trust in that profession. We have seen this recently in the case of the accounting profession. The widespread complicity of accountants in the accounting scandals of the early twenty-first century, including the scandal that brought down Enron, eroded trust in accountants and in the stock market (because would-be investors suspected that "creative accounting" would mislead them about the financial health of companies). There is no simple way to prevent abuses of power, however. Professionals are best qualified to monitor and evaluate the actions of others in their profession, but when professions fail to control their members' practice, such failure usually leads to greater regulation.

Given the history of professional codes of ethics, the question of whether the code of a particular professional society reflects the best moral reasoning of its

## Are "Codes of Ethics" Necessarily Ethical?

John Ladd has argued against the uncritical acceptance of codes of ethics as authoritative ethical guides in "The Quest for a Code of Professional Ethics: An Intellectual and Moral Confusion."[a] His main point is that ethical principles cannot be established by organizations or their members.

In this book, codes are used not as an authoritative source of ethical values or principles, but as a guide to the moral problems, temptations, and pitfalls common in engineering practice and guidance on how to respond well to them. As such, the provisions in codes of ethics are based upon a *prior* recognition of ethical values by those who formulate the codes of ethics.

The fact that something is called a "code of ethics" does not guarantee that its provisions are ethical in nature, much less that they are applicable to the domain of action for which they are proposed.

---

[a]This essay appeared in Rosemary Chalk, Mark S. Frankel, and Sallie B. Chafer, *AAAS Professional Ethics Project: Professional Ethics Activities in the Science and Engineering Societies* (Washington, DC: AAAS Press). It has been reprinted in *Ethical Issues in Engineering*, edited by Deborah Johnson (Englewood Cliffs, NJ: Prentice Hall, 1990), 130–136.

practitioners is empirical; one does not know that a code of ethics actually has any ethical content before examining the code. As we saw in the introduction, some codes of ethics list as one of their principles a provision that shows a primary concern for the well-being of the profession. The current ASCE code of ethics says, for example, "Engineers uphold and advance the integrity, honor, and dignity of the engineering profession by using their knowledge and skill for the enhancement of human welfare and the environment."[13] Enhancing human welfare is represented only as a means to that end of advancing "the integrity, honor, and dignity of the engineering profession." What reasons can you think of for and against the view that advancing "the integrity, honor, and dignity of the engineering profession" is itself an ethical consideration?

In engineering ethical codes and guidelines, and especially in some of the decisions by the NSPE's Board of Ethical Review, the treatment of one's fellow engineers and concern for the dignity of the profession receive strong, sometimes surprisingly strong, emphasis. For example, consider the following case.

---

### Information Due Potential Partners

Armandi, a principal in an engineering firm, submitted a statement of qualifications on behalf of her company to a governmental agency for a project. Armandi was notified that her firm was on the "short list" for consideration along with several other firms, but that it did not appear to have qualifications in some specialized aspects of the requirements, and that it might be advisable for the firm to consider a joint venture with another firm with such capabilities. Armandi then contacted Bent, a partner of a firm with the background required for the specialized requirements, and invited the Bent firm to join in a joint venture if Armandi was awarded the job. Bent agreed.

Thereafter, Engineer Chou, a principal in a firm that was also on the "short list," contacted Engineer Bent and also asked if the Bent firm would be willing to engage in a joint venture to supply the specialized services, if the Chou firm was selected for the assignment.

Bent wants to agree to work with either company if that company wins the contract. What, if anything, ought Bent do before making an agreement with Armandi and Chou?

*Source:* Adapted from NSPE Case 80–4

---

[13]This is the first of their four fundamental principles. (The ASCE Web site is at www.asce.org/.)

The 1980 NSPE Board of Ethical Review (BER) reviewed a similar case in 1980 and judged* that the engineer in question (whom we have called "Bent" in our open-ended version) behaved unethically in agreeing to participate in a joint venture arrangement with more than one other engineering firm without making a full disclosure to both the firms. The BER cited two sections of the then current version of the NSPE Code of Ethics as potentially relevant.

Section 1 – The Engineer will be guided in all his professional relations by the highest standards of integrity, and will act in professional matters for each client or employer as a faithful agent or trustee.[†]

Section 8 – The Engineer shall disclose all known or potential conflicts of interest to his employer or client by promptly informing them of any business connections, interests, or other circumstances which could influence his judgment or the quality of his services, or which might reasonably be construed by others as constituting a conflict of interest.

The BER reasoned that Bent has a relationship of trust with each company and that calls for the highest standards of integrity in Section 1 of its code of ethics. The board held that in this case *there is no potential or actual division of loyalty or conflict of interest*, because Bent's loyalty would be to the firm with which he worked after the contract was awarded. Therefore, the disclosure requirement of Section 8 does not *strictly* apply in this case, because at this point Bent "does not have a 'client,' as such." However, the BER maintained that the relationship of trust with each firm should not be diluted by establishing a similar and possibly competitive relationship without disclosure of this fact to all parties.

Sometimes a failure to show consideration is a failure of professional responsibility. This is true, for example, when one's clients can be expected to be frightened, confused, or in pain. In such a condition clients are unlikely to be effective advocates for their own needs and interests. However, there is a point at which being considerate, although *praiseworthy* and perhaps indicative of *moral virtue* (being a *good person*), goes beyond what is morally *required*.

My students have often judged that telling both firms of one's promise to form a joint venture with the other was a *good thing to do* (say, because it shows praiseworthy candor and honesty). Nevertheless, those engineers and engineering students with whom I have discussed this case often find it excessive to say that such behavior is ethically *required*. Among their *reasons* for disagreeing with the 1980 NSPE BER is that there is no implied understanding among engineering

---

*The question that the NSPE Board of Ethical Review (BER) considers for all of its cases is whether the engineers, or at least the engineers who did anything notable in the circumstances described, behaved ethically. The BER issues a judgment only about the actions of engineers in any of the cases it considers and bases its judgment solely on the then current provisions of the NSPE Code of Ethics for Engineers, and sometimes has bemoaned the lack of a provision that would forbid some action that it judges on independent ethical grounds to be wrong. Some of the revisions to the NSPE Code have been made in light of the experience of the Board of Ethical Review. This shows that its code of ethics is a "living document," that is, one that changes in response to changing circumstances.

†Notice that Section 1 of the NSPE code concerns dealings with clients (for those in private practice) or employers, rather than partners in joint ventures. It is clear that the BER exercises some discretion in applying the code to actual situations.

firms that each would refrain from making contingency plans and agreements to meet the conditions for a joint venture. Furthermore, entering into a second joint venture agreement does not harm the firm that does not get the contract.

How should we understand the NSPE's view that a high level of consideration of one's fellow engineers is ethically required? Neither the code nor the NSPE BER recommends that engineers *ignore* other ethically significant considerations to be considerate of fellow engineers. Rather, they demand that engineers go to some *extra effort* to show consideration. Mutual consideration is necessary for engineers to form *a cohesive community able to maintain standards of professional behavior*. To maintain standards of behavior, competition must be carried out within a framework of standards of decency and fair play. Striking an appropriate balance between promoting collegiality among engineers and supporting competition that encourages engineers to offer clients good value is a continuing challenge for professional ethics.

> Although a failure to show consideration is a failure of professional responsibility in certain circumstances, there is a point at which consideration of others, although praiseworthy, goes beyond what is morally *required*. Striking an appropriate balance between promoting collegiality among engineers and supporting competition that encourages engineers to offer clients good value is a continuing challenge for professional ethics.

### Revealing Wrongdoing

The fear of defending a lawsuit may inhibit university officials from reporting when a faculty member has been dismissed for cause, including research fraud. It also inhibits journal editors from printing retractions when not all the authors of an article agree that retraction is warranted. Professional societies such as the American Association for the Advancement of Science have urged similar legislation to protect university officials and journal editors who act in good faith in such cases. These considerations illustrate some of the factors that may inhibit peer control of professional conduct and therefore interfere with autonomy of professions.

On the one hand, some professions have suffered the loss of public trust when their members have put too high a priority on their loyalty to one another and sacrificed other morally significant aims for it. Physician peer review organizations charged with overseeing the work of physicians were found to be quite lenient with some physicians, notably those with drug and alcohol problems. Physicians who sat on the peer review organization committees (PSROs) once feared that they would be personally sued for removing the license of a negligent physician. This fear contributed to the pattern of excessive leniency. (Such suits were common and proved to be a great burden to the physicians who were sued, even if the plaintiff ultimately lost the suit.) To encourage better peer review, Congress enacted the Health Care Quality Improvement Act in 1986. This act gives legal immunity to the actions of peer review organizations if their judgment is based exclusively on the competence of a physician. (Peer judgments about such matters as membership or lack of membership in a professional organization or use of advertisements do not fall under this protection.) Similar legislation to govern review of the professional conduct and competence of engineers and scientists would strengthen the hand of their professional organizations in maintaining high standards of ethical behavior.

On the other hand, the failure to maintain standards of fairness and decency in the competition among members of some professions has led to failure in other duties to clients or to the public. This point has been missed by some writers who dismiss duties to fellow professionals as having no ethical content, and being matters merely of *etiquette*. Those who dismiss these matters as etiquette seem to be so concerned with the duties that a professional owes to the public that duties to members of the same profession seem insignificant. However, fairness and decency to members of one's profession are also ethical matters. Furthermore, when those ethical standards are not met, the conditions for provision of good services are often undermined as well. In Part 3, when we consider fair credit in research, we shall see the negative effect on the production of trustworthy research when standards of fairness and decency cease to govern the relations among research investigators.

*What is meant by professional autonomy? How does one know whether the provisions in some "code of ethics" actually have ethical significance? How does one know whether a provision in such a code is relevant to engineers' ethical practice?*

## Does Employee Status Prevent Acting as a Professional?

*How does being an employee affect the sorts of ethical problems an engineer may face?*

Discussions of professional ethics frequently start from an outdated model of professional practice in which the practitioner is assumed self-employed and in a one-to-one relationship with a client whose welfare is at stake. Indeed, as was mentioned, it has been argued that being an employee (rather than in private practice) makes a professional accountable to the employer and therefore less able to uphold professional values. Those who hold this view argue that the professional status of engineering is compromised by the fact that the majority of engineers are employees rather than in solo or group practice.[14] (This is true of the majority of mechanical, electrical, and chemical engineers, but not the majority of civil engineers. It is increasingly true of physicians. Although in the early 1900s most physicians were "in private practice," most are now employees [of HMOs, clinics, and other health care organizations]. Most lawyers are also employees, at least at the beginning of their careers. Most teachers are employees throughout their careers.)

The association of employee status with a compromise in professional status has some special relevance for engineers in the United States (but not, for example, in Canada[15]), because U.S. engineers employed in industry have an "industry exemption" from the requirement that they be licensed. As a result, the majority of engineers working for industry in the United States – unlike engineers in private practice, *any* lawyer, physician, or nurse – *have no license to lose* for incompetent or unethical behavior. However, as we saw in the case of the three

[14] See for example, Barber (1983).
[15] Unlike both Canada and the United States, Australia does not now have a general licensing procedure for its engineers although the National Professional Engineers Register (NPER) is a register, administered by the Institution of Engineers, Australia, for professional engineers who meet special qualifications.

chemical engineers who handled hazardous waste improperly at the Aberdeen Proving Ground, they are liable to other penalties.

Employee status of a profession can also change over time as can the scope of work. In the United States during the middle of the twentieth century most engineers were employees, but most physicians were not. Today both groups are primarily employees. Most of what in the writings of Hippocrates was considered the work of a physician would now be seen as nursing. Because nursing differentiated from medicine, most nurses have been employees. Therefore, it is anachronistic to continue to require self-employment as a mark of the "true" professional.

Being both a professional and an employee creates some special problems, even if most professionals are now becoming employees. First, professionals who are employees must answer to their employers, something they would not need to do if they were in private practice. They decide what departures from ideal professional behavior are important enough to object to and must figure out how to discharge their professional responsibilities in the face of minor departures from their personal standards. Second, they must figure out how to present their arguments on important issues so that others are most likely to appreciate their point. Third, if those within the organization continue to disregard an important matter, professionals must make a judgment about whether and to what extent they should either breach confidentiality or "make trouble" for the organization by taking matters outside the organization. Finally, they must decide where to take the matter and how best to raise the issue to get attention to the issue while being fair to those who disagree with them.

### Do Engineers Have a Right to Protest Shoddy Work and Cost Overruns?

Kim is an engineer who works for a large defense company. Part of Kim's job is to review the work of subcontractors on a large government contract to Kim's company. Kim discovers that certain subcontractors have turned in submissions with excessive costs, time delays, or deficient work, and advises management to reject these jobs and require the subcontractors to correct these problems.

After an extended period of disagreement with Kim over the subcontractor issue, management placed a warning in Kim's personnel file about insubordination and placed Kim on three months probation with a warning about the possibility of future termination. Kim continues to insist that the company has an obligation to ensure that subcontractors fulfill the specifications for their work and try to save unnecessary costs to the government. Finally, Kim requests an opinion from the NSPE Board of Ethical Review on the matter.

*Source:* Adapted from NSPE Board of Ethical Review Case 82–5

How should Kim treat the judgment of the NSPE Board of Ethical Review on the matter?

Is there any further information needed that would make a difference in your assessment? If so, what is that information and how would it affect your recommendation to Kim?

### Getting Started

What are Kim's professional responsibilities in this matter?

One piece of possibly relevant information that we lack is the likely consequences of the poor quality of the subcontractors' work. Safety is not mentioned, therefore, we may assume that the poor quality does not itself pose safety risks, but notice that if the system itself is a safety critical one, then failure in its performance will threaten safety. To resolve this situation, you would need to make some assumptions about just what effects the shoddy work would have and thus, whether there is any justification for the management of Kim's company deciding that Kim is being too picky.

In this case, the facts are presented as clear and unambiguous, perhaps more clear and unambiguous than one would typically encounter in actual practice. As it is, however, it displays the subtleties and difficulties in making a judgment on even a clear case. It also illustrates a situation in which the employee status of the engineer is a central feature.

> The decision of an employee engineer about whether and how to "blow the whistle" in specific circumstances has many parallels with the decision of an engineer in private practice about how to weigh other moral demands against the obligation to protect the interests of a client. The employee engineer is somewhat more vulnerable, however, than the engineer in private practice, in that the alienation of a client typically has less severe consequences than the alienation of one's employer, or even one's superiors within a company.

Although many civil and some mechanical (mostly structural) engineers are in private practice, most U.S. engineers are employees. The issue of **whistleblowing** (i.e., taking one's concerns outside of one's organization) arises for employee engineers. The decision of an employee engineer about whether and how to "blow the whistle" in specific circumstances has many parallels with the decision of an engineer in private practice about how to weigh other moral demands against the obligation to maintain client confidentiality or otherwise protect the client's interests. The employee engineer is somewhat more vulnerable, however, than the engineer in private practice, in that the alienation of a client typically has less severe consequences than the alienation of one's employer, or even one's superiors within a company. In discussing the central importance of public safety in codes of ethics, we have considered how that consideration might outweigh the obligation to preserve client confidentiality if the two were in irreducible conflict. (The NSPE BER judged that although if there are no safety implications an engineer does not have a duty to protest, an engineer has a *right* to do so (and so presumably should not be punished for doing so). The conflict with safety is clearer in the two cases that follow.

---

### The Responsibility for Safety and the Obligation to Preserve Client Confidentiality

The owners of an apartment building are sued by their tenants to force them to repair defects that result in many annoyances for the tenants. The owner's attorney hires Lyle, a structural engineer, to inspect the building and testify for the owner. Lyle discovers serious structural problems in the building that are an immediate threat to the tenants' safety. These problems were not mentioned in the tenants' suit. Lyle reports this information to the attorney who tells Lyle to keep this information confidential because it could affect the lawsuit.

*Source:* Adapted from NSPE Board of Ethical Review Case 90–5

What courses of action are open to Lyle?

Is there other information you would like to have about this case, and if so, how would having it affect your decision?

---

### Code Violations with Safety Implications

Lee, an engineer, is hired to confirm the structural integrity of an apartment building that Lee's client, Scotty, is going to sell. Through her agreement with Scotty, Lee will keep the report confidential. Scotty makes it clear to Lee that the building is being sold in its present condition without any further repairs or renovations. Lee determines that the building is structurally sound, but Scotty confides in Lee that electrical and mechanical code violations are also present. Although Lee is not an electrical or mechanical engineer, she realizes that the problems could result in injury and informs Scotty of this fact. In her report, Lee briefly mentions the conversation with Scotty about these deficiencies, but he does not report the violations to a third party.

What is your evaluation of Lee's actions? Of Scotty's? Is there any information not stated here that would make a difference to your judgment?

*Source:* Adapted from NSPE Board of Ethical Review Case 89–7

---

Professional education in engineering and science has only recently begun to give attention to questions of how to make fair and effective complaints. It has only developed in recent decades as a part of engineering ethics and since 1985 in research ethics. Until very recently medical education has also neglected teaching this moral skill, although for decades physicians have sent their patients to hospitals where those patients needed their physicians to be advocates for them.

Given the reliance of others upon the judgment of professionals, whether those professionals are employees or self-employed, the more important questions than whether certain professionals are employees are:

- how best to prepare professionals to cope with the potentially competing demands placed upon them so they behave responsibly;
- how to create supports within and without the employing institutions to support a high standard of ethical behavior; and
- how to create laws and policies that further rather than frustrate responsible practice.

*How does being an employee affect the sorts of ethical problems an engineer may face?*

## The Limits of Predictability and Responsibilities of the Engineering Profession

*What distinguishes a moral problem for the individual engineer from a problem that must be addressed by the engineering profession or society as a whole?*

In searching for criteria for responsible engineering practice, for example, there is no better place to begin than with the explicit criteria that reflective practitioners use to evaluate engineering practice. These criteria are not beyond criticism, but they are grounded in a critical appreciation of the realities of engineering.

As we discussed in the section on professional autonomy, society must rely on members of a profession to judge their peers' exercise of professional judgment. Therefore, as new consequences of some professional work are recognized, members of that profession must consider whether and how such consequences can be controlled. Criteria for responsible behavior regarding these potential consequences can then be proposed and discussed. Expectations need to be established about the consequences that a professional should take into account.

Bill Wulf, during his tenure as president of the National Academy of Engineering (NAE), from 1997 to 2007, proposed that the engineering *profession* itself has a moral responsibility to take up an emerging set of difficult moral problems resulting from the new complexity of technology. An example he gives is that complex technology such as digital technology cannot be checked in a feasible amount of time. Digital systems are not continuous, as physical systems are. In a physical system a small change in the system produces a small change in the behavior of the system. In a digital system, however, a small change (e.g., a change in one bit in the memory of a computer) can produce a radical change in what the system represents. As a result, some of the extraordinarily large number of bugs in software are not due to human error but are **emergent properties**, that is, properties that *could* not have been predicted.[16] The lack of continuity of a digital system also creates insuperable problems for testing. In an analog system one can pick test points that are spaced sufficiently close and trust that behavior at the intervening points will be similar; however, testing a digital system would require that one test *every configuration* of that digital system. Wulf points out that this is impossible, because even if

> every atom in the universe were a computer, and every [such] computer in the universe could test $10^{100}$ states per second, there wouldn't be enough time, even starting from the time of the Big Bang, to test all of the states in [Wulf's] laptop.[17]

This fact creates what Wulf calls a "macro problem," by which he means a problem for the *engineering profession* rather than a problem that the individual engineering/computer professional can address. That problem is: how to responsibly engineer technology when one knows in advance that there will be some behaviors of the resulting system that one cannot predict. We will discuss "macro problems" further in Chapter 11. It is not yet clear whether the engineering profession, or perhaps the leadership embodied in the NAE, will do more to address such problems.

*What distinguishes a moral problem for the individual engineer from a problem that must be addressed by the engineering profession or society as a whole?*

## Summary

Although everyone makes mistakes, we have seen that higher standards of care are applied to the behavior of professionals when operating in the *area of their expertise*. A profession has custody of a special body of knowledge. Those who have mastered that body of knowledge are in the position to recognize when it is being used competently and with due care. This is the basis of the

---

[16]Wulf, William A. 2004. Keynote Address, *Emerging Technologies and Ethical Issues in the Practice of Engineering*. Washington, DC: The National Academies Press, 1–6.
[17]*Ibid.*, 5.

claim of professions to be "autonomous," that is, self-governing. To live up to this considerable claim, many professional organizations issue codes of ethics specifying the norms of behavior required for ethical practice. Members of a profession do not always live up to the task of maintaining high standards of practice, however. When the public becomes aware of widespread dereliction by members of some profession, it generally clamors for more regulation of the behavior of members of that profession.

This chapter began the examination of the ethics of professions and the engineering profession in particular. It has focused on the norms of professional conduct that are expressed in the rules and obligations set forth by engineering professional societies. To fulfill such an obligation or follow such a rule one need only be able to perform (or refrain from performing) the specified act and be conscientious enough to follow through.

> A profession has custody of a special body of knowledge. Those who have mastered that body of knowledge are in the position to recognize when it is being used competently and with due care. This is the basis of the claim of professions to be "autonomous," that is, self-governing. To live up to this considerable claim, many professional organizations issue codes of ethics specifying the norms of behavior required for ethical practice.

This chapter has addressed what it means to be a professional, more specifically a professional engineer. It has examined some of the moral rules, rights, and obligations that pertain to fulfilling the trust that society places in engineers. One of the defining characteristics of a professional is the mastery of the particular body of expert knowledge and exercise of judgment in drawing on that knowledge to address problems that bear on other's well-being. Therefore, professional behavior requires more than following rules and fulfilling professional obligations. We have begun to see some of the more complex issues of responsible engineering practice that require much greater exercise of discretion and judgment than do obligations to perform (or refrain from performing) *particular acts*. To behave responsibly an agent must decide what acts are required to attain the desired state of affairs. In particular a responsible engineer must decide what to do to achieve desired ends. We will complete the picture of what it is for engineers to fulfill the trust that the public places in them by examining more thoroughly the engineer's professional responsibility to ensure some future state of affairs in Chapter 4.

# 2 Two Examples of Professional Behavior: Roger Boisjoly and William LeMessurier

What do you do when you realize that your work or your company's work has resulted in a serious threat to life and health, and how do you go about it?

## Section 1. Roger Boisjoly's Attempts to Avert the *Challenger* Disaster

What do safety problems look like to the engineer who encounters them? How do they develop over time? What are good ways of responding to such problems at each stage of their development? Much can be learned from the attempts of Roger Boisjoly, an engineer at Morton Thiokol, to avert the *Challenger* disaster of January 1986. His care and diligence in coping with the uncertainties about the nature and extent of the threat to the shuttle flights and his courageous persistence in raising issues exemplify responsible behavior.

Like others who have spent time with Roger Boisjoly, I have been impressed with his sincerity and forthrightness. These are matters of *moral character* over and above the particular *acts* he performed. Boisjoly's integrity and openness make his personal account of events especially illuminating, but at this point in our investigation we are concerned with his actions, what he did at various points in the unfolding story of the *Challenger* disaster, rather than with his character.

### *Moral Lessons from Actions Intended to Forestall the* Challenger *Explosion*
Does the fact that some disaster occurs, despite attempts to forestall it, show that the agents were negligent, incompetent, or in some other way blameworthy? Why or why not?

In hindsight, assigning blame for accidents and disasters based on the outcome is tempting. *Any* action that would have prevented the fatal *Challenger* flight, for instance, may seem justified; any failure to take an action to stop the flight may look like a mistake. This view is superficial, however. Judging actions *solely by a single feature of their outcome* omits consideration of the harmful side effects that might result from actions meant to bring about that single feature. Such superficial hindsight tells us nothing about how to act in situations in which we cannot perfectly foresee the outcome, which are the situations in which we usually find ourselves. The challenge is not merely to avoid *one* possible negative outcome, but to achieve a *generally good outcome*.

> Judging actions *solely by a single feature of their outcome* omits consideration of the harmful side effects that might result from actions meant to bring about that single feature. The challenge is not merely to act to avoid *one* possible negative outcome, but to achieve a *generally good outcome.*

To take Roger Boisjoly's actions as exemplary does *not* mean that they are above criticism, nor that they could not in any way be improved. Exemplary responses to moral problems, like excellent designs, may be improved, but they give us the shoulders of giants to stand on. Unfortunately, detailed accounts of exemplary responses to problems encountered in engineering are all too few. What is more common are stories of accidents, but accident stories only provide cautionary tales about what *not* to do.

Some have suggested that if Boisjoly had made a more effective graphical presentation of the data on leakage of hot exhaust gas through the O-ring seals and consequent erosion of those seals in previous shuttle flights, he could have made his case more convincing. If this is a valid criticism, it provides a lesson for engineering *educators*, because effective graphical presentation is a skill that receives little or no emphasis in most engineering education programs.

> Exemplary responses to moral problems, like excellent designs, may be improved, but they give us the shoulders of giants to stand on.

The point of examining Roger Boisjoly's actions is to learn from his example what sort of engineering responses are *good* responses to safety problems. Roger Boisjoly's story may help engineers recognize developing threats to safety and envision some actions to take at each stage of the developing threat. The causes of such threats to safety often lie in failures within organizations or in communications between organizations. Many factors that contributed to the *Challenger* explosion were of this sort. A good evaluation of the weaknesses of the shuttle program that led to the explosion of the *Challenger* is available in the report of the Presidential Commission[1] – commonly called the "Rogers' Commission" for William P. Rogers who chaired it – and in several books, including Malcolm McConnell's *Challenger: A Major Malfunction*, and articles.[2]

The hard truth is that many of the failures identified in the Rogers' Commission Report about the *Challenger* persisted and contributed to the explosion of the *Columbia* shuttle in 2003.[3] (In the *Columbia* accident, a 1.68-pound piece of insulating foam broke off during liftoff and hit the orbiter. The impact left a three-inch crack in the thermal protection system on the left wing of *Columbia*. When *Columbia* reentered the atmosphere, superheated air drawn through the crack melted the orbiter's aluminum structure, producing aerodynamic instability that shattered the orbiter, killing seven astronauts.)

---

[1] The updated URL for this report is available linked from the story of the *Challenger* in the Online Ethics Center. (Go to www.onlineethics.org, search for "*Challenger*" and near the bottom of any of the pages will be a link titled "Report of the Presidential Commission on the Space Shuttle *Challenger* Accident.")

[2] For example, T. E. Bell and K. Esch, 1987, "The Fatal Flaw in Flight 51-L," *IEEE Spectrum* 24(2): 36–51.

[3] *Columbia* Accident Investigation Board. 2003. *Report* Volume I, August 26, from http://caib. nasa.gov/.

The recurrence of organization and technical problems at NASA similar to those found in the *Challenger* case is evidence of the resistance to change commonly found in organizations, even those with failings that have been clearly identified.

Roger Boisjoly, for his effort to avert the *Challenger* explosion, received the AAAS award for Scientific Freedom and Responsibility. The implication of the AAAS award is that Boisjoly's actions were well-conceived and likely to have brought attention to the dangers he recognized to the *Challenger* flight. Acting well does not guarantee a good outcome, however, if one does not have complete control of the situation. Even the best practitioner rarely has control of all the factors that determine the outcome. As we shall see later in this chapter, William LeMessurier was successful in averting the collapse of the Citicorp Tower in New York City because he had more power to make his voice heard and because other figures also behaved responsibly. Roger Boisjoly's situation is most instructive for engineers who expect to work in large companies. For most such engineers, it is all the more important to find work situations in organizations that are responsive to their ethical concerns.

*Does the fact that some disaster occurs, despite attempts to forestall it, show that the agents were negligent, incompetent, or in some other way blameworthy? Why or why not?*

### The Post-Flight Inspection in January 1985

For Roger Boisjoly the story began in January 1985, a year before the flight of the *Challenger*. (The components of the space shuttle are shown in Figure 2.1.) Boisjoly was involved in the post-flight hardware inspection of another shuttle flight, Flight 51C.[4] During this inspection, he observed large amounts of blackened grease between the two O-ring seals, showing that the grease had been burned by escaping combustion gases. Gases from the rockets, which had been under immense pressure, created a blowhole through more than ten feet of zinc-chromate putty. Hot gas had blown by the primary O-ring seal as well. Were the gas to leak by the secondary seal, it might ignite the fuel tanks, causing them to explode.

Boisjoly reported his findings to his superiors, who asked him to go to the Marshall Space Flight Center to present his observations and explain the seal erosion. Boisjoly reported his hypothesis that lower than usual launch temperatures had compromised the resilience of the O-rings, and hence their capacity to seal.[5]

### The Significance of the O-Ring Seals and Escape of Hot Gas
Why was the escape of hot gas through the primary seal during shuttle flights a matter of grave concern?

[4]NASA had found severe erosion of O-ring seals from some flights before 1985, the starting point of Boisjoly's account.
[5]Report of the Presidential Commission on the Space Shuttle *Challenger* Accident (In compliance with Executive Order 12546 of February 3, 1986) (p. 112 in print version), accessed at http://science.ksc.nasa.gov/shuttle/missions/51-l/docs/rogers-commission/table-of-contents.html.

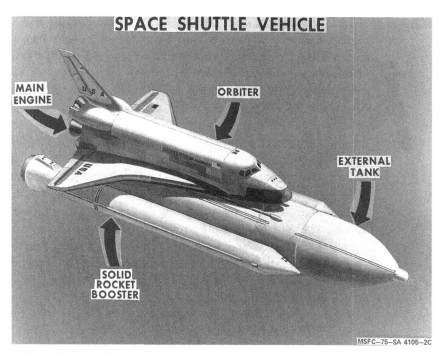

**Figure 2.1**    *Challenger* Shuttle: (1) the external tank, (2) two solid rocket boosters, (3) the *Challenger* orbiter, and (4) the main engine (Courtesy of NASA, *Report of the Presidential Commission on the Space Shuttle* Challenger *Accident*).

The two enormous solid rocket boosters (SRBs) attached to a space shuttle orbiter provide 80 percent of the thrust necessary to propel the shuttle into space. The aft field joint on the right SRB (shown in Figure 2.2 and labeled "aft segment with nozzle") is the one that failed in the *Challenger* flight, causing the orbiter and SRBs to explode. Each of the main segments is 12 feet in diameter and about 13 feet long. It is filled with solid fuel for the flight.

The SRBs are essential elements in the operation of the shuttle. An SRB is attached to each side of the external fuel tank. Each booster is 149 feet long and 12 feet in diameter. Before ignition, each booster weighs 2 million pounds. A few minutes after launch the SRBs are supposed to detach and parachute back to earth. Solid rockets in general produce much more thrust per pound than their liquid fuel counterparts. Once the solid rocket fuel has ignited, it cannot be controlled, however.

Seven cylindrical segments make up the SRB as shown. The joints where the segments are joined together are known as "field joints." (Figure 2.3 shows a field joint.) They are a tang and clevis connection with two large O-rings, positioned concentrically with the SRB. The O-rings are necessary to prevent hot gases from leaking through the joints of the SRB. Two are used for redundancy because leakage through the joints could ignite the external fuel tank. For this reason, Boisjoly's observation that hot gas had passed through the primary O-ring in Flight 51C alarmed him.

The tang fits in the clevis. The insulation protects the joint from the heat of the propellant. The O-rings are only a half-inch apart, but 12 feet in diameter. One hundred seventy-seven steel pins hold the joints in place. The O-rings shield the joint from 5800-degree combustion gases inside the booster. The diagram shows

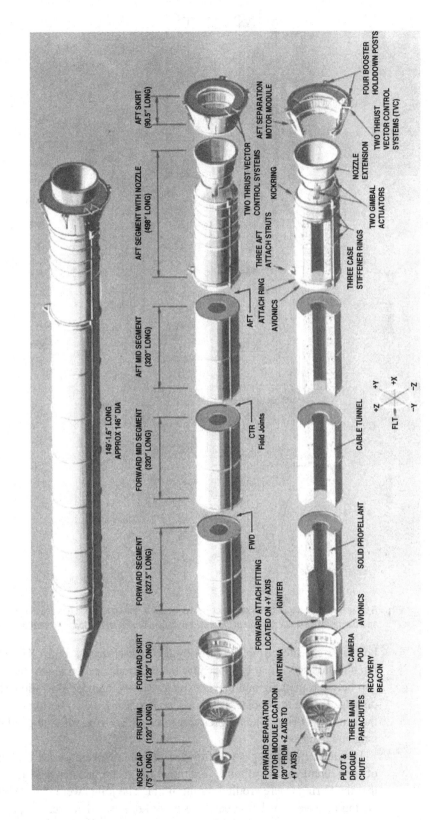

**Figure 2.2** A Solid Rocket Booster and Its Segments (Courtesy of NASA, *Report of the Presidential Commission on the Space Shuttle Challenger Accident*)

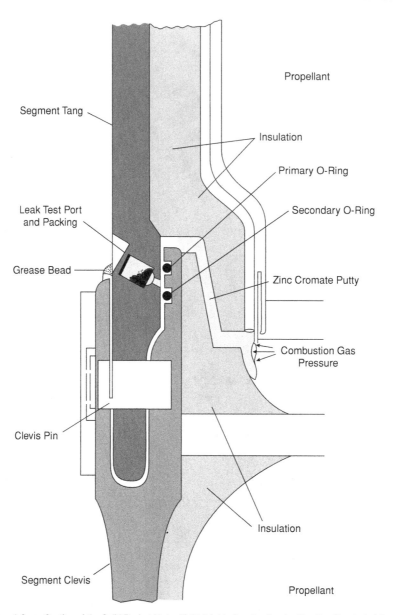

Segment Tang

Leak Test Port
and Packing

Grease Bead

Clevis Pin

Segment Clevis

Propellant

Insulation

Primary O-Ring

Secondary O-Ring

Zinc Cromate Putty

Combustion Gas
Pressure

Insulation

Propellant

**Figure 2.3**  A Cross Section of the Solid Rocket Motor Field Joint before the Combustion Gas Penetrated the Putty (based on a diagram from NASA's *Report of the Presidential Commission on the Space Shuttle* Challenger *Accident*)

hot gases shielded from the joint by the zinc-chromate putty. The O-rings are intended to "seat," that is, move into the positions needed to seal the joint, as the gap between the tang and clevis expands. In its intended operation, the putty is displaced when the booster is ignited, compressing the air between the putty and the primary O-ring. The air pressure forces the O-ring into the gap between the tang and clevis.

Leakage of exhaust gas was first discovered in November 1981, after the flight of the second shuttle mission. Examination of the booster field joints revealed that the O-rings were eroding during flight. The joints were still sealing, but the O-ring material was being eaten away by hot gases that escaped past the putty. Morton Thiokol studied different types of putty and its application to assess their effects on reducing O-ring erosion. The explosion on this particular shuttle flight occurred because hot combustion gases escaping from the rocket engines penetrated several meters of putty and passed the huge O-ring seals (rings 3 or 4 meters in diameter) to reach the external fuel tank. The record low temperatures experienced that night had made the O-rings less resilient, so they had not seated properly (i.e., moved into the proper place).

*Using the information about the shuttle design provided in this section, explain why the escape of hot gas through the primary seal during shuttle flights was a matter of grave concern.*

### Pursuing a Hypothesis about the Effect of Cold Temperature

When in your engineering work, you have a hypothesis about the cause of a life-threatening situation, what should you do?

NASA asked Morton Thiokol to give a more detailed presentation on the seal function as part of the flight readiness review for Flight 51E, scheduled for April 1985. Boisjoly presented his views at three successively higher-level review boards, but NASA management insisted that he soften his interpretation for the final review board.

The primary seal in Flight 51C, the January 1985 flight, had leaked gas in what was the worst temperature change in Florida history. That such conditions would soon recur seemed unlikely, certainly not for Flight 51E, which was scheduled for launch in Florida in April. Before pressing his hypothesis that low temperature had been a factor in the failure of the seals, Roger Boisjoly took the opportunity to test that hypothesis. He sought out his friend and colleague, Arnie Thompson, to discuss the blow-by (the leakage of hot exhaust gas) and the effect of cold temperature on O-rings' resilience.

---

### Lesson from Boisjoly's Action: Discussing Concerns with Peers

The situation in which Boisjoly found himself was not an emergency, and he had respected colleagues within his work group. In such circumstances, talking over the situation and one's own interpretation of it is a good idea because it enables one to:

- check one's own perceptions and interpretations.
- develop peer support for one's concerns.
- get practical suggestions and help in taking the next step in addressing those concerns.

---

Thompson proposed conducting tests of the effect of temperature on resiliency, which he and Boisjoly then carried out.

---

### Lesson from Boisjoly's Action: Conducting Tests

If the situation is not an emergency, define any risk as precisely as possible before carrying concerns outside one's immediate work group. This may require conducting experiments to improve one's estimate of the risk or to test the hypothesized causes of the risks.

However, if the test is expensive in time or equipment, securing approval to conduct the tests may first require making a good case for the hazard you perceive.

The resiliency testing showed that low temperature was a problem. Boisjoly and Thompson discussed the data with Morton Thiokol engineering managers, who considered the finding too sensitive to release.

*Suppose you are working in a large organization and have observed a serious problem with some system or device that you are working on. Describe what you think would be a good thing to do next. Give your reasons for your answer.*

### Stagnation in the Face of Mounting Evidence about Seal Erosion

What should you do, if, as an engineer working in a large company, you have good reason to believe there is a major safety problem in some system on which you are working and have presented your arguments and evidence through the usual channels, but nothing significant has been done to avert the danger?

Another post-flight inspection occurred in June 1985 at Morton Thiokol in Utah. This time a nozzle joint from Flight 51B, which flew on April 29, 1985, was found to have a primary seal eroded in three places over a 1.3-inch length. Inspectors postulated that the primary seal had never moved into place during the full two minutes of flight. The secondary seal in the same joint also showed signs of erosion. Boisjoly became even more concerned.

A flight readiness review presentation was prepared for Flight 51F, scheduled for launch on July 29, 1985. The status of the booster seals was the topic of a presentation to NASA at Marshall Space Flight Center on July 1, 1985, and of another the next day. The preliminary results of the O-ring resilience testing in March were presented for the first time during this meeting. Everyone in the program was by then aware of the influence of low temperature on the joint seals.

Management at Thiokol and NASA now had evidence that the seals did not prevent the escape of hot gases and that cold temperature was a factor in their failure to perform. Nonetheless, an attempt on July 19, 1985, to form a team to work on the seal erosion problem failed. Boisjoly recorded in his journal his frustration with management's failure to take steps to remedy the persistent failure of the O-rings.

### Boisjoly's Action: Keeping a Journal

By this time, Roger Boisjoly's heightening concern led him to begin keeping a journal of events pertaining to seal erosion. The journal would later become an important aid to him in giving testimony to the Presidential Commission investigating the *Challenger* explosion (the "Rogers Commission"). The journal had the immediate purpose of enabling Boisjoly to monitor events so that he could discover and remove roadblocks to fixing the problem.

Roger Boisjoly reports being influenced by the memory of an engineer who had been involved in another famous disaster: the 1974 crash of a Turkish Airline DC-10. For many years, this crash was the worst airline accident in terms of loss of life. Boisjoly remembered that the engineer whose designs had significantly contributed to this accident became almost totally dysfunctional for a long period after the crash and went through his workday under heavy sedation. This memory brought home to Roger Boisjoly that major safety problems might not be remedied through normal reviews. Therefore, he closely monitored all action taken on the problem of the O-ring seals.

In addition to the evidence from Flight 51B, Roger Boisjoly had his experimental results that showed seal erosion to be aggravated by cold temperature. Although no launches in cold temperature would occur in midsummer, Roger Boisjoly could see that the steps necessary to assure adequate sealing in all weather had not been taken. Therefore, he took action to force management's attention to the issue. He wrote a memo directly to the vice president of engineering, Bob Lund, stating his concern that failure to address the problem would mean an explosion of the shuttle. Roger Boisjoly's immediate superior immediately stamped the memo "company private," meaning that it was not to circulate beyond its addressee. (The AAAS award to Boisjoly for his efforts to avert the *Challenger* explosion was to specifically cite Boisjoly's letter to Lund as a praiseworthy attempt to address the safety issue of the behavior of the O-ring seals.) The persistent inability of management to take effective action on a major threat to safety justifies and even requires bringing the problem to the attention of responsible people in one's company, even if it requires going outside customary channels within that company. (Good companies generally set up alternate routes for raising concerns, and it is prudent to investigate these routes *before* joining a company.)

---

### Lesson from Boisjoly's Action: Expressing a Concern

Although the situation that Roger Boisjoly faced at this point was not an emergency, because no low-temperature launch was imminent, Boisjoly had good reason to think that management was not taking steps to properly address the hazard evidenced in seal erosion. He chose to communicate his concerns to Vice President Bob Lund in writing, although on other occasions Boisjoly had expressed his concerns by talking with Lund. Written communication ensures that a decision maker has a precise statement of the problem and that anyone else who is shown or who receives a copy of the communication sees the same statement, which they can each later review.

Some corporations, such as automobile manufacturers who have experienced lawsuits charging unsafe design in which employee memos have been subpoenaed, are reluctant to create records that might be used against them in future liability suits. They may discourage hard copy (paper and ink) or, more recently, email. However, these measures do attract attention.

---

### Lesson from Boisjoly's Action: Informing Others When Going to the Top

Inexperienced professionals who find that they are in a situation in which they need to go "over someone's head" or "to the top" with their concerns may neglect the question of whom they should inform or consult in doing so. Going to the top is less likely to offend those who are "leapt over" if they are at least informed, so that they are not caught unprepared for the actions that follow. Following his usual practice, Boisjoly showed his memo to his direct supervisor, who then countersigned it, although that supervisor signed it only as "concurred" rather than "approved" as that supervisor usually did with memos from subordinates.

---

Boisjoly's memo succeeded in getting the attention of top management, which then authorized the formation of a seal team.

*What should you do, if, as an engineer working in a large company, you have good reason to believe there is a major safety problem in some system on which you are working and have presented your arguments and evidence through the usual channels, but nothing significant has been done to avert the danger? Give reasons for your answer.*

### A Company's Concern about Its Image

Why are companies often reluctant to inform those outside their company of difficulties they are having? What factors are relevant in deciding how far to go in keeping one's company's problems confidential?

On August 19, 1985, Morton Thiokol personnel went to a meeting at the Marshall Space Flight Center on the problem with the seals on the booster rockets. In September 1985, NASA officials instructed Morton Thiokol to send a representative to the Society of Automotive Engineers (SAE) conference in October to discuss the seals and solicit help from others at the conference. Boisjoly was selected to make the presentation but NASA gave strict instructions that he was not to express the critical urgency of the joint problem, but rather to emphasize the progress on solving it that the company had made thus far.

Every organization has difficulties that it overcomes in the normal course of its operation. A company is usually reluctant to publicize its mistakes and failures. In September of 1985, NASA was receiving public criticism for promising commercial uses of shuttle flights that never materialized, for having cost overruns, and for being behind schedule. Morton Thiokol for its part was worried about losing its position as sole contractor for certain parts of the space shuttle effort.

NASA's unwillingness to reveal major malfunction in the solid rocket booster hampered Boisjoly's attempt to get expert advice about the seal problem. NASA's instruction was not so outlandish that Boisjoly felt morally obliged to defy it, however.

A situation in which a computer professional had concerns about a flaw that presented a serious threat is the 2005 case, popularly known in the information technology (IT) community as "Ciscogate" or the "Black Hat Bug." This case illustrates some differences between the IT community and other areas of engineering with respect to bringing at least security flaws to the attention of the whole community at professional meetings and on news Web sites. In the "Black Hat Bug" case, a computer professional (security researcher Mike Lynn) at Internet Security Systems (ISS) discovered a serious security flaw in Cisco IOS, the operating system powering ISS routers. He met with some reluctance from Cisco Systems to recognize the flaw, so he sought to bring this flaw to the attention of the IT community through a presentation at the 2005 Black Hat Conference. (This case is discussed in Chapter 5 in connection with free speech and whistleblowing. The subject of raising ethical concerns will be extensively examined in Chapter 7.) The story illustrates a feature of the IT community that is not common in the rest of engineering. Computer professionals who have despaired of having their concerns about security flaws heard within and remedied by the company with the flawed technology frequently take their concerns directly to the rest of the IT community, in part so that the members of that community can protect themselves against damage resulting from the flaw. When, unlike Mike Lynn, they wish to remain anonymous because they fear retaliation for "blowing the whistle," they can post their concerns under a pseudonym on one of the commonly visited news sites, such as Slashdot and The Register. Other areas of engineering do not have established Web sites for posting concerns.[6]

After his presentation, Boisjoly asked the audience for suggestions to improve the design, but received none. Boisjoly and another Morton Thiokol engineer, Bob Eberling, spent the remainder of the convention meeting with seal vendors whom they had previously contacted for help.

*Why are companies often reluctant to inform those outside their company of difficulties they are having? What factors are relevant in deciding how far to go in keeping one's company's problems confidential? Give reasons for your answer to the second question.*

### Working with Poor Management Support

If one is continually clamoring for attention, others are likely to stop listening. What factors are relevant in deciding whether and how to protest that one does not have the support to do one's job properly?

Although the seal task team had been formed in response to his July memo, Boisjoly reports that management did not give the effort much support and that the team lacked necessary resources and information. Many unanswered questions remained as the seal task team approached the end of 1985: Almost twenty flights had flown successfully and some of the cases of hot gas blow-by had occurred during warm as well as cold temperatures.

In attempting to bring attention to the problem, Boisjoly used normal channels to the fullest and had already presented his concerns directly to the vice president of engineering. He continued to keep his journal of progress, or lack of progress, on the seal problem and used activity reports to document the frustration of his efforts, including attempts to get more data on seal erosion. He received no response, however, and never knew if his comments went to upper management. The Presidential Commission investigating the shuttle disaster later cited frequent failures to pass along vital information as a principal pattern of errors that led to the shuttle disaster.

Contacting an ombudsman or a safety hotline at Morton Thiokol or at NASA might have been appropriate at this point, but Morton Thiokol and NASA had neither.

*If one is continually clamoring for attention others are likely to stop listening. What factors are relevant in deciding whether and how to protest that one does not have the support to do one's job properly?*

[6]In recent years the NSPE has created an "ethics hotline" for its members. (See http://www.nspe.org/Ethics/EthicsResources/index.html.) Although all registered engineers are eligible for membership in the NSPE, most of its members are civil or structural engineers. Therefore, this service could be regarded as a service primarily for civil and structural engineers. The IEEE sponsored an ethics hotline for a period in the 1990s. When they terminated that service, some of those who staffed it helped to initiate an ethics helpline at the Online Ethics Center (OEC) (http://www.onlineethics.org). When the Online Ethics Center moved to the National Academy of Engineering, the helpline format was changed to the Ethics Case Discussions, linked from the OEC or available at http://www.ethicscasediscussions.org/.

### The Day and Evening before the Challenger Flight

What factors have a major influence on your available options when a safety problem suddenly becomes urgent?

The *Challenger*, with a crew that included "the teacher in space," was scheduled to fly on January 28, 1986. The preceding day Boisjoly and his colleagues were shocked to learn that the overnight temperature at the launch site was predicted to be only 18 degrees Fahrenheit, lower than the record cold experienced the previous year. Boisjoly and several of his colleagues were firmly convinced that this extreme weather condition presented a major threat to the capacity of the O-ring seals to function, and thus to the survival of the flight crew.

With time running out Boisjoly and his colleagues went directly to the vice president of engineering to make their case for postponing the flight. They convinced him of the danger and secured his decision to recommend against flying. To make their point to NASA at the teleconference scheduled for that evening, they hurriedly prepared viewgraphs outlining their concerns about launching at such a low temperature.

The teleconference linked Morton Thiokol with Kennedy Space Center (KSC) in Florida and the Marshall Space Flight Center (MSFC) in Huntsville, Alabama. A manager colleague who had long shared Boisjoly's concerns, Al McDonald, was present at KSC for the teleconference ("telecon"). Discussion started with a history of O-ring erosion in field joints. Boisjoly reports "data was presented showing a major concern with seal resiliency and the change to the sealing timing function and the criticality of this on the ability to seal. I was asked several times during my portion of the presentation to quantify my concerns but I said I could not since the only data I had was what I had presented and that I had been trying to get more data since last October." When Boisjoly made this last comment, the general manager of Morton Thiokol glared at him.

This presentation ended with the recommendation not to launch below 53 degrees. NASA then asked Joe Kilminster, Vice President of Space Booster Programs at Morton Thiokol, for his launch decision. Kilminster said that because of the engineering judgment just presented he would recommend against launching. Then Larry Mulloy of NASA at KSC asked George Hardy of NASA at MSFC for his launch decision. George responded that he was appalled at Thiokol's recommendation against flying, but said he would not launch if Morton Thiokol objected. Mulloy then spent some time giving his interpretation of the data, arguing that these data were inconclusive.

The vehement reaction of NASA's George Hardy to the recommendation against launch surprised Boisjoly. Not only was Hardy usually moderate in speech, but in Boisjoly's experience, NASA had shown a great concern for safety consciousness. Nevertheless, the Rogers Commission was to find that, although failure of the seals on the field joints caused the explosion, many other flaws in the shuttle design and poor patterns of communication might also have resulted in a fatal crash. NASA's prior reputation for safety seems to have rested on its practice of placing the burden of proof on those who advocated launch. If there was any question of a risk, a flight was normally postponed. This time, however, NASA did not follow its established practice.

Joe Kilminster responded by asking for a five-minute off-line caucus to reevaluate the data. The mute button was pushed, so the two NASA groups could no longer hear Morton Thiokol's discussion. Immediately Thiokol's general manager, Jerry Mason, said in a soft voice: "We have to make a management decision." It would be a mistake to interpret Mason's remark to mean that he thought that from management's point of view the explosion of the *Challenger* would be an acceptable outcome. Mason intended to consider factors other than safety, factors that would weigh in favor of launching. To consider other factors meant downplaying the danger that Boisjoly knew to exist. Boisjoly reports that he became furious when he heard Mason's remark. He describes the subsequent discussion as follows:

> Some discussion had started between the managers when Arnie Thompson moved from his position down the table to a position in front of the managers and once again tried to explain our position by sketching the joint and discussing the problem with the seals at low temperature.

> Arnie stopped when he saw the unfriendly look in Mason's eyes and also realized that no one was listening to him. I then grabbed the photographic evidence showing the hot gas blow-by and placed it on the table and, somewhat angered, admonished them to look and not ignore what the photos were telling us. I, too, received the same cold stares as Arnie with looks as if to say, "Go away and don't bother us with the facts."

> At that moment, I felt totally helpless and that further argument was fruitless, so I, too, stopped pressing my case. . . .

> During the closed managers' discussion, Jerry Mason asked in a low voice if he was the only one who wanted to fly. The discussion continued, then Mason turned to Bob Lund, the vice president of engineering, and told him to take off his engineering hat and put on his management hat. The decision to launch resulted from the yes vote of only the four senior executives since the rest of us were excluded from both the final decision and the vote poll. The telecon resumed and Joe Kilminster read the launch support rationale from a handwritten list and recommended that the launch proceed.

> NASA promptly accepted the recommendation to launch without any probing discussion and asked Joe to send a signed copy of the chart.

> The change in decision so upset me that I do not remember Stanley Reinhartz of NASA asking if anyone had anything else to say over the telecon. The telecon was then disconnected so I immediately left the room feeling badly defeated.

In this instance, NASA responded in a way that was unprecedented in Boisjoly's experience. In the wake of the success of the Apollo project that had placed astronauts on the moon, NASA had conceived the shuttle project to retain public support. It had overpromised achievement. Having failed in many of its promises, NASA felt pressure to have the flight with "the teacher in space" as a visible success that could be mentioned in the State of the Union Address, which President Reagan was to give the next day. As Boisjoly was later to find out, Morton Thiokol

Figure 2.4 The Explosion of the *Challenger* (Courtesy of NASA, *Report of the Presidential Commission on the Space Shuttle Challenger Accident*)

was at this point negotiating a new contract with NASA and trying to remain the sole contractor for the solid rocket booster program. This negotiation undoubtedly influenced the top management's decision not to delay the *Challenger* flight, and so reversed its engineering decision.

Immediately after the launch of *Challenger* on January 28, 1986, the first signs of failure of a joint in the right SRB were visible in the form of puffs of black smoke that spewed out of that joint three to four times each second. Their color suggested that 5800-degree gases were eroding the O-rings; at the end of the first minute, a small but steady flame was evident. The explosion of the *Challenger* is shown in Figure 2.4.

Faced with a horrible outcome it is tempting to wish for a miraculous "rescue." Thus, it is often suggested that Boisjoly, Thompson, or McDonald ought to have done something more to see that the flight was stopped. Calling the press or notifying the astronauts are favorite suggestions. Calling the press would have been excessive before the teleconference because the vice president of engineering had already agreed to postpone the flight. After the teleconference, a media story would have come too late, even if the engineers had known a responsible journalist whom they could have readily contacted.[7]

The astronauts themselves are sequestered the night before each flight. Had someone called their families, it is not clear how they would have interpreted a call from a stranger claiming to be a Morton Thiokol engineer who thought the flight was unsafe, what they could have done with such information, or whom they could have convinced. Those like Boisjoly and Thompson who took many

[7] For a perspective from within the engineering profession on the appropriate use of the media in raising ethical concerns, see the NSPE Board of Engineering Review discussion of Case 88-7. This case, "Public Criticism of Bridge Safety," the Board's discussion of it, and three related cases are available in the Online Ethics Center at http://onlineethics.org. They originally appeared in *Opinions of the Board of Ethical Review* Volume VI, Alexandria, VA: National Society of Professional Engineers, 1989, pp. 117–119.

personal risks to bring their safety concerns to the foreground have regrets about the outcome but few about their own behavior. Boisjoly says he regrets only not vigorously advocating for a change the temperature criteria for launch to fifty degrees or more when he first became aware of the cold temperature threat. On the other hand Bob Eberling, one of the engineers who agreed with Boisjoly and Thompson but did not confront management the night of the teleconference, has said, in January of 1996 on the television program *60 Minutes*, that he regrets he did not do more that night.

The Presidential Commission found multiple instances in which safety concerns were not appropriately communicated to those with decision-making power. The failures to communicate safety problems turned up by the Presidential Commission were so grave that astronaut Sally Ride, who served on the Commission, refused to join any more shuttle flights thereafter. As Commission member Richard Feynman put it, "The guys at the top . . . didn't want to hear about the difficulties of the engineers. . . . It's better if they *don't* hear so they can be more 'honest' when trying to get Congress to OK their projects."[8] NASA's deputy administrator, Hans Mark, wrote in an article in the *IEEE Spectrum* that he had known and written a memo about the O-ring problem two years before, so it seemed incredible to him that Jesse Moore, the NASA decision maker in charge of the *Challenger* flight, did not know of the O-ring problem. Thus, even if Boisjoly *had* been able to find someone in authority at NASA to tell that evening, he would have been giving them information that they already showed resistance to receiving.

Later Boisjoly did take information to outside authorities – he became "a whistleblower." He gave documents to the Presidential Commission investigating the *Challenger* explosion without first giving them to Morton Thiokol to review and censor. His disclosure of the information to the Presidential Commission led to sanctions against him at Morton Thiokol and to his being ostracized in the little Utah town where he had previously been mayor. Boisjoly showed himself willing to give documents to the Presidential Commission despite the risk to his career and his desire to continue to reside in a town for which Morton Thiokol was the principal employer. Therefore, it is reasonable to assume that had he known of a more effective way of raising his concern about the effect of temperature on the seals he would have done so. To require that professionals behave responsibly cannot imply that they should be blamed for bad outcomes or that others are not responsible for supporting their efforts to safeguard the public's safety, health, or well-being.

In this instance, those in authority behaved not only unreasonably but also unpredictably. An engineer in Boisjoly's position could not have known in advance that going outside the company would be necessary. If engineers are prepared to raise safety issues as clearly, forthrightly, and persistently as Boisjoly and some of his colleagues, they will be meeting their responsibility for safety – although they may not be able to prevent every disaster. If corporations and government

---

[8] Unger, Stephen H. 1994. *Controlling Technology: Ethics and the Responsible Engineer*, second edition. New York: John Wiley & Sons, Inc.

agencies support engineers who raise safety concerns, a single bad decision will not create a disaster such as the explosion of the *Challenger*.

Accidents and safety problems are clearly not in the interest of management. When the general manager of Morton Thiokol recommended launching over the objections of his engineers, he made a very bad management decision.

*What factors have a major influence on your available options when a safety problem suddenly becomes urgent?*

### Preventing Accidents

What attention to safety problems would be reasonable to expect in a reputable technology company and why?

The case histories of technological accidents and related health hazards reveal few instances of flagrant disregard and suppression of evidence on health and safety risks: The behavior demonstrated by the asbestos industry and that by A. H. Robbins, makers of the Dalkon Shield, are the exception. Fewer companies even undertake the cynical comparison of the cost of legal liabilities for death and injury to the cost of preventing accidents (as Ford did with the Pinto gas tank). Cases such as those of the Dalkon Shield and the Pinto gas tank have received the greatest press attention precisely because of the outrageousness of the decisions involved. A review of case histories of accidents shows that the *Challenger* case is more typical of poor management decisions that result in accidents: Usually management does not flagrantly disregard safety, but rather fails to give sufficient attention to safety hazards because of the pressure of deadlines or financial exigencies.

A review of accident cases fails to pick up the many situations in which managers and corporations responded appropriately to the safety concerns of their engineers. Many corporations realize that it is in their interest to provide their employees with adequate opportunities to raise concerns about safety, and are taking steps to provide their employees with avenues to raise concerns about safety and other ethical matters. Attention to how organizations support the timely expression of employees' ethical concerns is at least as important to preparing for professional responsibility as reviewing disasters. Therefore, this book discusses good ways that universities, departments, companies, and organizations foster responsible action by their members. This information is meant to help readers assess the ethical climate of an organization before they join it.

*What attention to safety problems would be reasonable to expect in a reputable technology company and why?*

### A Note on the Challenger Disaster as a Formative Experience for Many Engineers and for Popular Culture

For the currently employed generation of engineers the *Challenger* accident functions as a so-called flash-bulb memory, much as the memory of the Kennedy assassination did for the previous generation. Many engineers can remember where they were when they heard about, or saw, the *Challenger* explosion. Often it was watching the launch with "the teacher in space." Many testify that when the explosion occurred, killing the seven astronauts, including that teacher, their own teachers (perhaps because they were so shocked) were unable to talk about what had occurred and simply ushered them back to their classes. Some students were lucky enough to have teachers or parents who could talk to them about what had occurred, but many older students recall it as an experience of confusion or disillusionment with authority figures.

The case has provided not only another famous accident for study but also a personal experience of people being reluctant to deal forthrightly with bad news. This formative experience for engineers born between the mid-1950s and the mid-1970s is likely to affect the culture of engineering for years to come.

Roger Boisjoly's attempt to avert the *Challenger* disaster deserves careful study for what it reveals about the commonly occurring situations in which engineers in large companies are challenged to fulfill their professional responsibility for safety. Boisjoly's problem situation and the responses he made to it at various stages of its development illustrate the recognition of safety problems and actions that are likely to be effective in addressing those problems.[9]

## Section 2. William LeMessurier's Handling of the "Fifty-Nine Story Crisis"

Technology is always advancing. How should one respond when one learns that previous regulations failed to bring to light serious safety problems with one's previous engineering work?

The Citicorp Tower* in New York City was completed in 1977.[10] William J. LeMessurier designed the supporting structure for this unusual skyscraper. Shortly after its completion, LeMessurier discovered that the building was more vulnerable than expected to toppling over in strong winds such as hurricanes. Hurricanes of the strength that New York City experiences about every sixteen years could bring down the building. Stresses caused by quartering winds (winds that hit the building on the corner and so acted on two sides at once) were a special threat that was compounded by the substitution (without LeMessurier's knowledge) of bolts for welds on the diagonal supports of the structure, and two few bolts at that! The harrowing story of LeMessurier's discovery of the defect and correction of the flaw in his own design is another instructive example of a person fulfilling the moral responsibilities that go with being an

---

[9]The description of these attempts given here draws heavily on the account that Roger Boisjoly gave in January 1987 in an address at MIT, on another address that he gave in September 1989, and on answers to questions he gave to those audiences. The text of the first talk is published in several places, including *Books and Religion*, 1987, 15 (March/April): 3–4, and Johnson, 1991, pp. 6–14. It may be found in the Online Ethics Center at http://www.onlineethics.org/CMS/profpractice/exempindex/RB-intro.aspx.

*In 2008, its official name was changed to "611 Lexington Avenue."

[10]The story of William LeMessurier's resolution of the crisis with the Citicorp Tower appears in the Online Ethics Center. Included in the site are slides showing Chicago's Hancock Building and other innovative skyscrapers as well as the Citicorp Tower. These are from an address LeMessurier gave in 1995. A videotape of this talk is available through the Online Ethics Center.

engineer. Only in the 1990s was the story made public.[11] LeMessurier's timely revelation of the danger to authorities, coupled with appropriate action of other key decision makers, averted what would have been a disaster of astounding proportions.[12]

LeMessurier drew on his experience with the Citicorp building when he later consulted on the equally serious problems with I. M. Pei's John Hancock Tower in Boston – and there is much to be learned from how LeMessurier went about marshaling the resources to eliminate the threat.

When Citibank began planning for a new headquarters tower in midtown New York, the art of designing and building a structurally safe skyscraper seemed nearly perfected. After the development of steel frame buildings and Elisha Otis's successful introduction of the safety-brake-equipped elevator in the 1850s, architects began to design ever-taller buildings. The Home Insurance Building constructed in Chicago in 1885 was the first multistoried building to have a complete structural frame supporting its masonry walls.

By the 1930s, when the 102-story Empire State Building was completed, skyscrapers had begun to appear in cities all over the world. Creative architects and engineers introduced further innovations in the design and construction of tall structures that called for lighter materials and columnar supports. Chicago's Hancock Building, for instance, incorporated an innovative system of diagonal bracing that allowed the building to be much leaner and lighter than would have been possible with the more customary structural supports.

*Technology is always advancing. How should one respond when one learns that previous regulations failed to bring to light serious safety problems with one's previous engineering work?*

### LeMessurier's Innovative Design for the Citicorp Tower
Do innovative designs present special dangers?

Before his firm was engaged as consultant on a new corporate headquarters for Citibank, William LeMessurier had already distinguished himself as a preeminent structural engineer with extensive experience with skyscrapers. In the first skyscraper that he designed, Boston's State Street Bank, he incorporated an inventive cantilever girder system. In the design of the famous Boston Federal Reserve Bank, he created an opening at the base of the building that proved large enough for an airplane to fly through.

The Citibank headquarters needed an innovative design because of special constraints of the site. A church congregation owned part of the block on which

---

[11] It was told in the May 29, 1995, issue of the *New Yorker*, 45–53.

[12] The account given here is primarily based both on Joe Morgenstern's May 29, 1995, *New Yorker* article "Fifty-Nine Story Crisis" (pp. 45–53) – which is currently available at http://www.duke.edu/~hpgavin/ce131/citicorp1.htm – and on an address by William J. LeMessurier at the Massachusetts Institute of Technology on November 17, 1995. From 1995–2007, copies of LeMessurier's lecture (which included many technical details not repeated here) were widely distributed through the Online Ethics Center as a videotape, and then as a CD, both titled "The 59 Story Crisis: A Lesson in Professional Behavior."

Figure 2.5 Citicorp Tower* (Photo courtesy of Caroline Whitbeck)

Citicorp planned to build and needed a new church building. Citicorp agreed to replace the old church building with a new structure. In return, Citicorp gained the air rights above the new building. Leaving more open space at ground level also allowed the Citicorp Tower to be taller than zoning laws would otherwise have allowed.[13]

To allow for the church underneath, the Citicorp Tower was constructed on four nine-story stilts and a central column in which the elevators were to be located. The church site was at a corner of the lot, which was one reason the stilts had to be under the middle of each of Citicorp Tower's outer walls, rather than under its corners (see Figure 2.5). This posed a challenging structural engineering problem. LeMessurier's solution was to use large diagonal girders throughout the building; these would transfer the Tower's great weight to four huge columns that would run the height of the building on each side and anchor the structure to the ground (see Figure 2.6). The new church could then be constructed underneath one corner of the Citicorp Tower.

LeMessurier's innovative diagonal bracing for the Citicorp Tower greatly reduced the weight of the structure. The reduced density of the Tower also makes it more dynamically excitable: it would have had an unpleasant tendency to sway in the wind had it not been for a tuned-mass damper. This damper, installed at the top of the building, consists of a 400-ton concrete block floating on pressurized oil bearings attached to two horizontal springs at right angles to one another. The

---

*Other photos of the Tower, including dramatic shots taken at the base of one of the stilts, may be found at http://www.thecityreview.com/citicorp.html.

[13] For this last constraint on the design of the Tower, I am indebted to Diane Hartley, personal communication, June 15, 2010.

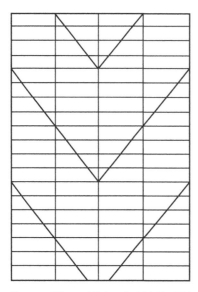

**Figure 2.6**    Diagram of Diagonal Girders Each Linking Eight Floors

innovation of the damper made this the first building to have a mechanical aid *as part of its design.*

*Why might innovative designs present special dangers?*

### The Discovery of the Change from Welds to Bolts

When one creates a design with the expectation that it will be constructed or manufactured in a particularly expensive way, what possible changes to one's design are relevant to consider?

In May 1978, LeMessurier, acting as structural consultant to another building being planned in Pittsburgh, again considered using diagonal bracing in his design. As in the Citicorp Tower, his plans called for full-penetration welds to hold the sections of the braces together. (Welded joints are extremely strong, but as a potential contractor for the Pittsburgh construction job pointed out, they are time-consuming to make and much more expensive than bolted connections.) At this point LeMessurier learned that the substitution of bolts for welds had been suggested during the Citicorp Tower's construction, and the Citicorp contractors had decided to put the braces together with bolted joints to save the cost of welding. Those steel contractors for the Citicorp Tower had judged that bolts would be strong enough to make the structure safe.

When LeMessurier referred the Pittsburgh contractor, who was concerned about the cost of welding, to the successful Citicorp job, he (LeMessurier) learned of the substitution of bolts for welds. Initially LeMessurier was not alarmed, because the substitution was reasonable from an engineering standpoint. Because LeMessurier had served as a consultant on the Citicorp project, he saw no reason that they should have informed him earlier. His assessment of the effect of the substitution was to change, however, when another threat came to light, a threat that was compounded by the substitution of bolts for welds.

*When one creates a design with the expectation that it will be constructed or manufactured in a particularly expensive way, what possible changes to one's design are relevant to consider?*

### Investigating the Effects of Quartering Winds

Once an engineering assignment has been completed are there any reasons for thinking that an engineer involved in the original work would have a special obligation to warn of any dangers he finds in the resulting construction of product?

In June 1978, a month after LeMessurier was told of the switch from welds to bolts, he says he received a telephone call from a student from New Jersey.* LeMessurier recounts that the student's professor had been studying LeMessurier's design for the Citicorp Tower and had concluded that LeMessurier had erred in placing the building's nine-story stilts in the middle of the four walls rather than at the Tower's corners.

LeMessurier's account is that the professor had misunderstood the constraints of the design problem that faced LeMessurier. So he said he called the student back (when he had more time) to explain his reasons for putting the Tower's supports at the building's midpoints. He adds that he thought his unique design, including the supports and the diagonal-brace system, made the building particularly resistant to quartering winds.

Shortly thereafter, LeMessurier says that he decided that the subject of the Citicorp Tower and quartering winds would make an interesting topic for the structural engineering class he taught at the time. Because the then current requirements of the New York City building code (like other building codes of the time) extended only to the effects of perpendicular winds, LeMessurier did not know how his design would fare in quartering winds.

LeMessurier says that he was interested to see if the building's diagonal braces would be as strong in quartering winds as they had been calculated to be in perpendicular winds and did some computations (shown diagrammatically). He found to his dismay that in a quartering wind, although some of the stresses would vanish, in four of the eight chevrons, stresses would increase by 40 percent.[14]

Then he became concerned about the replacement of welds with bolts. Had the New York contractors considered quartering winds when they replaced the welds with bolts? Had they used enough bolts? The second question was particularly important, because the 40 percent increase in stress on certain structural members resulted in a 160 percent increase in stress on the building's joints. Therefore, it was vital that enough bolts be used to ensure that each joint was sufficiently strong.

---

*That student has been reported to me by Professor Donna Riley (email 2007) as having been Diane Hartley, who was at the time a Princeton engineering student studying with David Billington. Diane Hartley's story is explored in the next section of this chapter.

[14]There is an inconsistency about just how much the stress on some members increased between the account in the *New Yorker* and LeMessurier's MIT lecture, although the first as well as the second is presumably based on LeMessurier's calculations. Morgenstern's *New Yorker* article has the figure of 40 percent, but in his MIT talk, LeMessurier (speaking without notes) said that some "doubled." Partly because LeMessurier was speaking without notes and partly because Diane Hartley's thesis also arrives at the figure of 40 percent, I have used 40 percent here.

LeMessurier reports being disturbed by what he next discovered. The New York contractors had not considered quartering winds when they substituted bolted joints for welded ones. Furthermore, the contractors had considered many of the Tower's diagonal braces as *trusses* (rather than as supporting *columns*) and so exempted them from load-bearing calculations. As a result, they had used far *too few* bolts.

Shaken, LeMessurier reviewed old wind-tunnel tests of the building's design. When he compared these against his new quartering-wind calculations, he discovered that under adverse weather conditions, the Tower's bracing system would be put under even further stress.

*Once an engineering assignment has been completed are there any reasons for thinking that an engineer involved in the original work would have a special obligation to warn others of any dangers he finds in the resulting construction of product? If so, what would be the basis for such a special obligation?*

### Wind Tunnel Evidence of the Danger

How do LeMessurier's actions in asking Alan Davenport to repeat the wind tunnel tests compare, ethically, with Roger Boisjoly's conducting bench testing and discussing his concerns with his colleague, Arnie Thompson?

LeMessurier now believed there might be grave danger. He turned to Alan Davenport, a Canadian consultant during the building's design phases, for further confirmation. Davenport, who had run the original wind tunnel tests, ran the tests again using new calculations to reflect quartering winds and the change from welds to bolts.

The results, when compared with the building's original testing, confirmed LeMessurier's suspicion that stress in some of the building's structural members would increase. His concern grew, for the results indicated that a 40 percent theoretical increase in a member's structural stress would be much greater under real-world conditions. During a storm, the whole building could shake, causing all of the structural members to vibrate synchronously.

A "sixteen year storm" in New York City (one that occurs on average every sixteen years) would have the strength to cause total structural failure, if the storm also knocked out electrical power necessary to run the building's tuned-mass damper. (The tuned-mass damper keeps the building from swaying in a wind and discomforting the occupants. It has an enormous steadying effect.)

The engineering remedy was straightforward; heavy steel-welded "band-aids" over the joints would give the building more strength than its original design. The repairs needed to be finished before a hurricane hit, however. It was then the last day of July, and hurricane season was just beginning. To accomplish the repair, LeMessurier would have to reveal the building's vulnerability. Doing so could cost him his career and reputation as a structural engineer. He could not predict the reception of his news by Citibank leadership, city officials, or the public.

*How do LeMessurier's actions in asking Alan Davenport to repeat the wind tunnel tests compare with Roger Boisjoly's conducting bench testing and discussing his concerns with his colleague, Arnie Thompson? (What is ethically important is how these sorts of actions might have been expected to further safety, rather than details of differences in the tests performed.)*

### *Informing Those Who Need to Know and Mobilizing Support*
How comparable was LeMessurier's mobilization of support for fixing the joints on the Citicorp Tower with Roger Boisjoly's mobilization support among his peers at Morton Thiokol for preventing launch of the shuttle in cold weather?

On July 31, LeMessurier contacted the lawyer of the architectural firm that had retained him as its structural consultant for the Citicorp Tower. He then contacted the firm's insurance company. As a result, a meeting was arranged for the following day with several lawyers for the insurers, to whom LeMessurier related the entire story. The lawyers soon decided to bring in a special consultant, Les Robertson, a respected structural engineer. Robertson listened to LeMessurier's description of the situation, and then took a more pessimistic view than even LeMessurier. Robertson did not believe that, for instance, the tuned-mass damper would serve as a safety device despite LeMessurier's assurances that generators could keep the dampers running during an electrical power loss.

Citicorp had to be informed of the danger, so LeMessurier and his partner tried to contact Citicorp's chairperson, Walter Wriston. Initially, Wriston was not available to them, but LeMessurier's partner was able to arrange a meeting with Citicorp's executive vice president, John Reed, who had engineering experience and played a part in the construction of the Tower. Once more LeMessurier detailed the situation. When prompted for a cost estimate, he guessed that $1 million would be sufficient. He also explained that the repairs could be done without inconvenience to the tenants by isolating the bolted joints within plywood "houses" and doing the necessary work at night within those houses.

Reed appreciated the gravity of the situation, and arranged a meeting with Walter Wriston on August 2. LeMessurier again told his story. Much to his relief, Wriston recognized the importance of the Tower as Citicorp's new corporate emblem, and so readily agreed to the repair proposal. He approved a plan to install emergency generators as a backup power supply for the tuned-mass dampers, and oversaw much of the relations with the public as well as with the building tenants.

The next day, LeMessurier met with two engineers from the construction company that was to perform the repairs. After examining the joints, these engineers approved LeMessurier's plan to reinforce the bolted joints with welded band-aids.

Before undertaking the repairs, several steps were necessary. LeMessurier contacted the company that had constructed the tuned-mass damper to help ensure the device's continuous operation. Meteorological experts were retained to give advance warning of any storm that could cause the building's destruction. LeMessurier reluctantly agreed with Robertson that, as a further precaution, an emergency evacuation plan for the building and the ten-block-diameter surrounding neighborhood be drafted. In its final form, the plan was to involve up to 2,000 emergency workers provided by the Red Cross.

LeMessurier explained the situation to city officials both to secure their cooperation with the evacuation plan and to comply with the building code.[15] They

[15] The NSPE BER created a case, 98–9, that resembles the Citicorp story in some but not all respects. It crucially ignores William LeMessurier's notification of city officials and that he did return the phone calls from the *New York Times* (only to find that the paper was shut down due to a strike). The NSPE BER judges that the engineer in its story failed to act ethically in failing to do

responded with approval and encouragement, rather than with the cynicism that LeMessurier expected. They too recognized both the seriousness of the problem and the immediate need to solve it. Energy was not wasted on rancor or placing blame.

The final task and the one LeMessurier most dreaded was informing the press of what was going to be a major undertaking on the brand-new Citicorp Tower. An initial press release was issued. It indicated that the building was being refitted to withstand slightly higher winds. Indeed the meteorological data suggested that the winds for that year were going to be somewhat higher than normal. The *New York Times* (*NYT*), for one, was sure to express further interest in what could be a very juicy story. LeMessurier did return the phone call from a *NYT* reporter, but found an unexpected reprieve from the interview when the paper shut down in a citywide press strike.

> *How would you compare LeMessurier's mobilizing support for fixing the joints on the Citicorp Tower with Roger Boisjoly's mobilizing support among his peers at Morton Thiokol for preventing launch of the shuttle in cold weather? (Specify ethically relevant similarities or differences and their significance.)*

### Accomplishing the Repair without Causing Panic

Consider the lack of awareness of danger by the occupants of the Citicorp Tower and surrounding buildings during the repairs. What reasons can you give for thinking they should or should not have been told of the danger? If you believe they had a right to have been told, identify who should have told them and why it was that person's or persons' moral obligation to do so.

Repairs to the Citicorp building commenced immediately. The plan of action was to complete the repair welding at night when the tenants were not in the building, so as not to inconvenience them. Each bolted joint in the building was to be exposed by ripping away the flooring and walls around it and each was to be covered with a plywood house to minimize any visible signs that things were awry with the building's structure.

The pace of the work was fast. Parts of the interior around the bolted joints were torn up at night and put back together in the morning. LeMessurier occupied himself with repair process calculations. Les Robertson calculated how to repair the joints and, suspecting that other components of the building could be vulnerable, investigated the floors, columns, and braces for weakness.

The repair work was in full swing on the first of September, when a hurricane moving toward New York was detected. The news was met with alarm. The partial repairs – along with the tuned-mass damper – greatly improved the building strength, but no one wanted to see it tested. There was great relief when the hurricane moved out over the ocean.

Two weeks later, repairs had progressed to the point that, with no storms predicted, the elaborate evacuation plans could be scrapped. The next month

so. Presumably, the NSPE BER's primary concern was to emphasize the importance of notifying the proper authorities, rather than making a judgment on William LeMessurier's actions, whom they did *not* name.

repairs were complete. On completion, even if the tuned-mass damper were to fail, a 700-year storm – a storm so strong that it is expected to occur in New York City only about once in a 700-year period – would not pose a threat to the Citicorp Tower.

The engineering problem had been solved, and the building now exceeds even its originally intended safety factor.

> *Consider the lack of awareness of danger by the occupants of the Citicorp Tower and surrounding buildings during the repairs. What reasons can you give for thinking they should or should not have been told of the danger? If you believe they had a right to have been told, identify who should have told them and why it was that person's or persons' moral obligation to do so.*

### The Insurer's Response: LeMessurier's Good Name

What sort of insurance risk do you think LeMessurier's actions showed him to be?

LeMessurier feared for his career but did not allow any worries or self-protective impulses to distract him from carrying out the repairs. In the middle of September, when work was almost complete, Citicorp notified LeMessurier and his partner that it expected to be reimbursed for the cost of the repairs.

The cost for the building's repair ranged from a high estimate of $8 million for the structural work alone, given by one of the construction companies involved, to $4 million, which, according to LeMessurier, was the Citicorp estimate. (Citicorp did not make public its estimate of the cost of repairs.)

LeMessurier's liability insurance company had agreed to pay $2 million, and LeMessurier brought that figure to the negotiating table. The Citicorp officials eventually agreed to accept the $2 million, to find no fault with LeMessurier's firm, and to close the matter.

A relieved LeMessurier nevertheless expected his insurance company to raise the premiums on his liability insurance. He would, he reasoned, appear as an engineer who had bungled an expensive job and caused the insurer to pay a large settlement.

At a meeting with officials from the insurance company LeMessurier's secretary was able to convince them that LeMessurier had "prevented one of the worst insurance disasters of all time!" Far from behaving in an incompetent or devious manner, LeMessurier had acted in a commendable way: He had discovered an unforeseen problem; acted immediately, appropriately, and efficiently; and solved it. LeMessurier's handling of the Citicorp situation increased his reputation as a competent and honest structural engineer. It also prompted his liability insurers to lower his premium.

> *What actions of LeMessurier would lead an insurance company to think that LeMessurier's actions showed him to be a particularly good risk?*

## Section 3. **The Mystery of the Misidentified Student**

> If the history of the Citicorp Tower redesign is as Diane Hartley recalls, what, if any, credit for detection of the hazard posed by quartering winds do the student and her advisor deserve?

In 1978 Diane Hartley was an engineering student at Princeton, studying with David Billington who was offering a course on structures and their scientific, social, and symbolic implications (subsequently titled, "Structure and the Urban Environment"). This course interested Diane Hartley early in her engineering studies and led her to pursue her undergraduate thesis with Billington, a thesis titled "Implications of a Major Office Complex: Scientific, Social, and Symbolic Implications."

In her thesis, Hartley looked into the Citicorp Tower, which had been recently built and was interesting for a number of reasons, including its innovative design. That design not only allowed a preexisting church to remain at ground level, but, because it left more open space at ground level, was permitted to be taller than zoning laws would otherwise have allowed.

When she contacted William LeMessurier's firm, they put her in touch with Joel S. Weinstein in their New York office, at the time a junior engineer with the firm. Mr. Weinstein sent her the architectural plans for the Citicorp Tower and many of his engineering calculations for the building. She reports that at the time she thought it odd that she did not see initials of another person beside those calculations, because the usual practice was for such work to be checked and initialed by a second engineer.

When Diane Hartley calculated the stresses due to quartering winds (winds hitting one of the corners of the building and so hitting two sides of the building at once), these made her concerned that they produced stresses that were significantly greater than those produced by a single side. Although calculation of quartering winds was not *required* by the then current building code, she assumed those calculations would have been done for a building with such an innovative design and asked Joel Weinstein for his calculations of the effects of quartering winds.* He said he would send them, but she did not receive them. When she told Joel Weinstein of the increased stresses that her calculations showed for quartering winds, he reassured her that the building was safe and its design was, indeed, "more efficient." Being an undergraduate, Diane Hartley reports that she deferred to Weinstein and quoted his words in his thesis, although his judgment was inconsistent with her calculations of quartering winds, which are also in that thesis. (David Billington, in his comments on Hartley's thesis, questioned this inconsistency.[16])

In recent years, when LeMessurier was asked by a coworker (who was acquainted with David Billington) whether the student might have been a woman, LeMessurier reportedly responded that he did not know because *he did not actually speak with the student*.[17] Was the reason that the (unnamed) student from an engineering school in New Jersey whom LeMessurier reports having prompted his examination of the effects of quartering winds on the Citicorp Tower was

---

*David Billington, Diane Hartley's undergraduate thesis advisor, reports that because the columns or "legs" of the Citicorp Tower were in the middle of each side, rather than at the building's corners, he had specific concerns about the effects of quartering winds. (Telephone interview, June 30, 2010.)

[16] Hartley, Diane. 1978. *Implications of a Major Office Complex,* senior thesis, Princeton University, 377.

[17] For this observation, I am indebted to Diane Hartley, personal communication, June 15, 2010.

represented as a male was that LeMessurier, having never spoken with Hartley, *assumed* the student was male?

> The story of the Citicorp Tower is at least a story of detection and remedying of hazards to the public safety and of how new engineers may lack confidence in their own engineering reasoning to press their recognition of safety problems.

When I first heard about Diane Hartley, I thought it was a case of inadequate credit, but although Hartley raised the issue with the New York office of LeMessurier's firm, she does not claim to have *consistently* pressed the issue of stresses due to quartering winds. What I do not know and cannot know is whether the load bearing calculations for the Citicorp Tower were done by Weinstein and these went unchecked, or whether they failed to include calculations for quartering winds, or even whether such calculations, though not required by the building code of the time, would have been expected for such an innovative design, as Diane Hartley believes, or would have been unusual, as LeMessurier says he was prepared to argue had Citicorp sued him or his firm for negligence for failing to consider quartering winds and the matter gone to trial. The story of the Citicorp Tower is at least a story of evaluating previously overlooked hazards to the public safety and marshaling resources to remedying them. It is also a cautionary tale about how new engineers may lack confidence in their own engineering reasoning to press their recognition of safety problems, and how readily (in the United States[18] at least) females in engineering are overlooked.

*If the history of the Citicorp Tower redesign is as Diane Hartley recalls, what, if any, credit for detection of the hazard posed by quartering winds do the student and her advisor deserve and why?*

## Section 4. Comparison of the Stories of Boisjoly and LeMessurier

> Consider the work situation that you expect to be in when you finish your final engineering degree. How would you act to bring appropriate attention to a safety problem, if you were the first to notice it?

It is emotionally satisfying to close with a happy ending, but life is not always so obliging, as we saw in the case of the *Challenger* flight. Despite the dramatically different outcomes of the two cases, the responses of Boisjoly and LeMessurier to the problems they faced have much in common. The principal differences between the two were their positions, their ability to influence the outcome, and the intelligence with which their communications were received.

After becoming aware of the possibility of danger, both Roger Boisjoly and William LeMessurier sought further information from testing and from the advice of others. Boisjoly was embedded in the organizational structures of Morton Thiokol and NASA, and worked with colleagues at every stage of his efforts. He enlisted the support of those who would listen and made appropriate use of

[18] The sex stereotyping of engineers and engineering is not found in all countries. I recall in particular a talented engineering student from Mauritius telling me that it was only when she came to the United States that she heard that engineering was a male field.

the organizational channels open to him. Although more solitary in the initial discovery of a threat, LeMessurier showed a similar resolve to test his concerns with peers and pursue a course that would safeguard life.

The inadequate response of Morton Thiokol and of NASA to evidence of the danger that the seals would not prevent hot gas from reaching the fuel tank contrasts strongly with the cooperation that LeMessurier received from Citicorp and from city officials. Boisjoly recognized Morton Thiokol's slowness in addressing the malfunction of the seals and was able to overcome it by taking the highly unusual action of writing directly to Morton Thiokol's vice president of engineering about the danger. Both Boisjoly and LeMessurier took appropriate actions that neither had witnessed anyone else take. Despite the profound difference in the ultimate outcomes, the actions of both Boisjoly and LeMessurier demonstrate some ways in which a concern for safety can be implemented in engineering practice.

*Consider the work situation that you expect to be in when you finish your final engineering degree. How would you act to bring appropriate attention to a safety problem, if you were the first to notice it?*

# PART 2

# ENGINEERING RESPONSIBILITY

# 3 Ethics as Design – Doing Justice to Moral Problems

How does one go about addressing an actual moral problem?

## The Perspectives of the Judge and the Agent

Deliberation about what to do is a subject on which, as philosopher Stuart Hampshire observed in 1949, philosophical ethics has had little to say. Hampshire made his point by saying that courses in ethics only teach students to critique moral actions rather than to resolve ethical problems. Writing *Innocence and Experience* some forty years later, he found the situation no better.[a]

As Hampshire pointed out, an agent (i.e., the person who confronts the problem) needs the skills of a judge in weighing alternative courses of action once these are formulated. The skills of a judge are only part of the skills an agent needs to respond to an ethical problem. The rest of the task is a constructive or synthetic one of devising and refining candidate responses.[b]

---

[a] Hampshire, Stuart. 1949. "Fallacies in Moral Philosophy," *Mind* 58: 466–482; reprinted in *Revisions: Changing Perspectives in Moral Philosophy*, edited by Stanley Hauerwas and Alasdair MacIntyre (Notre Dame, IN: Notre Dame University Press, 1983); and Hampshire, 1989, *Innocence and Experience*, Cambridge, MA: Harvard University Press.

[b] Hampshire does at one point speak of imagination in connection with the task of thinking up responses, and Patricia Werhane has picked up on this way of speaking about the task. "Imagination" may suggest a purely mental activity, however, which I find misleading. The engineering design experience, which in *concretely* exploring some responses often reveals others, seems to me a much more instructive guide to moral problem solving.

People confronted with ethical problems must do more than simply evaluate alternatives; they must also come up with those alternative responses: they must figure out what to do and *devise* a plan of action.

Ethical evaluations do have a role in devising responses to ethical problems, of course. These evaluations come in many forms, from "What is being proposed is morally wrong" to "This margin of safety is sufficient for the circumstances in which this device will operate." This book is concerned with devising good responses, which includes, but is not confined to, making ethical evaluations.

Suppose my supervisor tells me to dispose of some regulated toxic substance by dumping it down the drain. In this case, part of my problem is that what I have been ordered to do is potentially injurious to human health and illegal.

Assuming my supervisor knows that the substance is a regulated toxic substance – an assumption I should verify – then my supervisor is knowingly ordering me to act illegally. This evaluative judgment is one that I make in describing the situation.

In this case, the question is what can and should I do? It is not enough to say that I should not dump the waste down the drain. My problem is not the simple choice of answering yes or no to the question of whether I should follow the order. I need to figure out what to do about the supervisor's order. Shall I ignore it? Refuse it? Report it to someone? To whom could I report it? Someone else in the company? The

Environmental Protection Agency? Should I follow another course of action altogether? Is there any place I can go for advice about my options in a situation like this? How or where can I find out about the likely consequences of each course of action? Also, what do I do with that toxic waste, at least for the time being? These are questions with important implications for human well-being, for fairness to others, and for the environment, as well as for my relationship with my supervisor and future with this company. Any answer to the question of what to do will depend on a variety of factors. Learning what factors to consider and how to assess them are components of responsible professional behavior.

The importance of finding good ways of acting (and not merely the ability to come up with "the right answer" to a question of whether or not to do some particular act) may be brought home by an example from daily life. Did you ever pour paint, gas, acetone (nail polish remover), motor oil, garden pesticides, or other household hazardous waste down the drain or put disposable batteries in the trash? Did you do so before you knew that "the right answer" to the question of whether the liquid should be disposed of in this way is "No"? If you did know that such disposal was harmful to the environment, did you do so because you did not know what else to do with the refuse? This example illustrates how the construction of options by a society, options such as collection procedures for hazardous household waste, as well as an agent's good or bad intentions affect conduct.

> People confronted with ethical problems must do more than simply decide whether to perform some particular act. Those agents must do more than evaluate alternatives; they must *devise* possible responses: they must formulate a plan of action.

The need for a response in the form of action is what makes ethical problems practical problems. The similarities between ethical problems and a specific class of practical problems, design problems, are instructive for thinking about the resolution of ethical problems and correcting some common fallacies about ethical problems.

Practical problems may or may not have solutions. Of those practical problems that are ethically significant – which are what we have been calling "ethical problems" – some call for coping rather than for solution. The perennial problems of human vulnerability, suffering, and mortality are such problems. Both ethical problems that call for solution and those that call for coping have their counterpart in design problems, although good ways of coping are also called "solutions" in the case of design problems. For example, a system of drainage ditches might be designed to prevent damage from periodic flooding of a river. The design would count as a solution to the problem of how to cope with periodic flooding, although the ditch system would not prevent the floods and would not solve the problem of flooding itself.

Design problems are problems of making or repairing objects and processes to satisfy human wants and needs. The analogy that I draw between ethical problems and design problems holds for a variety of design problems, from designing or repairing a bookshelf to devising a rotating work schedule, to designing or redesigning an experiment. The analogy between ethical problems and problems of *engineering design* is especially instructive, however. Because engineering design is a subject in the university curriculum (and a subject required by the

Accreditation Board for Engineering and Technology [ABET] in accredited engineering programs), the design process in engineering has been widely studied and discussed. In contrast, craft skills are often transmitted by apprenticeship and articulated only in ways peculiar to a specific craft. Furthermore, engineering design stands out among college subjects in giving sustained attention to the synthetic reasoning necessary to construct good responses to practical problems. Engineering appreciates the importance of practical as well as theoretical problems, of engineering design as well as engineering theory, and of synthetic as well as analytic reasoning. Devising a good response requires synthetic reasoning. Philosophical ethics has paid more attention to analytic reasoning and the analysis of ethical problems and possible answers to them. Analysis is important, but it is not sufficient to devise responses.

*What is involved in addressing an actual moral problem?*

## Design Problems

How do problems of engineering design differ from numerical problems that teach and test the application of scientific and engineering theory?

Engineering educators recognize that, although the ability to analyze the designs of others is a useful skill for designers to possess, it is not sufficient to make a person a good designer. For this reason, most engineering schools offer courses in engineering design that are markedly different from the engineering theory courses that teach students how to apply theory to the solution of numerical problems. Unlike design problems, the problems used to teach and test knowledge of how to apply engineering and scientific theory typically have unique mathematically exact solutions and are stripped of any realistic details that are not immediately relevant to the application of the relevant theory.

> Typically, numerical problems differ from problems of engineering design in being shorn of details that are not immediately relevant to the application of the relevant theory and in having unique mathematically exact solutions.

Engineers design a range of objects, systems, and processes. These range from large constructions (e.g., a bridge at a given site), to small consumer products (e.g., a new type of handheld computer-telephone), to complex public systems (e.g., a traffic control or subway system), to manufacturing processes (e.g., a more cost-effective way of making newsprint from recycled newspapers or a process for making nontoxic weather-resistant paint).

Design problems in engineering (and problems of experimental design) are typically highly constrained, as are challenging ethical problems. The design process, especially in the ways in which it differs from merely analyzing the designs of others, highlights the very aspects of the agent's response to ethical problems that philosophy and applied ethics, in their preoccupation with ethical evaluation, have had difficulty illuminating.

Design problems in engineering (and problems of experimental design) are typically highly constrained and require one to take account of multiple factors, as do challenging ethical problems.

To develop a good response to an ethical problem one must typically take into account a variety of considerations. In situations like that of being instructed to dump toxic waste that raise questions of blame – for either negligence or intentional wrongdoing – fairness (in assigning blame) is a prominent consideration. Some tension or conflict may exist between the moral demands or values associated with considerations of fairness, but it is often possible at least partially to satisfy many of these demands simultaneously. Indeed, doing so is a mark of wisdom. This seemingly commonsense observation about ethical problems has been obscured in recent years by a tendency to represent ethical problems as irresolvable conflicts between opposing principles or obligations. Although ethical conflicts are occasionally irresolvable, starting from the assumption that a conflict is irresolvable is misguided because it defeats any attempt to do what design engineers often do so well, namely, to satisfy potentially conflicting considerations simultaneously and find a resolution.

*How do problems of engineering design differ from numerical problems that teach and test the application of scientific and engineering theory?*

## The Design Analogy

What characteristics of problems of engineering or experimental design apply also to ethical problems? Name as many as you are able.

To illustrate the characteristics of a design problem, consider the design of a mechanically simple object: a child seat to fit on the top of wheeled suitcases designed to be "carry-on" luggage. When removed from the suitcase, the seat must double as a child seat that will strap into an available airline seat, and the child seat itself must fit easily into the overhead compartment. Several manufacturers make such suitcases. Most have similar features, making it possible to design a child seat that fits most of the suitcases in use. I directed three mechanical engineering students in this design project. One student, Colleen, investigated what the potential user would require in such a device – such as ease of cleaning and having a place in the seat to carry feeding bottles, pacifiers, and similar paraphernalia. Two other students, Lisa and Kimberly, investigated standards and safety requirements and built rough prototypes. The students demonstrated solutions to the design problem and developed some of the features of such solutions.

Lisa and Kimberly's designs are significantly different solutions to the child seat design problem. For example, in Lisa's design, the horizontal crossbar that holds the child in place pivots around its permanent attachment to the end of the right armrest and its other end secures into the other armrest. In Kimberly's, the crossbar and armrest form a single U-shaped piece that lifts overhead like the tray of an old-fashioned highchair. (Both designs have the advantage that they do not detach from the rest of the chair, so they will not become lost.) Kimberly's design has larger dimensions. A larger seat might better suit a heavier child, but would be more expensive to manufacture. Lisa's seat would accommodate most children under two years old, the age at which infants in arms fly free with an adult.

These differences illustrate the first point about design problems that is significant for ethical problems: For interesting or substantive engineering design problems, there is rarely, if ever, a uniquely correct solution or response, or indeed, any predetermined number of correct responses. This is in contrast to puzzles, math problems, and most of the problems that engineering students typically do in problem sets. In the same way, ethical problems, may have a variety of good solutions or ways of coping.

> *The First Point about Design Problems that is Significant for Understanding Ethical Problems*: For interesting or substantive engineering design problems, there is rarely, if ever, a uniquely correct solution or response, or indeed, any predetermined number of correct responses.

## The Role of Imagination in Problem Solving

Stuart Hampshire himself does speak of imagination in the previously cited article. Imagination does seem to be deficient if an agent can think only of responses that the agent has seen others perform in like circumstances. The construction of a response is commonly a matter of both thinking and acting. Therefore, "imagination" is not a good name for the relevant skills of a designer. Action stimulates thinking as much as thinking stimulates action. This consideration may be seen to apply to moral problems as will be further illuminated when we consider the often-neglected *dynamic character* of moral problems.

People sometimes speak about "doing the right thing," but speaking of *the* right thing should not be taken to imply that ethical problems have *uniquely* correct solutions or responses. The assumption that possible responses to ethical problems are determined in advance would make the view that ethical problems have unique correct solutions more plausible. That may be true for very simple ethical problems: in such a case, the possible responses may be evident. That would transform an ethical problem into a type of multiple choice problem that might have a unique best answer. Some engineering design problems and some ethical problems may be trivial in that the specific action of the problem leaves little leeway in an acceptable solution. The question of what to do about a promise that one has freely made, in circumstances where no morally compelling counterclaims exist, is trivial in this sense: Obviously, one should keep the promise. So in the design of a bolt to fasten the housing of the radar for a large commercial aircraft only a few questions would be open, such as whether the bolt should be made of corrosion-resistant material and whether this bolt should be interchangeable with many other bolts used in the aircraft. In both the cases of keeping a promise and designing a bolt, devising an appropriate response is not demanding, so the principal moral question is whether one is sufficiently conscientious in acting to do what is required.

Experience shows that responses to challenging ethical problems are not obvious, however. That is what makes an ethical problem challenging. Responses to some ethical problems may seem obvious, if an agent mistakenly assumes that the only solutions are ones that the agent has seen others give to such problems. Part of what ethics education is about is increasing the store of good responses that people can readily think of in the face of a moral problem.

There may be no solution to a given design problem at a given time – no way of making a thing that answers a certain set of specifications. (Perhaps no design of the child seat could be light enough to be supported by a soft-sided suitcase, strong enough to meet safety requirements, and affordable to manufacture, for example.) However, if one solution to a design problem exists, others usually do as well.

It is for nontrivial ethical problems that the analogy with problems of engineering design is most important. The fulfillment of moral responsibilities in general and professional responsibilities in particular provides many examples of ethical problems that resemble interesting design problems. Where ethical problems do not take the form of multiple choice problems, it is not surprising that where there is one course of action that provides an ethically responsible resolution of an ethical problem, other responses may also be ethically acceptable.

The initial problem about the toxic waste is an interesting ethical problem with several acceptable responses. It may be possible to change the supervisor's mind, perhaps by detailing the potential health effects or the legal liability to the company, or by simply stating that in conscience, one cannot dump the waste. If the supervisor is adamant, it may be possible to get others in the company – the ethics or environmental office, if any, or the legal department, if any – to countermand her order. The character of one's organization makes a difference to one's response as well. Although some organizations have a strict chain of command, others, including most universities, make a point of having "multiple channels" for working through problems. There may be several ways of properly disposing of the waste while not embarrassing the company or coworkers more than necessary.

> *The Second Point about Design Problems that is Significant for Understanding Ethical Problems*: Although there is not a uniquely correct solution, nonetheless, some possible responses are clearly unacceptable. There are wrong answers even if there is not a unique right answer, and some solutions are better than others.

One commonly hears the assertion that for some or all ethical problems, "there are no right and wrong answers." Those who say this may be attempting to acknowledge that there are no uniquely correct solutions to ethical problems or they may be espousing an extreme subjectivism about ethical matters. Some possible responses to moral and design problems are so poor as to be clearly wrong, however. "Intimidate vulnerable parties into acquiescing to whatever we want" or "make it with a safety factor of 1" – which means that there is no safety margin – *are wrong answers*. The first violates basic moral standards and the second violates basic safety standards.

Returning to the original student design project, a child seat that could not recline when in the airline seat would be more of an irritation than a comfort to the child. A suitcase child seat lacking appropriate safeguards should the handle slip out of the adult's hand, allowing the seat (and the back of the youngster's head) to fall to the floor would be prohibitively dangerous. (As these examples illustrate, some design questions are *also* questions of ethical responsibility.) In our toxic waste problem, dumping the waste down the drain is one of many wrong answers. (Dumping the waste under the supervisor's hedge is another.)

Table 3.1  Comparison of Designs

| | Lisa's Child Seat | Kimberly's Child Seat |
|---|---|---|
| Size of child | Most children from 6 mos. to 2 yrs | Most children 1 to 4 years old |
| Crossbar design | Armrests are stationary. Crossbar attaches at the end of one to latch on the other | Armrests and crossbar form a single U-shaped piece that lifts over the child's head |
| Additional relevant design criteria in which the two designs do not differ | Comfortable for the child<br>Means to protect the child if the suitcase handle slips from the adult's hand<br>Manufacturable at a reasonable cost<br>Easy to install<br>Robust<br>Easy to clean and maintain<br>Does not interfere with maneuverability of suitcase | |

> *A refinement of the First Point about Design Problems that is Significant for Understanding Ethical Problems*: Although for interesting or substantive engineering design problems there is rarely, if ever, a uniquely correct solution, two solutions may each have advantages of different sorts. Therefore, it is not necessarily true that, for any two candidate solutions, one must be incontrovertibly better than the other.

In the case of Lisa's and Kimberly's child seat designs, one is not clearly better than the other, although some features of one are clearly better than the corresponding features of the other. If no design feature were constrained by the design of some other feature, it might be possible to put the best features into one best design. However, some features are so constrained. For example, the design of the security strap that fits between the child's legs and runs between the crossbar and the seat depends on the design of the crossbar. Furthermore, even a given feature may be better in some respects (easier to keep clean, more comfortable for the youngster, less expensive to manufacture) and worse in others (more cumbersome for the adult to operate, more likely to break). Such a feature may be an overall advantage for some users and a disadvantage for others. The two designs are compared in Table 3.1.

For ethical problems, too, different courses of action that satisfy all basic constraints may have different advantages. Suppose, for example, in the case of the disposal of the toxic waste, my supervisor is acting on habits established in the 1970s and 1980s when such dumping of waste was prevalent. My supervisor may not appreciate what is wrong with dumping, or know he is violating the law in doing so. Significant changes in knowledge, regulation, and company attitudes have taken place in the last three decades. My supervisor may then be open to arguments that things have changed, especially if they come from her boss or from our environmental or legal department. Before I go to those offices, I could tell her that I think we should get their view and that I intend to do so. That forewarning may prevent her feeling undercut when I take the concern further, and it may even convince him to consult those offices himself.[1]

---

[1] An opinion by the NSPE's Board of Ethical Review on a case that concerns going "over the head" of one's supervisor when one believes the supervisor to be in the wrong is Case No. 82–7. That case was originally published in *Opinions of the Board of Ethical Review* Volume VI, Alexandria, VA: National Society of Professional Engineers, 1989, pp. 27–29.

Suppose, instead, I take the approach of saying that in conscience I cannot dump the waste. This has the advantage that it does not raise the specter of my continually going over her head – an effect that may be more of a danger if I am new to the job. It also leaves less opportunity for him to find out that standards have changed, however. Suppose that in response to my conscientious objection he says, "Well if you are so squeamish, I will do it myself." What, if anything, do I say then? What if he then proceeds to dump the waste? Each of the two avenues that I have outlined has advantages and disadvantages. It may not be possible for me to find out ahead of time which one is more likely to work well with my particular supervisor.

Notice that many of my subsidiary judgments concern *how far to go* to convince my supervisor of the error of her order. I must first think through the possible responses and further actions I might take before deciding whether to take those actions. This point is worth emphasizing because, as I mentioned, ethical problems have often been misrepresented as choices between two (or more) preset options. (The formal "decision problems" considered in decision analysis are problems of deciding among *prescribed alternatives* and thus are species of *multiple choice problems*.) Many common ethical disagreements are about how far to go in trying to achieve some end that both sides agree is desirable or trying to avoid violation of some ethical norm that both sides agree upon. Relatively few disagreements take the form of one side thinking that some ethical value is a noble one and the other side thinking it is harmful or of absolutely no importance. More commonly, disagreement is over what else should be sacrificed or risked to achieve some desirable outcome.

Nothing in the argument for a multiplicity of acceptable ethical responses requires ethical relativism. The variety of acceptable solutions to complex ethical problems does not require that agents hold different moral beliefs or commitments. Even if all agents have exactly the same moral beliefs and commitments, a variety of good responses may exist. Advantages and disadvantages of acceptable solutions may differently suit the life circumstances of different individuals independent of their moral beliefs, even as the loss of two fingers would have very different implications for a singer and for a pianist.

---

The *Third and Final Point about Design Problems that is Significant for Understanding Ethical Problems*: Their solutions must satisfy all of the following requirements:

- Achieve the desired performance or end – e.g., create a child seat that fits on a wheel-on-board-suitcase, or fulfill one's responsibility for environmental safety.
- Conform to specifications or explicit criteria for this act – e.g., the seat must fit inside the overhead rack and be a comfortable booster seat that straps into an airline seat; resolving the toxic waste issue should not take so much time that one fails in other major responsibilities.
- Be reasonably secure against accidents and other miscarriages that might have severe negative consequences.
- Be consistent with existing background constraints – e.g., for the child seat: be manufacturable without scarce or hazardous materials. For any ethical problem background constraints include the requirement to avoid violating anyone's human rights.

*What characteristics of problems of engineering or experimental design apply also to ethical problems? Give as many as you can think of. You need not limit yourself to those described in this section.*

## Four Moral Lessons from Design Problems

How does one go about addressing design problems, and what features of moral problems and approaches to designing them does that illuminate?

The analogy between ethical and design problems suggests some strategies for addressing moral problems.

First, consider the examination of the situation and definition of the problem. An initial assessment is needed just to *name* the problem. In the case of design problems, the ambiguity is typically limited to a lack of knowledge of what potential users might require in such a device (and hence the constellation of features in such a device) and what solutions are already available. Often it is not clear how far one can go in meeting some requirements and still satisfy others. For example, in the case of a child seat, it would be desirable to accommodate large toddlers and three-year-olds, as well as average size two-year olds, but the seat must not be too heavy to be supported by commonly used suitcases.

It is important to recognize unknowns, ambiguities, and uncertainties at early stages of problem solving. The following problem of engineering design is one in which there were many unknowns: Engineers sought to design a device to automate testing for a variety of immune factors. This complex device was so novel that when it was designed, there were no industry standards for the characteristics of such a device. At the initial stage designers had to decide such questions as how constant the temperature at which the device maintains chemical reactions must be: Should the specifications be for a temperature of $37°$ C $+ 1°$ or $37°$ C $+.1°$? Once such specifications were decided upon, the designers built a feasibility model – that is, a model that meets the specifications and embodies the core features of the technology. Such a feasibility model demonstrates that it is possible to create the device in question but typically leaves open many questions about the device that ultimately will be manufactured and sold. (In the case of a device as complex as this one, "engineering models" are then built. These include some user interfaces (i.e., some of the controls that a lab technician running the assays would use) similar in fit, form, and function to the device that will eventually be manufactured and sold. After the engineering model, the next step is an "engineering prototype," which is an economically and technically manufacturable version of the device. Next, "manufacturing models" are built on the manufacturing floor, with detailed documentation to catch any problems that arise when the device is actually manufactured. Finally come production units that can be sold.

The initial phase of the design of the immuno-assay device illustrates the task of problem definition. Engineers recognize the importance of allowing for as much flexibility as possible in the definition of the problem to avoid foreclosing options to change or add features in successive models to improve safety, performance, reliability, or manufacturability. Comprehensive foresight prevents difficult or costly changes when far along in the process. For example, changing the manufacturing process typically requires retooling for manufacture, which is very expensive.

The first lesson from design problems for ethical problems is to begin by considering the unknowns, ambiguities, and uncertainties in the situation. In the case of ethical problems, the situation may even be fundamentally ambiguous, creating even more of a challenge for foresight. At least with a design problem it never turns out that what seemed to be a problem of designing an airplane turns out to be a problem of designing a coffee pot. In contrast, if one hears from one person that another is doing something wrong, it may be that the second is doing wrong, or that the first is slandering the second, or something entirely different. All that is certain at the beginning is that something is not as it should be.

> The first lesson from design problems for ethical problems is to begin by considering the unknowns, ambiguities, and uncertainties in the situation.

Appreciating ambiguities and uncertainties is important. These are often underemphasized in professional ethics. The original edition (1989) of *On Being a Scientist* – the handbook on research and research ethics for young scientists put out by the National Academy of Sciences (NAS), the National Academy of Engineering (NAE), and the Institute of Medicine (IoM) – contains advice that fails to take account of ambiguity. That handbook recommends that when one believes one has witnessed research misconduct, one should talk it over with a trusted experienced colleague and "[o]nce sure of the facts, the person suspected of misconduct should be contacted privately and given a chance to explain or rectify the situation."[2] It is often *not possible to wait for certainty before responding*. The advice to act only when one is sure of the facts is advice to avoid action.* That advice has not been repeated in subsequent editions of *On Being a Scientist*.

> It is often not possible to wait for certainty before responding to a situation.

What is needed are ways of acting that will prove prudent and fair no matter how uncertainties are resolved. In cases in which crucial ambiguities cannot be fully resolved early in the situation, the ambiguity should be understood as a defining characteristic of the situation – for example, ambiguously either malfeasance or slander. Faced with an ambiguous problem, agents typically need to figure out: whether to gather more evidence, how to raise the issue (or gather more evidence) without being unfair to others, and how to best elicit support for their concern to achieve a fair resolution. When reporting an ambiguous situation to others, a good ethical rule of thumb is to clearly state the facts as one knows them with as little interpretation as possible.

---

[2] Committee on the Conduct of Science. 1989. *On Being a Scientist,* first edition. Washington, DC: National Academies Press, 16.

*Actually, two things are wrong with the advice. As is now widely acknowledged, confronting persons who have committed research misconduct frequently leads to data destruction or other methods of concealment. Therefore, in the face of strong evidence that misconduct has occurred, and having talked over the matter with a knowledgeable and trusted person or institutional ombudsperson, the matter is best turned over to one's institutional research standards officer.

The second lesson from engineering design for ethical problems is that the development of possible solutions is distinct from definition of the problem and may require more information. This is a difference between ethical problems (or design problems) and formal "decision problems." "Decision problems" or problems in decision analysis include specification of the alternatives among which one is to decide. Therefore, a fully defined decision problem is a type of multiple choice problem. The need to develop possible solutions in real life shows that open-ended statements of ethical problems do more justice to them than do representations of them as multiple choices.

> The second lesson from engineering design for ethical problems is that the development of possible solutions is distinct from definition of the problem and may require more information.

Furthermore, before proposing solutions, agents must frequently clarify the problem. Although open-ended statements do more justice to ethical problems than do multiple choice statements, even open-ended statements are only outlines of ethical problems. If one had an actual ethical problem, there would be real details to examine. For example, if I really had the problem about the disposal of toxic waste, there would be a particular person who would be my supervisor whose character I might learn more about. My organization (a company, a university) would have particular policies that I could investigate.

One of the important characteristics of a responsible or wise response to a practical problem is appropriate investigation of a problem before attempting to solve it. Investigation of the requirements of potential users by engineering designers was already mentioned. This is especially important for a novel device. If engineers are seeking to improve the design of a currently available product, say a mousetrap, they engage in "benchmarking," that is, they gather information about the mousetraps already available. Just as important, they do an investigation of the demand for features not currently available in mousetraps and of the relative importance of all features in the mind of the user. In U.S. engineering, the demands of the user are often reflected in what is called "the voice of the customer;" however, the user is not always the buyer. In Sweden a fuller range of users are often consulted. For example, the workers who will use a new medical device are consulted on its design.

Too often when statements of ethical problems are presented, students' attempts to interrogate the problem are cut off. Answering problems without seeking to investigate them is poor preparation for understanding and addressing actual ethical problems.

From the place of brainstorming in the practice of engineering design, we learn more about how an agent goes about developing responses. Brainstorming requires an uncritical atmosphere in which people can present "half-baked" ideas that may be later refined or combined. Articulation of any half-baked ideas is discouraged in the many ethics classes where adversarial debate is the primary method used. Although an adversarial debate format may provide some useful pre-law training, it does not help develop the ability to think constructively about resolving ethical problems.

A child, known as "Amy," who was asked to respond to the notorious "Heinz dilemma" demonstrates a rather heroic capacity to brainstorm in the face of critical response. When Amy is asked if a man, Heinz, should steal a drug he cannot afford to save the life of his wife, she proposes new alternatives to either stealing or letting Heinz's wife die:

> Well, I don't think so. I think there might be other ways besides stealing it, like if he could borrow the money or make a loan or something, but he really shouldn't steal the drug – but his wife shouldn't die either.

Asked why he should not steal the drug she replies:

> If he stole the drug, he might save his wife then, but if he did, he might have to go to jail, and then his wife might get sicker again, and he couldn't get more of the drug, and it might not be good. So, they should really just talk it out and find some other way to make the money.[3]

The brainstorming (in this case yielding the idea of "borrowing the money" or, as Amy elsewhere suggests, persuading the druggist to lower the price) and interrogation of the problem are not entirely separable activities. In addressing design problems suggestions from potential users about their needs frequently stimulate new ideas, and ideas for approaches to the design may stimulate new questions for potential users. For ethical problems, additional information gained through interrogating the problem frequently changes the desirability of possible responses.

The point is illustrated by consideration of the following situation: A highway safety engineer is allocating resources for safety improvements and considers two intersections. Both have the same number of fatal accidents per year. However, one is in a rural setting and the other is an urban setting. The urban intersection handles on average four times the number of cars as the rural intersection and has a higher rate of minor injuries and property damage than does the rural intersection. There is just enough money in the budget to improve one intersection. Which one should it be?

The choice of improving the urban intersection is often justified on the ground that there improvements will have the greatest overall reduction of injury, and this choice is cited as illustrating a utilitarian choice of "the greatest good for the greatest number." The choice of the rural intersection is justified on the ground that it is a more dangerous intersection in the sense that the likelihood of a fatal accident for a given use of the intersection is four times higher. This consideration is taken to represent concern for fairness (presumably equal distribution of the risk of fatal injury associated with going through any given intersection) or even respect for individual rights.

What is relevant here is not how well this story illustrates the philosophical distinctions between utilitarian and competing rationalist foundationalist schools of thought in ethics, but the danger that this example will be misunderstood as an example of problem solving. Notice first that the problem is presented as a *forced choice* between spending all the remaining resources on one intersection and spending it all on the other. In fact, there would likely be many other choices.

[3] Gilligan, Carol. 1982. *In a Different Voice: Psychological Theory and Women's Development.* Cambridge, MA: Harvard University Press, 27–28.

For example, putting up traffic signs at both intersections may be an alternative to installing traffic lights at one. However, even accepting the multiple choice character of the problem as stated, there is a great deal of potentially relevant information that the example does not give us about the accidents. For example, suppose that at one intersection, but not the other, in all serious accidents at least one of the drivers involved was drunk (or fell asleep or had a heart attack, etc.). Such information might show that the most crucial variable for reducing serious accidents at one site is reducing driver impairment, whereas at the other it is the physical characteristics of the intersection; only the latter would be best remedied by changing the intersection itself.

A third lesson from design problems concerns acting under time pressure. It is often important to begin by pursuing several possible solutions simultaneously, so that one will not be at a loss if one meets insuperable obstacles, but also to avoid spreading one's energies *too* broadly. This admonition applies both to the design of individual features of the product and to approaches to revising the design when obstacles are encountered at later stages. For the immuno-assay device discussed earlier, the possible design corrections were of three general kinds: mechanical modification, modification of the chemical procedure, and modification of the software. Because modification of the software was generally the cheapest modification, where it could provide the requisite fix software modification was best.

> A third lesson from design problems concerns acting under time pressure. It is often important to begin by pursuing several possible solutions simultaneously, so that one will not be at a loss if one meets insuperable obstacles, but also to avoid spreading one's energies too broadly.

The need to act under pressure of time is also a common feature of ethical problems. In the face of time pressure, it is reasonable to pursue several possibilities simultaneously in case one fails to prove practicable. Consider the ideas proposed by Amy, the child who rejects the forced choice of the Heinz dilemma and "brainstorms" a variety of possible courses of action. (Some of these have to do with relationships, such as remonstrating with the druggist; others, such as "taking out a loan," do not.) The simultaneous pursuit of several options is a mark of good design strategy when there is any danger that one line of development may prove unfeasible. Pursuing several options contrasts with the representation of an ethical problem as a static situation with static solutions. If the situation were static, the problem would become a simpler one of selecting the right alternative and steadfastly pursuing it.

Fourth and finally, the dynamic character of problem situations has further implications. Both the problem situation and one's understanding of it are likely to change and develop over the course of time.

> Fourth and finally, the dynamic character of problem situations has further implications. Both the problem situation and one's understanding of it are likely to change and develop over the course of time.

For example, in attempting to avert the *Challenger* accident discussed in Chapter 2, engineer Roger Boisjoly's problem situation began with evidence, in the form of blackened grease, that hot gas was escaping through the joints. The problem then became one of conducting experiments to test the effect of temperature on the seals, and then one of getting a seal team formed to redesign the seals and of getting resources to do so. The final problem became one of stopping the flight in view of predicted record cold temperature.

> *How does one go about addressing design problems, and what features of moral problems and approaches to designing them does that illuminate?*

## Implications of the Dynamic Character of Ethical Problems

> Why should it make a difference whether actual moral problems are dynamic or whether they are static (unchanging)? How does the static or dynamic character affect how one best approaches such a problem?

If the dynamic character of the ethical situation is neglected, it is easy to confuse doing the wrong thing (which creates a changed situation) and then making the best of the resulting bad (changed) situation on the one hand, with taking an action that is justified in some circumstance or the other. For example, a colleague with whom I was working to formulate some criteria for research ethics raised the question of whether gift authorship is ever ethically justified. (Gift authorship in a research context is the listing as an author a person who has not contributed substantially to the research reported in the paper. An overview of fair credit and authorship will be discussed in Chapter 9.) My colleague was recalling an incident from her own research experience. She had had an idea for a collaborative project and proposed it to a second researcher who had established some of the groundwork for my colleague's new effort. The second researcher at first expressed interest, but then failed to respond when my colleague actually proposed to start the work. After several communications brought no response, my colleague undertook the work with members of her own lab only. There was some delay because my colleague's group had to recreate some research materials that would have been on hand at the second researcher's lab. In due course, my colleague and one of her post-doctoral fellows completed the research and wrote a manuscript reporting the work. As a courtesy, because the work was built in part upon the earlier work of the second researcher, my colleague sent a "preprint" (i.e., a copy of the unpublished manuscript) to the second researcher. That researcher replied that my colleague could not publish the paper because a virus that my colleague's group had used had been obtained from the second researcher's lab for a different purpose and the second researcher had not given permission for the new use. (The sharing of research materials, or the means for making them, is encouraged in science, although no lab is expected to take on great burdens to supply others with materials. Some science and engineering journals require that those who publish articles in their journal furnish to others the reagents and similar materials necessary to replicate the work. In this case, however, my colleague had agreed to use the virus only for a single purpose. That purpose did not include making materials for the project described in the manuscript.)

Hoping to shame the second researcher into desisting from her complaints, my colleague wrote back asking if there was someone from her team whom

## Implications of the Dynamic Character of Problems

Practical problems, and ethically significant practical problems in particular, are typically dynamic situations. In responding to such a problem, therefore, one's initial response may change the character of the problem or create new problems. It is important to recognize and think through that potential before taking action.

This feature of practical problems is one that illustrates why solving problems requires more than both judgment skills and imagination. In thinking through a situation, one needs to consider eventualities that have some likelihood of occurring, rather than every possibility that one can imagine happening. Thinking through a situation requires guidance, relevant experience, or access to others with relevant experience in order to estimate the likely consequences of one's actions.

## Aristotle's Concern to Warn against Pitfalls

As Edmund Pincoffs has argued, much of Aristotle's ethics consists in warning against moral pitfalls.[a] Philosophical ethics in the Enlightenment period (and in rationalist thought generally) largely neglected this aspect of ethics in favor of developing ethical systems that purported to answer the question of what reason (alone) tells us ethics is all about. Sometimes such systems have been "applied" to yield ethical tests of particular acts.

---

[a] Pincoffs, Edmund. 1971. "Quandary Ethics," *Mind*, 80: 552–71; reprinted in Stanley Hauerwas and Alasdair MacIntyre (Eds.), *Revisions: Changing Perspectives in Moral Philosophy*, Notre Dame, IN: Notre Dame University Press, 1983.

the researcher thought should be added as an author on the manuscript. To my colleague's dismay, the second researcher sent a letter back nominating both herself and a post-doc in her lab as coauthors. My colleague's post-doc was about to take a job in proximity to the second colleague. Because this post-doc, who had done nothing wrong, would be vulnerable to retaliation, my colleague decided to go ahead with gift authorship. She added the names of the second researcher and her post-doc to the list of authors on the manuscript. She did so despite her firm conviction that gift authorship is a corrupt practice.

One might agree that my colleague "made the best of a bad situation." Her story was not one in which gift authorship is *justified*, however, because the situation itself was one that was partly her own doing and one that, as an ethical matter, she ought not to have brought on herself. Hers is a **cautionary tale**. Cautionary tales help others avoid the same pitfalls. In the future, she would take care not to get into this situation; she will be more careful to check the conditions under which she receives research materials and would not again make the mistake of offering gift authorship as a backhanded form of moral criticism. (It is hard to imagine a bad situation, in which one could make the best of it *only* by violating some norm of responsible conduct, *unless* one had acted unwisely.)

In this book, we will often be concerned with learning how to anticipate the possible consequences of actions and how to seek reliable information about likely consequences of actions to avoid ethical pitfalls that place a person in a position of having to choose the lesser evil.

*Why should it make a difference whether actual moral problems are dynamic or whether they are static (unchanging)? How does the static or dynamic character affect how one best approaches such a problem or understands its moral dangers?*

## Problems as Experienced by Agents

What would be a good way to respond to the situation in which wrong may have been done, but you are not certain of what was done or by whom?

Because the situation may not be what it first seems, the previously stated rule of thumb, to clearly state the facts with a minimum of interpretation, has special ethical significance when there is a question of another's negligence or malfeasance. It is ethically important to be fair to others and avoid spreading false rumors about them. Minimizing your interpretation will minimize the possibility that you have interpreted the situation incorrectly. (The same rule is also simply *prudent*, in that sticking to the facts makes it less likely that raising the issue will get you into trouble.) As with all rules of thumb, the advice is not always applicable. If a larger overall pattern seems apparent – for example, the person whose actions are in question has a pattern of cheating in some way or harassing others – then the best idea may be to raise the possibility of that interpretation of some current observations, stating the facts that lead to that interpretation. However, the question of that overall pattern is often best raised with some designated neutral (like an ombudsperson) or a person who has official responsibility and some experience looking into such matters.

> It is ethically important to be fair to others and avoid spreading false rumors about them. Clearly stating the facts with a minimum of interpretation minimizes the possibility that you have interpreted the situation incorrectly.

Here are two problem situations that illustrate some of the points we have been discussing, especially coping with unknowns, ambiguity, and lack of certainty in responding to an ethical problem and responding to its dynamic character.

### Scenario: Is It Plagiarism?

You find that two academic publications have remarkably similar text expressing an idea that is not part of common knowledge in your field.
   What if anything can/should you do, and how ought you go about it?

The similarity between the texts in this scenario requires explanation. There may be intentional wrongdoing, namely: plagiarism – the first author of the second or the second author of the first – but this is a matter that cannot be confidently decided from publication dates alone. The plagiarist may have seen the work of the plagiarized, but published first. Furthermore, especially if the idea came up in conversation, the person who received the idea may no longer remember where he first heard it, or even that he did hear it from another person, so the fault may be one of negligence or recklessness rather than intentional wrongdoing.

The situation may not be plagiarism at all. Perhaps the ideas were original to a third party who was the teacher of both and who gave each the mistaken impression that those views were common knowledge in the field. Therefore, in this case too, it is important for the person who raises the issue to do so in a way that does not prejudge the issue.

> ### Scenario: What about My Contribution?
>
> You are a new graduate student. You started graduate research with Prof. One in Great Lab working on the Fantastic project. By the end of the first year you had not only become proficient at many of the more routine tasks of the project, but had made one small but notable refinement to the approach to the segment assigned to you. At the end of the first year, Prof. One went on leave for a semester and you started working with Prof. Two in the same lab but on a different project. Prof. One returned for the spring semester and took up the Fantastic project, among others. The following fall, the beginning of your third year, you learned from another student who was working on Fantastic that Prof. One is publishing a paper on some aspects of Fantastic with this student, a paper that contained your refinement.
>
> What, if anything, can and should you do?
>
> Are there any ambiguities in the situation? If so, how can you fashion a response that will be appropriate however the ambiguities are resolved? (Ambiguities in ethical problems are frequently greater than the unknowns in a conceptual design problem.)
>
> Does Prof. One remember your involvement and contribution? Does he remember it but judges your contribution to be insignificant? Does he judge it significant, but worthy of an acknowledgment rather than joint authorship? Might he be planning to add you as a third author, but has not gotten around to telling you? In resolving the situation, it is important to attend to alternative possibilities.

*Describe how best to respond to the situation in the scenario, What about My Contribution? Identify areas of uncertainty or ambiguity and describe how to cope with them.*

## Making and Assessing Ethical Judgments

What is the role of making ethical judgments in designing or devising responses to ethically significant moral problems?

Understanding, assessing, and making ethical judgments are a significant part of learning how to respond well to ethical problems. Considering the judgments of experienced individuals and authoritative organizations is a good way to discover both what factors these individuals and organizations find most ethically significant and what they think are the most likely realities underlying the apparent situation. It is also prudent to discover the priorities of individuals and groups whose views are influential in the sphere in which one is working because those priorities will influence the reception your own actions receive.

> The ethical judgments of experienced individuals and authoritative organizations reveal both the situational factors these individuals and organizations find most ethically significant and what they think are the most likely realities underlying the apparent situation.

In this book, you will find references to or summaries of cases, decisions, and ethical opinions issued by a variety of sources. These include the National Society of Professional Engineers' (NSPE's) Board of Ethical Review (BER), ethical guidelines of other ethically active professional societies such as the Institute of Electrical and Electronic Engineers (IEEE) or the American Chemical Society (ACS), the National Academy of Engineering (NAE), and the National Academy of Sciences (NAS), and from the U.S. Supreme Court. You will also find opinions from ethics offices of companies that employ large numbers of engineers. The reasoning offered in these judgments and the recognition of morally relevant features of the situation exemplify complex moral reasoning.

> The ethical judgments quoted in this book are not to be accepted uncritically. Notice the different priorities of the various groups issuing statements on ethics.

Ethical judgments by different authoritative bodies *do* reflect the differing additional value priorities of the organizations they represent. Ensuring that lower courts correctly interpret law and follow legal procedure as well as that laws are consistent with the U.S. Constitution are the concerns of the Supreme Court but not for some other bodies making moral judgments, for example. In contrast, the NSPE is concerned with preserving the cohesiveness of the engineering profession. Therefore, although it affirms an engineer's responsibilities and obligations concerning public welfare, clients, and employers, the NSPE tells engineers to go to considerable lengths to show consideration to fellow engineers in fulfilling these responsibilities and obligations. Not surprisingly, although they expect engineers to fulfill their professional obligations and responsibilities, the ethics offices of reputable companies encourage their engineers to do so in ways that will minimize the likelihood of damage to the company. The ethical judgments quoted in this book are included to be instructive, but *not* to be accepted *uncritically*. Notice the different priorities of the various groups issuing statements on ethics.

The steps in responding to an ethically significant problem are shown in Figure 3.1.

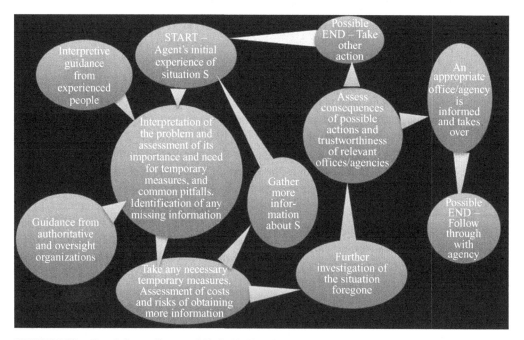

Steps in Responding to an Ethically Significant Problem Situation, S

Following is a problem situation that lies outside the domain of situations that NSPE BER, the U.S. Supreme Court, or corporate ethics offices would normally consider, although many of the ethical values that those bodies affirm are relevant to it. Unlike the cases presented by the NSPE, this scenario, and most in this book, have some realistic uncertainties that may be familiar to undergraduate engineers. As we saw earlier, these uncertainties may make the nature of the ethical problem ambiguous.

---

### Scenario: Risky Racing[a]

You are a new member of your university's solar car team. In just two months, a prestigious race will take place in Australia and you are frantically trying to get your car to run properly. Sixty entries from all around the world will compete for the glory, prestige, and sponsorship a good performance brings. Media coverage will be extensive.

One of the main selling points you used to attract sponsors was the light weight of your small car in comparison to other entrants. Now it appears that the lightness of your car may be a liability. Last week during a test run, the car spun out of control after a moderate 45 mph turn and came to rest 100 feet into someone's front lawn. The driver was not injured, but was so shaken that she refused to ever drive the car again. Just two days later, the car slid off the road and flipped over after passing through some rain puddles. The new driver got away relatively unscathed, but might have been killed had the car veered into the lane of oncoming traffic.

Some of the team members quietly admit that the car is an overpowered 3-wheel torpedo, but much time and money have already been invested in the project. The plane tickets to Australia and the entry fee for the race are nonrefundable. The race is held only once every three years. For most of the team, this is the only major race in which they will ever compete.

The team leaders reason that because the roads in Australia are likely to be long, flat stretches, the car should probably make it through the race without a serious accident. You are not convinced. To be competitive, the car will have to travel at 60 mph for most of the race. Very large 80-wheeled trucks known as "land trains" commonly barrel down the race route at 70 mph. You are a relatively new member and have not put as much time into the project as some of the others, so your opinions do not carry much weight.

What should you do and how ought you to go about it?

---

[a] Based on a scenario by Mike Wittig (MIT '95).

*What is the role of making ethical judgments in designing or devising responses to ethically significant moral problems? What else is involved?*

## Summary and Conclusion: Improving on Excellence

If, in some respects, a person has behaved well from a moral point of view in a certain situation, does that mean that in a similar situation, everyone should copy that person's actions?

Neglect of the perspective of the moral agent accounts for the misunderstanding and misrepresentation of ethical problems in much of recent ethics (although attention to the perspective of the agent has increased since the publication of the first edition of this book). Where ethics exclusively emphasizes the perspective of the judge or that of a disengaged critic who views the problem from "nowhere" and treats it as a "math problem with human beings," it does not illuminate the real tasks of moral life. For the agent facing an ethical problem, not only are possible responses undefined, but the nature of the problem situation itself is often ambiguous. As a result, the agent faces a whole series of smaller problems about what to do next in the face of multiple ambiguities and uncertainties. To be responsible requires consideration of how to treat others, and what becomes of others and oneself in addressing intermediary problems, as well as in the outcome of the larger story in which the smaller problems are located.

The understanding of the design activity in engineering, especially in the ways in which it differs from merely analyzing existing designs, highlights the aspects of the agent's response to ethical problems that philosophy and applied ethics have had difficulty illuminating. The *multiply-constrained* nature of many problems in engineering design provides an instructive analog of challenging ethical problems, which involve many types of moral considerations, all of which must be taken into account. Many ethical problems that are represented as conflicts are better understood as problems with multiple constraints, constraints that may or may not turn out to be simultaneously satisfiable.

> Many pressing problems, both problems for individuals and social problems, are multiply-constrained problems that require continuing input from many individuals and organizations.

The analogy with design problems implies that we should expect that even excellent responses to a problem might be improved upon. I embrace this implication. To frame ethical problems exclusively from the vantage point of the judge or the moral critic, to the neglect of that of someone facing the problem, associates ethics with judgment and criticism and creates incentives for people to insulate themselves from criticism, either by narrowing the scope of the problems they address or by developing ready rationalizations for their behavior. However, pressing problems, both problems for individuals, such as how to be a good engineer, teacher, parent, or friend, and social problems, such as how to provide for public health or protect the environment, are multiply-constrained problems that require continuing input and oversight by many individuals and organizations. Recognizing that good resolutions of ethical problems often can be improved upon should have the salutary effect of promoting open, constructive, and nondefensive discussion of ethical problems.

*If some person, group, or organization has behaved well, from a moral point of view, in a certain situation, does that mean that when in a similar situation, everyone should copy that action? Give reasons for your answer.*

# 4 Central Professional Responsibilities of Engineers

## The Centrality of Responsibility in Professional Ethics

What characteristics or behavior on the part of the professionals on whose work your own welfare depends would qualify them as trust*worthy*?

In today's era of specialized knowledge, we all must depend on professionals for our safety, health, and well-being. What a person needs from an engineer, a health care provider, or any other professional is more than that the professional obey simple rules of practice and ethics. What each of us needs of professionals is that they exercise their professional *judgment* to devise a plan for securing us a good outcome in our specific situation. Exercising **professional judgment** typically requires more than following simple rules. It requires taking into account a range of factors, marshaling relevant parts of the body of knowledge specific to one's profession, and devising a course of action that achieves a good (or even "the best") outcome in the circumstances. Because exercising judgment (rather than simply following a rule) requires higher cognitive functions and some intellectual maturity, the subject of professional judgment and the moral responsibility that goes with it has been left for this and later sections of this book, sections that are addressed to juniors, seniors, and graduate students.

> To understand the ethics of one's profession one must understand not only what *acts* are ethically required, permitted, or forbidden but also what is entrusted to one as a member of one's profession, that is, what good outcomes one is expected to marshal one's professional knowledge to achieve. The pursuit of those good outcomes is a **moral responsibility**, specifically, a **professional responsibility**. Fulfilling that responsibility may require considering *many potentially competing factors.*

To understand the ethics of one's profession one must understand not only what *acts* are ethically required, permitted, or forbidden but also what is entrusted to one as a member of one's profession, that is, what good outcomes one is expected to marshal one's professional knowledge to achieve. The pursuit of those good outcomes is a **moral responsibility**, specifically, a **professional responsibility**. Fulfilling that responsibility may require considering *many potentially competing factors.*

### An Exemplary Professional Response: Landing a Disabled Plane

A handy example of someone who succeeded in taking account of a wide range of relevant factors in the situation and provided an exceptionally *good* outcome is the pilot, Chesley "Sully" Sullenberger, who landed his disabled Airbus A320 airplane in the Hudson River with no loss of life.

The basic facts of the case are that within *90 seconds* after takeoff on January 15, 2009, US Air flight 1549 collided with a flock of birds. That collision disabled *both* of the plane's engines. Captain Sullenberger then took over the flight controls from the copilot, who had been at the helm and who then turned his attention to the complex task of trying to restart the engines.

Sullenberger's judgment under pressure has been consistently praised. Some commentators have pointed out the extensive training that pilots receive. The unique features of the situation, including the location of the plane, could not have been joint features of any of the pilot's training situations, however.

Sullenberger's experience is commonly mentioned. His familiarity in making a host of previous flight decisions would have been an asset to him in noticing and promptly taking account of relevant features of his situation. Not only did he need to decide whether there was time to return to the airport from which he had taken off (or reach another) or to risk ditching the plane in the river but also how to go about landing in the river in a way that made it possible to evacuate the plane. To accomplish this he had to maneuver the plane so that it would float for at least a few minutes. He also chose a location in the river close to an active ferry terminal, where boats could quickly reach the plane and remove its passengers and crew before the plane's wings sank.

Sullenberger demonstrated many virtues, including courage, steadfastness, and what, following Aristotle, is called "practical wisdom" in this emergency. Unlike a test pilot, Sullenberger did not have an ejection seat, so he could not have ejected himself from the aircraft to save only himself. Therefore, acting in the interests of others coincided with his interests in handling the crash. He exited the plane last, however.

This example illustrates the development of judgment to which education in the ethics of a profession such as engineering is intended to contribute. In this case, Captain Sullenberger exercised professional judgment that brought together his theoretical and practical knowledge as a pilot to bear on the unique circumstances that faced him. His first professional responsibility was for the safety of those on board. His goal was to save their lives, a goal that he achieved.

The critical thinking and moral reflection you engage in now will provide you with practice and some background for recognizing and fulfilling your professional responsibilities and maintaining your moral integrity.

In Part 1, we took the first step, which was to understand why certain moral rules are seen as especially important for engineers and appear in influential codes of ethics. In the remaining three parts of this book, we will employ the more complex notion of (forward-looking) moral responsibility and examine the responsibilities that are characteristic of the engineering profession.

As we saw in Chapter 3, devising good responses to challenging moral problems (and therefore fulfilling an engineer's professional responsibilities) requires problem-solving skills much like those used to address problems of engineering design and experimental design.

As philosopher John Ladd and sociologist Bernard Barber have both argued,[1] **responsible professional practice** combines two elements: *proficiency in the*

---

[1] Ladd (1978) and Barber (1983).

*knowledge and skills* of the profession in question and *concern* for the well-being of others – **"due care"** as it is often called in legal discussions. Responsible practice is practice that is conducive to producing desirable *results* in the area entrusted to the profession in question. For example, responsible engineering produces technology that *does not do* what it *is not* supposed to do – the safety requirement – and *does do* what it *is* supposed to do – the performance requirement. (Performance requirements of *safety critical* systems, such as artificial hearts or traffic control systems, have additional safety implications.) Unlike rules of conduct, rights, or obligations, the specification of responsibilities specifies the *outcome to be achieved* rather than the particular *act* to be performed or to be avoided.

Busy lives tempt people to take shortcuts. The example of Timothy Geithner (Treasury Secretary in the Obama administration) shows that a candidate for Treasury Secretary may neglect paying more than $34,000 in federal taxes! Greater moral lapses (and incompetence) lie at the heart of the 2007–2009 banking and credit crisis. Negligence, incompetence, and some deliberate malfeasance was involved in the behavior of Enron accountants that came to light in 2001–2002. Bernard Madoff, in promoting a pyramid or "Ponzi" scheme revealed in 2008, exhibited deliberate malfeasance.

We cannot be our own experts on all subjects and so we are vulnerable to the incompetence, carelessness, negligence, and malfeasance of others.

We saw in Part 1 that professions recognize certain standards of behavior as incumbent upon members of that profession, if those members are to use their specialized knowledge in a manner consistent with the public trust. Those standards are usually set out in documents with names like a "code of ethics," which are either phrased as a set of broad aspirations or a set of rules of behavior that establish a minimum for professional conduct.

Both competence and moral responsibility require more than rule following, however. The example of Captain Sullenberger illustrates what we all need of professionals and other experts: We need them to recognize what is relevant to meeting the needs of their clients and the public, to integrate that information into a plan of action, and to carry it out.

Fulfilling responsibilities typically requires both creativity and more exercise of judgment than do fulfilling obligations, respecting others' rights, or fulfilling any other moral requirement that can be expressed as a moral of the form "Do X" or "Do not do Y." (Additional judgment is required both to think up possible responses to a situation and to assess those responses.) Unlike moral rules or statements of obligations that specify what acts to perform or refrain from committing, (forward-looking) responsibilities specify the *ends (i.e., the good result)* that are the goals of the professional practice in question, such as "the responsibility for *the integrity of the research record.*" One must figure out what to do or avoid doing to achieve the specified ends while respecting other ethical norms.[2]

---

[2] Although "responsibility" can be used as a synonym for "obligation" – "the responsibility *to do* something" just means "the obligation to do that thing" – some responsibilities are not captured by the language of "obligations." There is no expression "obligation for." Achievement of ends requires judgments about *what actions* will best achieve those ends without causing major negative side effects in the research context under consideration.

Revelations that members of some profession have practiced incompetently or without due regard for that aspect of others' well-being entrusted to their profession are frequently followed by calls for greater regulation. The United States saw a call for greater regulation after failures in the accounting profession resulted in the Enron scandal in 2001. (The Department of Justice, the Securities and Exchange Commission, and the Public Utility Commission of Texas found Enron to have inflated its profit statements and hidden debts by improperly using off-the-books partnerships, manipulated the Texas power market and the California energy markets, and bribed foreign governments to win contracts abroad.[3]) Although the public was relatively enthusiastic about oil drilling immediately before the 2010 explosion of the BP offshore oil rig, now that that accident has proven to be the worst environmental disaster in U.S. history, the segments of the public who do not work in the oil industry call for greater regulation (and stricter enforcement of regulation) of oil drilling.

Professional groups often resist regulation and oversight by those outside their profession unless faced with overwhelming evidence that self-regulation by the profession has failed. Regulation does burden professionals by requiring more documentation at the least. Professional groups typically argue for their members' freedom from outside control, arguing that only members of their profession know how to evaluate the work of their profession.

> Some professionals may not in fact be competent or may not show proper concern for the welfare of their clients (or the public in general). From the point of view of both the professionals and the public at large, regulation, like other forms of legal redress, does not provide a *good* alternative to having professionals behave responsibly.

On the one hand, it is important to avoid a repeat of the mistakes and malfeasance that threaten the general welfare. On the other, regulation has drawbacks. Regulation may provide an interim solution in the present crisis of confidence, somewhat like a FEMA tarpaulin. In the long-term, however, regulations add to the tasks people are tempted to ignore (even with new penalties for doing so). May human watchdogs and regulators be assumed to be more conscientious than those humans they watch and regulate? If not, who watches them? Some professionals may not in fact be competent or may not show proper concern for the welfare of their clients or the public in general. From the point of view of both the professionals and the public at large, regulation, like other forms of legal redress, does not provide a *good* alternative to having professionals behave responsibly. Trustworthiness of professionals and other experts is a better situation. It requires that professions and other bodies of experts devote attention and effort to developing and transmitting high standards of professional practice.

*What characteristics or behavior on the part of professionals on whose work your own welfare depends would qualify them as trustworthy?*

[3] Patsuris, Penelope. 2002. "The Corporate Scandal Sheet" (updated through September 2002), *Forbes.* Accessed at http://www.forbes.com/2002/07/25/accountingtracker.html.

### *Ethical Responsibility and Official Responsibility*

What is the relationship between an engineer's professional responsibilities and the duties delineated in the job description for that engineer's position?

## Accountability and "Responsible To"

Because at a certain age (typically 18 or 21 years old for legal maturity in the United States) people are assumed to have the intellectual and emotional maturity to fulfill responsibilities, at that age they are held **accountable,** that is, answerable to others, including the legal system, for doing so.

Although "responsible" is sometimes used as a synonym for "accountable," this derivative sense of "responsible" is easily distinguished from being responsible *for* some desirable outcome. When "responsible" is used as a synonym for "accountable" it is paired with "to" (followed by the name of the party to whom one is answerable) rather than "for" (the desirable outcome).

For someone to have a **moral responsibility** *for* something means that the person must exercise judgment and care to achieve or maintain a desirable state of affairs with regard to whatever is in that person's care. For example, accountants are responsible for the accuracy of financial reports, physicians are responsible for health outcomes (and certain aspects of public health), and engineers are responsible for safety and performance in the design, manufacture, and operation of technology. Notice that we speak of people reaching "an age of responsibility" or "age of discretion," indicating that although children may follow moral rules, greater intellectual abilities are required to exercise responsibility appropriately.

The moral sense of responsibility, in which one undertakes to achieve some future state of affairs or maintain some present one, should not be confused with the *causal sense* **of responsibility** for some existing or past state of affairs. (Consider the example of the storm that is described as being "responsible for" deaths and property damage. Causal responsibility, not moral responsibility, is attributed to the storm. Storms do not have moral responsibilities and are neither responsible nor irresponsible in the moral sense. They are causal but not moral agents, so their actions are not subject to moral evaluation.)

> For someone to have a **moral responsibility** for something means that the person must exercise judgment and care to achieve or maintain a desirable state of affairs.

Moral responsibilities *of a moral agent* may derive from their causal responsibilities, however. If a person has caused a difficulty, that is one reason to think that the person has a moral responsibility for remedying the resulting situation. If a person breaks something, that person has some responsibility for fixing it or for cleaning it up and replacing it. However, people often find themselves faced with a responsibility not of their own making. If an infant or young child breaks something, someone else must clean it up. If pollution of the environment is not adequately addressed in one generation, subsequent generations find themselves responsible for cleaning up the contaminants that another has left.

The moral sense of responsibility is related to the **virtue sense of "responsible,"** the sense which names a character trait of being a person who regularly recognizes and fulfills her responsibilities and hence, is trustworthy.

Characteristically, the achievement of the desired outcome involves some exercise of discretion or judgment. This is what distinguishes a responsibility from

## The Concepts, Obligation and Responsibility

The relation between the concepts of obligation and responsibility is actually more complex, and the two terms sometimes overlap. For example, notice that the Code of Ethics of the National Society of Professional Engineers (NSPE), after saying that engineers shall "Hold paramount [that is, take as their primary concern] the safety, health and welfare of the public in the performance of their professional duties;" goes on to say, "Engineers shall at all times recognize that their primary obligation is to protect the safety, health, property and welfare of the public." Consider the obligation stated in this last passage. It is stated in the form of an obligation to *do* something (in this case protect the safety). As philosophers point out, "protect" is an achievement word in that it specifies not what action one performs, but what one *succeeds* in doing. Contrast the word "look" and the word "see." "See" is an achievement word. "Look" is not. So one might sensibly say, "I looked, but I did not see the bird that was singing." When an obligation is stated in terms of an achievement word, the goal of the action is specified. This makes the obligation statement *equivalent* to a statement that specifies a responsibility for achieving that goal. In neither case is one told precisely what acts to perform or refrain from performing. In contrast, the obligation to refrain from taking bribes specifies what *acts* are forbidden, namely the offering of payments or inducements to someone to motivate them to do something for the bribe payer to which the bribe payer is not entitled.

other moral requirements and makes moral responsibility a concept best taken up in later college years,[4] as it requires some intellectual and, perhaps, emotional maturity to fully grasp. An obligation specifies what acts a person is required to perform or refrain from performing. Notice that this difference is reflected in the difference between the expression "responsible for (some end)" – such as responsible for the safety of some device, responsible for the welfare of some person – as contrasted with being "obligated to *do* (or refrain from doing) certain things." Generally, the statement of an obligation specifies the acts one is expected to perform or refrain from performing. Contrast a professional's responsibility for the well-being of her clients with a professional's duty or obligation to be truthful about her qualifications or anyone's obligation to refrain from assaulting others.

Confusion may be caused by the fact that the term "responsibility" is sometimes used narrowly as a synonym for obligation, so that one may say, for example, "It is your responsibility to back up the computer files before you leave." To avoid confusion in this book the term "responsibility" will never be used in that sense, that is, as a synonym for obligation – that is, it will never be used in the form "responsibility to perform some act." It will be used exclusively in the form "responsibility for some outcome or state of affairs to be achieved."

Moral responsibilities derive from either one's relationship to a person whose welfare is in question, or from the special knowledge one possesses, such as professional knowledge that is crucial to an aspect of another's well-being. Examples of the first sort include the responsibility of one friend for another and of a parent for a child. Notice that a person can have this first kind of responsibility without having any particular knowledge that helps him fulfill the responsibility. Examples of the

[4]In my teaching experience, only a minority of incoming freshmen understand the notion of moral responsibility, whereas all freshmen understand the notions of moral obligation, moral rule, and moral right. After several years of college most students do grasp the notion of moral responsibility. I don't know what causal mix of brain development, experience of living apart from parents, and anticipation of life after graduation produces the change, however, and individual students vary greatly in the readiness with which they learn the concept.

second sort are the responsibility of a health practitioner to stop and give aid to an injured person who may be a stranger, and the responsibility of an engineer to ensure public safety and thus safeguard many individuals whom the engineer will never meet. One person's responsibility for another's welfare may combine both elements. For example, a health care practitioner may have a significant personal relationship with a patient who also is dependent on the practitioner's knowledge for adequate care. Little knowledge and few relationships are shared by everyone. Therefore, most moral responsibilities are *special* moral responsibilities, that is, they belong to some people and not others. No generally accepted group of "human responsibilities" is analogous to the catalog of commonly accepted human rights or (by derivation from human rights) universal obligations.

**Professional responsibility** is the most familiar type of moral responsibility that arises from the special knowledge a person possesses. Mastery of a body of advanced knowledge, especially knowledge that bears directly on the well-being of others, distinguishes professions from other occupations. Today, it is not possible for one person to master all the knowledge that is relevant even to her own well-being. Because society looks to members of a given profession to master and develop knowledge in a particular area, the members of a profession bear special moral responsibilities in the use of the special knowledge vested in them. For example, a state environmental protection division would employ an environmental engineer to decide whether plans for construction of a power plant meet the regulation requirements of the Clean Air Act (specifically, whether the plans provide sufficiently for the reduction of such pollutants as sulfur dioxide and nitrous oxides), and thus whether a building permit should be issued. Engineering knowledge is required to make this assessment. Neither the public nor administrators can make that assessment without such knowledge.

Although some moral demands on professionals are adequately expressible in rules of conduct that specify what acts are permissible, obligatory, or prohibited, there is more to acting responsibly. A good consulting engineer not only shuns bribery, checks plans before signing off on them, and the like but also must exercise judgment and discretion to provide a design or product that is safe and of high quality. Moral agents in general and professionals in particular must decide what to do to best achieve good outcomes in matters entrusted to their care.

## When Official Responsibilities Become Moral

Because there is a prima facie moral obligation to keep one's promises, arguably, when one accepts a job or office, one implicitly promises to take on the obligations and responsibilities that go with it. Thus, if one freely takes on a job or office, official obligations and responsibilities may also be moral responsibilities and moral obligations.

Not only does responsible behavior require more judgment than does the performance of specified acts, but the person with the responsibility may need only see to it that someone else does what is needed. Thus the question "Who will be responsible for the lead screening program?" does not ask who will do the screening tests, but rather who will see that the program is implemented.

Now consider the differences between a moral responsibility and an **official responsibility** – that is, a responsibility that someone is charged to carry out as part of her assigned duties. The description of a job or office specifies some of its official responsibilities. Moral responsibility does not reduce to official responsibility. Some official responsibility or obligation may even be immoral. "I was just doing my job" or "I was just doing what I was told" is not a generally valid excuse for unethical behavior of an adult.

Corresponding to the notion of moral responsibility is the notion of **legal responsibility**. A legal responsibility may arise in either of two ways: as a moral responsibility that is legally recognized and enforced, such as a responsibility of a parent, or as a legally mandated official responsibility. An example of the latter is the legal responsibility for deciding whether to move some community members from their homes to prevent their further exposure to toxic contamination. A legal responsibility may be part of the job of a public official.

### Are Workers Interchangeable? – Simon's Model

According to Herbert A. Simon's model of organizational behavior, people in formal organizations ideally make the decisions delegated to them on the basis of the organization's interests and values rather than on the basis of the values they themselves hold. Simon presents this model, not as a description of how administrators do make decisions, but of how they *ought* to. That makes Simon's model what we call a "normative" theory, rather than a descriptive theory. Many others in addition to Ladd have challenged it. This model of organizational behavior treats one competent person as completely substitutable by any other who comes to occupy the same position in the organization; that is, any agent in a given position would have the same official responsibilities, and any competent person in that position would make essentially the same decision.

The notion of official responsibility is central to the attribution of decisions to organizations rather than to the people in them. For example, people may say that the *Ford Motor Company* decided to rush the Pinto into production, rather than that particular people, such as Lee Iacocca, then president of Ford, made the decision. This way of thinking about decisions turns on the idea that an organization is a "decision-making structure" and that the actual person or people who make a decision carry out their official responsibilities and obligations according to the values and criteria attributable to the organization. Organizational values determine the goals to be achieved. The technical skills and scope of authority are held to specify the scope of actions that the agent is to take in achieving those organizational goals. The agent's own values or the values of the agent's profession, religion, or culture are all assumed to be irrelevant to what the agent will do in "doing her job." Therefore, on this model doing one's job is unaffected by the character and values of the person doing the job.

Any decisions that a person makes in her official capacity are attributable to the organization rather than the individual.

As John Ladd has argued, *official* responsibilities differ significantly from moral responsibilities, in that they attach to job categories and impersonal roles rather than to particular people in particular circumstances, with histories and human relationships that are unique to them.[5]

The scope of one's official responsibilities is specified by one's position and one's job description, apart from one's larger insights into the situation. One person's official responsibilities exclude another's. This exclusionary feature makes

---

[5]Ladd, John. 1970. "Morality and the Ideal of Rationality in Formal Organizations," *The Monist* 54(4): 488–516.

official responsibility quite unlike moral responsibility. Two friends of the same person may *both* have a *moral* responsibility to see that he does not drive while intoxicated, for example.

If a supervisor were to say to an engineer, "It is not your job to think about safety questions," this might be true as a statement about official responsibilities but would not mean that the engineer lacked any moral responsibility for raising safety concerns. Although a person's job description may not include some matter, she may have a moral responsibility in that matter, especially if it is a responsibility of her profession and hence one for which she has professional knowledge to fulfill.

Moral responsibility, unlike official responsibility, cannot be simply transferred to someone else. This feature of moral responsibility is expressed by saying that it is not "alienable." Suppose an engineer in charge of a project assigns the responsibility to make certain safety checks to another member of the team and the subordinate fails to do so. The engineer in charge will bear some responsibility for the failure, especially if the engineer in charge had reason to know that the subordinate was not reliable or did not have the relevant competence.

Consider the following case based on real-life events and reviewed by the Board of Ethical Review [BER] of the National Society of Professional Engineers [NSPE]:

---

### The Responsibility for Safety and the Obligation to Preserve Client Confidentiality

Tenants of an apartment building, annoyed by many building defects, sue the owners to force them to repair those defects. The owner's attorney hires Lyle, a structural engineer, to inspect the building and testify for the owner. Lyle discovers serious structural problems in the building that are an immediate threat to the tenants' safety. These problems were not mentioned in the tenants' suit. Lyle reports this information to the attorney who tells Lyle to keep this information confidential because it could affect the lawsuit. Lyle complies with the attorney's decision.

*Source*: Adapted from NSPE Board of Ethical Review (BER) Case 90–5

What, if anything, might Lyle do other than keep this information confidential? Which, if any, of those actions would have better fulfilled Lyle's responsibilities as an engineer?

What other information may be needed to make this decision?

---

The question that the BER explicitly addresses in its discussion of cases is whether certain actions of engineers described in its version of the case are "ethical or unethical," and BER decisions are based solely on applicable provisions in the NSPE code of ethics. The reasoning behind these simple binary judgments is what makes them interesting. The discussion of these cases reflects the norms of ethical practice put forward by this professional society. Most of the cases that come to the attention of the NSPE's Board of Ethical Review are based on complaints of one licensed engineer about the behavior of another, although a few are based on news stories.

There is some danger that in emphasizing the professional responsibility to work for the well-being of a client – rather than just emphasizing the rights of the client – we encourage **paternalism** on the part of the professional. **Paternalism** derives from the Latin word for father (*pater*). Acting like a parent toward those who are not your children may or may not be justified in particular circumstances. An act of paternalism may be roughly defined (following Gert and Culver) as infringing on a moral rule of conduct toward someone or infringing on that person's rights (such as the right of self-determination) for what the agent believes is that person's own benefit.

The question of paternalism often arises in medicine and health care with respect to the treatment of patients. Because many engineers in industry must protect the safety and health of anonymous members of the public rather than identified clients, and usually do not occupy positions of greater power than the clients they do have, paternalism is not a frequently discussed topic for engineers in industry. However even for such engineers in industry, the issue of paternalism can arise in connection with "idiot-proofing" as we shall see later in this chapter. Issues of paternalism often do arise for engineers and scientists in connection with relationships among coworkers and students.

### Paternalism in the Supplying of First Aid Supplies

The first aid kits in some of the teaching laboratories at a major university contain only small bandages. When some members of the engineering faculty tried to have more adequate supplies put in the kits, they were told that if the kits contained more supplies, those supplies might be misused in a way that would cause injury. Anyone who needs more than bandages, they were told, should go to the health service for treatment.

This example illustrates that if one is determined not to put anything in people's hands with which they might harm themselves, they will not be able to do themselves much good either.

To say that some act counts as paternalism does not yet tell us whether it is justified or unjustified paternalism. However, acts of paternalism do *need* justification, because they involve infringement of moral rules; the burden of proof is on the side of those who claim that a given infringement, in this case, a given act of paternalism, is morally acceptable. If the rule infringed in a given case was not an absolute moral rule, then other moral considerations may show that the act of paternalism was, on balance, *justified* – that is, it was right to do it in those circumstances. The responsibilities of a professional to look out for a client's welfare in the area of the professional's expertise do not *necessarily* conflict with any of the client's rights, especially if the professional explains the pros and cons of the situation to the client rather than simply making a judgment that is left unexplained to the client.

*What is the relationship between an engineer's professional responsibilities and the official responsibilities and obligations delineated in the job description for that engineer's position?*

### Trust and Responsibility

What is trust? What is the relationship between being a responsible professional and being a trust*worthy* professional?

Trust of various sorts is necessary for carrying out many tasks in ordinary life: trust of technology, trust of institutions, and trust of other individuals.[6] Without trust there can be no cooperative activities and thus no life in a community or society. (Cooperative activities include many that are *also* competitive. Competitive sports is a handy example of a competitive activity in which there are standards of "fair competition" to be mutually upheld.) **Trust** is confident reliance.

Confidence and reliance do not always go together. We may rely on someone or something, trusting that the thing or person in question will perform as needed and expected. However, we may also rely on people or things even where we have good reason to *dis*trust them. If I am told that my well may have been contaminated with toxic substances, then I will stop using water from the well only if, or to the extent that, I have another source of water available. Conversely, we may have great confidence in something – say that the automobile of the president of Ford Motor Company is in good repair – without relying on that fact. Unless we in some way rely on this fact, we do not *trust* it.

As Annette Baier has argued, trust relationships do not always have an ethically sound basis. Someone may trust another whom she has successfully threatened or otherwise coerced into doing her bidding. Baier's general account of the morality of trust illuminates the strong relation between the trust*worthy* and the true. A trust relationship according to Baier is decent if, or to the extent that, it stands the test of disclosure of the basis for each party's trust. For example, suppose one party trusts the other to perform as needed only because the truster believes the trusted to be too timid or unimaginative to do otherwise. Alternatively, suppose the trusted fulfills the truster's expectations only because he fears detection and punishment. Disclosure of these premises will undermine the trust relationship. Knowing the truth will give the trusted person an incentive to prove the truster wrong, or give the truster the knowledge that if undetected defection or betrayal becomes feasible, the trusted will likely defect or betray. Telling the truth about the basis for trust is an operational test of whether the trust is rooted in trustworthiness and a confidence in the other's trustworthiness. (That does not mean that it is always appropriate to speak everything one knows to be true, of course.) If the trust relationship cannot withstand having the truth told about it, it is corrupt by Baier's criterion.

Although explicit philosophical examination and discussion of moral trust and trustworthiness are relatively recent, both professional ethics and the philosophy of technology have given considerable attention to the concept of responsibility. Behaving responsibly in certain contexts, or being a responsible person in those contexts, means being willing and able to take responsibility for one's own actions. Acting responsibly in a professional capacity or being a responsible professional makes one a trustworthy professional. Therefore, the literature on responsibility, which has been extensive in recent discussions of professional ethics, provides at least an implicit discussion of many aspects of the morality of trust in professional practice.

---

[6] Portions of this section are adapted from "Trust and the Future of Research," *Physics Today,* November 2004, 48–53.

*What is trust? What is the relationship between being a responsible professional and being a trustworthy professional?*

### Trustworthy Engineers/Trustworthy Professional Practice

What factors do engineers need to consider in order to be trustworthy/responsible in their practice?

> For professionals or their professional practice to be trustworthy is a matter of both ethics and competence.

For professionals or their professional practice to be trustworthy is a matter of both ethics and competence. Trustworthy practice requires sustained attention to relevant aspects of others' well-being and the knowledge and wisdom to promote or safeguard that well-being. The well-being of many parties may be at stake in a given situation. It is important to consider all of them and, as far as possible, promote the well-being of all.

As an example, a trustworthy structural engineer who is building a bridge or a tunnel needs all of the following:

- A concern for public safety, public convenience, and environmental protection
- Proficiency in structural design; an understanding of the characteristics of building materials
- An understanding of traffic demands
- An understanding of the environmental implications of the work
- An estimate of the likelihood and severity of earthquakes, hurricanes, and other natural threats to the integrity of the bridge.

The engineer might also need to consider such factors as:

- Other technologies that might influence the use of the bridge (e.g., the characteristics of any ships or vehicles that might go under the bridge, or collide with its supports)
- An estimate of any likely intentional human threats to the bridge (sabotage or terrorism)

As another example, in overseeing the storage of chemicals, a chemical engineer would need to consider how the chemicals react with one another in case they leak or spill, as well as how the chemicals will interact with containers of various types. In designing software, a computer professional needs not only up-to-date knowledge about software and its potential but also an understanding and commitment to meeting the explicit design constraints (criteria) of the client/job, plus possibly unstated background criteria of security and reliability.

Ethical and technical considerations frequently become inextricably intermingled in the exercise of professional discretion and judgment.

### Values in Diagnostic Decisions

For a period, another area of practical and professional ethics, bioethics, discussed diagnosis as though it were an ethically neutral technical inquiry. Ethically interesting issues were assumed to enter only with the selection of treatment. It is now recognized that the question of whether to obtain more information in pursuit of a firm diagnosis raises its own ethical issues. Suppose a patient is seriously ill with a disease that will kill her within a year, and she is having symptoms of what may be a passing condition or may be the early stages of another disease with a slow progression. Are the pain, risk, disruption of life, and expense to the patient of tests to diagnose the new symptoms justified by the difference the knowledge will make to her treatment? In other cases, one should consider whether patients are prepared to cope with unexpected information that may turn up – for example, information about abnormalities with largely unknown prognoses that may turn up in prenatal tests such as amniocentesis.

Unless design engineers consider the safety of future operators of the object/ system they design, they might never think to ask some technical question about the behavior of some material under unusual temperature or humidity conditions, conditions under which the object/system may be operated. It is only after noting the potential for one device to influence another, as microwave ovens can influence pacemakers, that the question of how best to warn users of the potential risk becomes an issue. When the complex intermingling of ethical and technical issues is ignored and professional responsibility is treated as having separable ethical and technical components, the exercise of responsibility is distorted. Of course, one can distinguish between reservations about people's technical competence and concerns about their moral character when considering whether to trust their professional judgment. When they come to exercise that judgment, however, technical and ethical components will usually be inextricably involved. Recognition of and concern about ethical questions commonly lead to new technical questions, and technical insights or breakthroughs often raise new ethical questions.

Ethical and technical questions are closely connected in the engineer's responsibility for safety.

---

### Technical Disagreement and Ethical Responsibility

Hilary is an engineer working for the state environmental protection division. Hilary's supervisor, Pat, tells Hilary to quickly draw up a building permit for a power plant and to avoid any delays. Hilary believes that the plans are inadequate to meet clean air regulations, but Pat thinks that these problems are fixable. Hilary considers whether to ask the state engineering registration board about the consequences of issuing a permit that goes against environmental regulations.

What values, obligations, and responsibilities are at stake in Hilary's deliberations about what to do? Should Hilary consult the state registration board? If so, how ought the information from the state board affect Hilary's decision about what to do after that? (Consider all likely responses of the registration board.)

What, if anything, can and should Hilary do if Hilary's department authorizes the building permit over Hilary's objections?

Is there any other information you would like to have to help you answer these questions, and what difference would it make to your assessments?

*Source*: Adapted from NSPE BER Case 92–4[a]

---

[a]The NSPE Board of Ethical Review (BER) cases with judgments offered by the BER based on application of the then current NSPE Code of Ethics are available in hard copy in the volumes V–VIII of *Opinions of the Board of Ethical Review*, Alexandria, VA: National Society of Professional Engineers.

---

Moral problems are sometimes treated as questions of whether to do something; the question of how to go about it is then treated as merely pragmatic. However, questions about *how* to do things often raise ethical questions of *fairness*, and questions of *how far to go*, say in protecting safety, are at the core of professional responsibility.

*What do engineers need to consider in order to be trustworthy/responsible in their practice?*

## Character and Responsibility

What, if any, virtues or vices have you observed to be especially likely for engineers to possess? Are these in any way related to the nature of engineering work – say as necessary to do the work well or tolerated because of temptations in the work? For example, it is sometimes argued that arrogance often develops and is tolerated in surgeons, because they head teams whose work has momentous consequences for patients' lives and they must make quick decisions with incomplete information about what action they and those teams should take.

### Concept of Responsibility and Human Rights

Previously we noted the argument by John Ladd that taking the concept of responsibility and the virtue of being a responsible person as central creates a necessity for human rights. That argument can now be briefly rephrased: because being a responsible person means being able to take responsibility for one's own actions, one assumes the rights of self-determination. If people's control of themselves is disrupted – for example, if people were to be regularly drugged – it would effectively undercut their fulfillment of their moral responsibilities, personal and professional, and so undermine moral life.

The implication is that taking responsibility as the central ethical notion does *not* imply that human rights are irrelevant, but rather provides a "warrant" or justification of those rights.

In the last section we briefly considered the virtue sense of "responsible," in the notion of a responsible person. In the discussion of Roger Boisjoly, passing reference was made to some of the virtues, such as honesty and courage, with which he is widely credited, even though the emphasis in that section was on the *actions* he took rather than his *moral character*. This section provides an overview of concepts of moral character and, hence, virtue and vice as they apply to engineering ethics.

In contrast to moral rights, moral rules, and moral obligations, traits of **moral character**, or **virtues** and **vices**, such as honesty, kindness, courage, responsibility, insincerity, hardheartedness, cowardice, and recklessness, are *characteristics of people*, rather than features of acts or the consequences of those acts.

Those entering engineering generally value honesty in themselves and others. Honesty (encompassing truthfulness, which is also called "veracity"), candor, and sincerity do support engineering accomplishment and the advance of engineering knowledge. Because people entering engineering are commonly people who enjoy solving technical problems, and stealth and guile are no aid to success in doing that work, engineers as a group may be more honest than those entering other professions.

Character traits that are considered desirable or undesirable (and are therefore called "virtues" and "vices," respectively) vary somewhat with sphere of activity and the relative importance accorded to specific activities in particular cultures and subcultures (including professional subcultures). For example, the intellectual self-discipline required to rigorously test hypotheses in engineering and other scientific fields may not be crucial in parenting an infant. A degree of arrogance is tolerated in skillful surgeons that would not be acceptable in many other professions, probably because they must make immediate decisions that may have life-and-death implications and lead a team of others in carrying out those decisions. (Skills are something one can choose to apply or not; character traits are part of what a person is. Although people do sometimes act "out of character," such behavior is remarkable and a more complex phenomenon than deciding when to stop exercising a skill.)

As philosopher Alasdair MacIntyre has argued, the development of certain traits of moral character is essential in the development of complex cooperative activities, including social practices. Among such social practices is the maintenance of a profession and of (ethical and technical) professional standards.

### MacIntyre on the Development of Virtues

Social practices achieve ends and produce results for which their practitioners may receive what MacIntyre calls "external" rewards: pay, fame, or career advancement. For example, engineering research produces new knowledge, for which researchers may receive external rewards in addition to knowing that they have succeeded in advancing knowledge. In addition to developing certain skills, engaging in a practice also develops certain virtues in its practitioners. Practicing engineering research, for example, develops not only research skills but also virtues such as patience, thoroughness, and diligence. MacIntyre calls these virtues "internal" goods, or rewards of the practice, because they are achieved quite apart from whether the research yields any particularly notable results, or advances researchers' careers.

Nonetheless, many scholars agree that some virtues, such as honesty and courage, are necessary to the successful conduct of all or most social practices.

To understand a person's character one must understand the configuration of ethically relevant considerations that influenced her actions. Knowing that a person has often broken the law might lead one to conclude the person was dishonest. However, if the individual habitually hid people from unjust persecution by a tyrannical government, then the person could well have been an honest person in circumstances that justified lying to law enforcement officials. If a person's apparent bravery and willingness to risk her life in battle derived mainly from an obsession with killing and maiming people, then the quality would not be the virtue of bravery, but merely the successful redirection of a character defect in a socially acceptable way.

The concept of **moral integrity** is central to the assessments of character, but it is not one more character trait. Roughly, "moral integrity" is the ethical coherence of a person's life and actions. Honesty and consistency characterized by the absence of hypocrisy or betrayal are part of the notion of moral integrity. People's values may be expected to develop over the course of their life, so moral integrity is not simple persistence in maintaining value commitments, however. The coherence of people's lives is a narrative coherence, the coherence of their life stories. Understanding a person's life story in turn requires understanding the place of values or ideals as they develop throughout a person's life.

A loss of integrity can be forced upon a person. One example is Sophie, in the book and film *Sophie's Choice*. Nazi guards force a true dilemma on Sophie: She is required to choose, in the presence of her two children, which of the two is to be killed on pain of having them both killed. Being forced to send one child to her death is fatal to Sophie's moral integrity and sense of self. This is an extreme case but it illustrates that circumstances as well as personal resolve are factors in maintaining moral integrity.

A more common situation is one in which all of the responses that are *obvious* to a person are ones that threaten to betray some relationship or trust. A common circumstance in which this can happen is when someone is called upon to make a grave health care decision on behalf of a family member that the decision maker feels unprepared to make in a way that the ill person would have wanted.

Mismatches between preparation and needed skills and virtues are important to remedy. Even well-meaning people may respond badly when they have not thought through how they will fulfill potentially conflicting responsibilities simultaneously. In developing or reviewing policies and practices in a work situation, it is also important to be alert to mismatches and conflicts between the skills and virtues of key actors and the skills and virtues that others *need* in those key actors. For example, a devotion to the progress of scientific research might interfere with health care providers' fulfillment of the responsibility to secure the best health outcome for their patients, or with an engineering faculty member's oversight of their graduate student's education.

*What, if any, virtues or vices have you observed to be especially likely for engineers to possess? Are these in any way related to the nature of engineering work – say as necessary to do the work well or tolerated because of temptations in the work?*

### The Specific Professional Responsibilities of Engineers

In Chapter 1, we saw some examples of the ways in which the moral rules contained in codes of ethics vary with different professions. What are the specific responsibilities of engineers and how do these responsibilities differ from, say, the professional responsibilities of physicians?

The body of knowledge that characterizes a profession enables its practitioners to foresee possibilities, to devise ways to achieve desirable results and to avoid undesirable side effects. Specialized knowledge enables engineers and scientists to design interventions, devices, processes, or constructions and to foresee how those products, processes, and constructions will act or interact. The designers of these products are uniquely qualified to foresee and modify many consequences of their use or misuse. Prominent among such consequences are hazards to the safety of manufacturing workers, operators, consumers, and the public.

*What are the specific responsibilities of engineers and how do these responsibilities differ from, say, the professional responsibilities of physicians? Explain the differences that you have identified.*

## The Emerging Consensus on the Responsibility for Safety among Engineers

Why would anyone think that engineers have a special responsibility for safety?

The engineer's responsibility for safety is a useful place to begin discussion of specific professional responsibilities, because it is both familiar and generally agreed upon. It provides an undisputed area of responsibility that illustrates more general features of professional responsibility. Examination of the general features of professional responsibility for a generally agreed upon area of responsibility will provide a basis for clarifying responsibilities that are newer or more controversial.

Engineering students are often taught that safety is their responsibility. "First make sure the system doesn't do what you don't want it to do – that's the safety issue; then make sure it does do what you want it to do – that's the performance issue." This admonition is remarkably similar to the admonition to physicians: "First, do no harm."[7]

Emphasis on the engineer's responsibility for safety is also found in the codes of ethics or ethical guidelines of many engineering societies. These codes specify that it is the engineer's responsibility to protect public health and safety. Indeed, five of these societies – American Society of Civil Engineers (ASCE), American Society of Mechanical Engineers (ASME), American Institute for Chemical Engineering (AIChE), National Society of Professional Engineers (NSPE), and National Council for Engineering Examiners and Surveyors (NCEES) – continue to say in the latest revisions of their codes of ethics:

> Engineers, in the fulfillment of their professional duties, shall [h]old paramount the safety, health, and welfare of the public.

To "hold something paramount" is to say that something is the foremost concern.

Seven societies enjoin the engineer to report or otherwise speak out on risks to health and safety.[8]

Although recognition of the engineer's responsibility for health and safety is well-established, the range of factors that an engineer is expected to consider is ever-expanding. Lessons about the consequences of previous design decisions accumulate and technological innovation continues.

---

### Unanticipated Factor, Auto Safety

You are a new engineer working as part of a design team for a large automobile manufacturer. The company is doing a major redesign of one of its product lines.

Your team is responsible for designing part of the frame of the new car. As part of the company's drive to make cars lighter and more efficient, your team is directed to make some of the structural members out of carbon fiber composites. The cross member that holds the rails of the frame apart was ideally suited for composite replacement.

You test several different composite materials and lay-ups, and finally choose one that you have reason to believe will work. Several prototypes of the car are built, which you checked carefully. Your design is then approved and is about to go into production.

Just today you found a problem with your cross member. A few inches of the cross member from a car that was winter tested showed extensive cracking. After looking at the design, you realize that the cracked portion is in proximity to the exhaust system. You conclude the hot pipe in cold weather created thermal stresses and caused cracking.

What can and should you do and how do you go about it?

*Source*: Adapted from a scenario by Dan Dunn, Chris Minekime, and John Van Houten (MIT '93)

---

[7] For a discussion of the (non-Hippocratic) origins of this admonition, see Albert R. Jonsen, "Do No Harm: Axiom of Medical Ethics," in *Philosophical Medical Ethics: Its Nature and Significance*, edited by S. F. Spicker and H. T. Engelhardt, Jr. (Dordrecht, Holland: D. Reidel Publishing Co., 1977).

[8] Middleton, William W. "Ethical Process Enforcement and Sanctions – The Engineering and Physical Science Societies," delivered at the AAAS-IIT Workshop on Professional Societies and Professional Ethics, May 23, 1986.

*Why would anyone think that engineers have a special responsibility for safety?*

## Lessons from the 1979 American Airlines DC-10 Crash and the Kansas City Hyatt Regency Walkway Collapse

How did the experience of the 1979 DC-10 crash and the 1981 collapse of the walkway at the Kansas City Hyatt Regency expand the *scope* of safety considerations?

Two famous accidents in the last decades of the twentieth century – the 1979 DC-10 crash and the 1981 collapse of the walkway at the Kansas City Hyatt Regency – expanded the *scope* of safety considerations still further. They dramatically illustrated that designs that may be safe when constructed or maintained as specified may nonetheless create hazards indirectly by creating temptations for others to take unsafe shortcuts. In these two instances, the unsafe shortcuts were in maintenance, fabrication of connections, and construction.

### The 1979 Crash of an American Airlines DC-10 in Chicago

The 1979 crash of a DC-10 occurred just after takeoff. The left engine ripped off its mounting just before liftoff. The 1979 DC-10 crash was the worst disaster in U.S. aviation history before 9/11. The cause of the crash was found to be a ten-and-a-half-inch crack in the rear bulkhead of the pylon that attaches the engine to the wing. Pylons of other DC-10s were found to have similar cracks.

FAA investigation found the cracks to have been caused by improper maintenance at American Airlines and Continental Airlines. Rather than separating the engine and the pylons during maintenance, as recommended by the manufacturer, McDonnell Douglas, the crews had been removing and reinstalling the engine and pylons as a unit using a forklift. The case is complicated by the fact that the FAA had approved the use of the forklift in this way, on the condition that leather straps were provided for cushioning. The heavy aircraft components were liable to be misaligned during the forklift maneuver, resulting in the cracking of the rear bulkhead's flange. The maintenance shortcut saved approximately 200 person hours per engine, however, which provided a great incentive to use it. The designs of comparable aircraft by competing airline manufacturers, Boeing and Lockheed, did not present the same temptations for unsafe maintenance.

This accident made aircraft designers aware of the ways in which their designs *may create temptations to take unsafe shortcuts at other stages of manufacture or maintenance.* Therefore, this case added a new element in the evaluation of a design's safety.[9]

[9] For a fuller discussion of this case, including complexities that there is not space to discuss here, see Martin Curd and Larry May, *Responsibility for Harmful Actions.* Dubuque, IA: Kendall/Hunt Publishing Co., 16–21; Paul Eddy, Elaine Potter, and Bruce Page, 1979, "Is the DC-10 a Lemon?" *New Republic* (June 9): 7–9. For a discussion of the economic causes of design decisions that contributed to DC-10 crashes (and other airline accidents), see John Newhouse, 1982, "A Reporter at Large: The Airlines Industry," *The New Yorker*, (June 21): 46–93; and Newhouse's book from which this piece was taken, *The Sporty Game* (New York: Alfred Knopf, Inc., 1982).

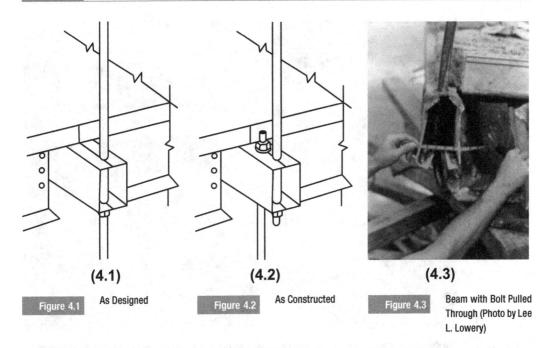

| (4.1) | (4.2) | (4.3) |
|-------|-------|-------|

| | | |
|---|---|---|
| Figure 4.1 | As Designed | |
| Figure 4.2 | As Constructed | |
| Figure 4.3 | Beam with Bolt Pulled Through (Photo by Lee L. Lowery) | |

## Hyatt Regency Hotel Walkway Collapse

On July 17, 1981, walkways in the Hyatt Regency Hotel in Kansas City, Missouri, collapsed, leaving 114 dead and more than 200 injured. Many of the dead and injured had been attending a tea-dance party in the atrium lobby at the time of the accident. Some had been standing and dancing on the walkways that were suspended above the lobby floor at the levels of the second, third, and fourth floors. Connections supporting the ceiling rods that held up the second- and fourth-floor walkways failed. The fourth-floor walkway collapsed onto the second-floor walkway directly below. The third-floor walkway, which was offset from the other two, remained intact. It was the worst structural failure in U.S. history in terms of injury and loss of life.

Jack D. Gillum & Associates, Ltd. had been subcontracted to perform all structural engineering services for the project. Jack D. Gillum was president of that firm and the professional engineer for the project. He was also one of the principals of Gillum-Colaco, the consulting structural engineering firm for the project from which Jack D. Gillum & Associates subcontracted their work. Eldridge Construction Company was the general contractor on the project. Havens Steel Company was the fabricator for the connections, working under a subcontract to Eldridge Construction. During January and February 1979, more than a year before the collapse, Havens Steel Company changed the design of the rod connections that hung two of the walkways, one above the other, from a floor above. The original plans had a single rod used at each point of connection, passing through the first walkway, fastened with a bolt underneath and extending to suspend the mezzanine level. Havens proposed to use two separate rods to simplify the assembly task and to eliminate the need to thread the entire length of the rods. (See Figures 4.1–4.4.)

This change doubled the load on the lower bolts, which now supported the weight of *two* walkways. The excessive load ultimately caused a lower bolt to pull through the beam so that one walkway collapsed upon the one below, causing it, too, to collapse.[a] (As originally designed, the walkways were barely capable of holding up the expected load and would have failed to meet the requirements of the Kansas City Building Code.[b])

The fabricator, in his testimony, claimed that his company had telephoned the engineering firm Gillum-Colaco, Inc. for change approval. Gillum-Colaco, Inc. denied ever receiving such a call

**Figure 4.4** The Kansas City Hyatt Regency during the First Day of Investigation. The second and fourth floor walkways lie on the lobby floor. They were moved from their original positions to extricate people from the wreckage. The third floor walkway is still suspended at left. (Photo by Lee L. Lowery).

from Havens.[c] Yet, Jack D. Gillum had affixed his seal of approval to the revised engineering design drawings.

On October 14, 1979, while the hotel was still under construction, more than 2,700 square feet of the atrium roof had collapsed because one of the roof connections at the north end of the atrium failed.[d] The engineering firm testified that the owner, Crown Center Redevelopment Corporation, had, on three separate occasions, refused its request for on-site project representation to check all fabrication during construction because of the expense.[e]

---

[a] *Missouri Board for Architects, Professional Engineers and Land Surveyors vs. Daniel M. Duncan, Jack D. Gillum and G.C.E. International, Inc.*, before the Administrative Hearing Commission, State of Missouri, Case No.AR-84–0239; Statement of the Case, Findings of Fact, Conclusions of Law, and Decision rendered by Judge James B. Deutsch, November 14, 1985, pp. 54–63.

[b] *Ibid.*, pp. 423–425. See also Edward O. Pfrang and Richard Marshall, "Collapse of the Kansas City Hyatt Regency Walkways," *Civil Engineering-ASCE*, July 1982, pp. 65–68. This article contains the official findings of the failure investigation conducted by the National Bureau of Standards, U.S. Department of Commerce.

[c] Administrative Hearing Commission, State of Missouri, Case No. AR-84–0239, pp. 63–66.

[d] *Ibid.*, p. 384.

[e] The synopsis given here is primarily derived from W. M. Kim Roddis, 1993, "Structural Failures and Engineering Ethics" in the *Journal of Structural Engineering ASCE* (May), and from the Hyatt Regency case in the case materials in *Engineering Ethics*, edited by R. W. Flumerfelt, C. E. Harris, M. J. Rabins, and C. H. Samson (Texas A&M), final report to the NSF on Grant Number DIR-9012252.

---

Having learned from such accidents, engineers now consider how their designs may indirectly increase risk. Following the Hyatt Regency disaster, the American Society of Civil Engineers (ASCE) urged civil engineers to accept design work *only if they oversee the subsequent fabrication and construction as well.*

Examples of other mistakes that threatened even graver consequences are those of the design and construction of the John Hancock Tower in Boston and the Citicorp Tower in New York City. Both structures were designed in ways that met the standards of engineering practice of the day, but each turned out to be endangered by effects that were not considered at the time. The Hancock Tower case is summarized in "Unexamined Influences." The Citicorp Tower case appeared in Chapter 2.

---

### Unexamined Influences: Boston's John Hancock Building

The Hancock Tower manifests several unrelated problems. The excessive swaying of the Tower, which architect William LeMessurier sought to remedy with a tuned-mass damper like the one he had pioneered for the Citicorp Tower, proved symptomatic of a serious structural problem. Structural engineering of the day did not consider certain second-order effects that were significant for the John Hancock Tower. In particular, the increase due to gravity of the displacement caused by the wind hitting the building on its short side had been overlooked. This effect was greatly magnified because of the 300-foot length of the long side of the Hancock building. The correction of the problem with the Hancock Tower was a more elaborate and expensive operation than the remedy of placing heavy steel-welded reinforcement for the bolted connections on the diagonal supports on the Citicorp Tower, which was discussed in Chapter 2.

---

This case provides a stark lesson in the possible importance of effects and relationships that engineers may not know to examine.

The point that accidents are frequently the way in which a society learns to foresee dangers is thoroughly discussed in Henry Petroski's elegant book, *To Engineer Is Human*. The cases discussed in the present chapter illustrate Petroski's point, because all led to changes in what engineers are expected to do in safeguarding the public.

### Foresight and Responsibility

To be morally responsible for outcomes, people must have some ability to foresee and influence them. I draw attention to this seemingly obvious point because some commentators have sought to blame technology (and engineers/computer professionals) for everything that is objectionable in modern life. Certainly technology has had harmful unintended consequences, as have political decisions. However, there is no possibility of doing without all technology any more than there is a possibility of doing without political institutions and decisions.

The history of engineering reveals at least two major considerations governing the evolution of requirements for engineers to fulfill their responsibility for safety. First, engineers' professional responsibility for safety extends only as far as the possible outcomes that an engineer can foresee at a given point in the development of engineering knowledge. Therefore, so-called end uses of technology (i.e., the uses to which some technology is ultimately put) and social byproducts of technology are the engineer's professional responsibility only if the competent engineer or scientist can foresee them. Second, the engineer has a professional responsibility to examine those determinants or results that she might control or influence.

## Foresight and Consequences

The view in philosophical ethics known as "Utilitarianism" maintains that what one ought to do in a situation is a matter of what, in those circumstances, would achieve "the greatest good for the greatest number." To *know* what one ought to do in a given situation would require a reliable estimate of the net amount of good produced by alternative actions. However, as we saw in Chapter 1, alternatives are rarely given, and instead must be devised. Indeed, for a problem that is technical as well as moral, the responses that an agent can devise are strongly influenced by the agent's disciplinary background. Therefore, even if an agent were somehow able to know *all* the consequences of particular responses, the set of "possible responses" would vary from one agent to another.

Although Utilitarianism has severe limitations as a guide to action, it may still be understood as one of several theories about what makes an action the best action, using the criterion that the action provides the greatest good to the greatest number, even if it is impossible to know whether a given action meets the criterion.

In professional life, the responses one is expected to consider and the foreseeable consequences of those responses are a function of the state of knowledge at the time. Furthermore, the possibilities that a member of one profession will consider are very different from those that a member of another profession will consider when faced with the same circumstances. For example, faced with the threat of an epidemic of cholera, a civil engineer will think of sanitation improvements as a way to control the spread of disease. A physician will think of medical means. This is regularly described as "**disciplinary bias**," but calling it "bias" is not to fault the agent who has it, because it is *a form of bias that cannot be eliminated*. (What can be eliminated is the illusion that *one's own expertise* is all that a situation requires.) In a democracy all citizens have some responsibility for the policies undertaken or allowed by their government, including policies on the development and use of technology. This responsibility is not a professional responsibility peculiar to engineers, but one they share with all citizens. As we will see in the next section, the engineering profession and even individual engineers may have some special professional responsibility to educate the rest of the public about new technology, however.

*How did the experience of the 1979 DC-10 crash and the 1981 collapse of the walkway at the Kansas City Hyatt Regency expand the scope of considerations that engineers are expected to consider in meeting their responsibility for safety?*

## "Bugs," Glitches, and Errors as Central Concerns in Software Engineering

Why are bugs and glitches more commonly the focus of attention for software and computer professionals, rather than safety problems *per se*?

The central problem for software engineers (and others who design and test software) is that of creating bug-free software. Sometimes the bugs clearly threaten human health and safety. As we saw in Chapter 1, a bug in the software was one of many design flaws in the Therac-25 X-ray machine that allowed it to deliver a potentially lethal dose of X-ray to patients when the operator attempted to correct a typographical error using the delete key. Bugs are especially likely to threaten safety in safety-critical systems, such as traffic control systems, but whether bugs threaten life and limb may depend on other circumstances that have nothing to do with the software and the technology it immediately affects. An example is

**Figure 4.5**  Two Air Force F-22 Raptors Shown Landing at Kadena Air Base, Japan, after almost Being Brought down by a Software Bug (U.S. Air Force photo: Sheila deVera).

the software glitch that threatened the U.S. Air Force's F-22 "Raptor" airplane, a case that we will examine here.

## A Software Bug That Threatened the U.S. Air Force's Superfighters

This example concerns the U.S. Air Force's latest superfighter, the F-22 "Raptor." The Raptors cost more than $300,000,000 each, but for a while, a software bug caused havoc in these pricey planes when they crossed the international dateline. The glitch came to light when, in February 2007, a group of ten Raptors headed across the Pacific for exercises in Japan. These Raptors suffered simultaneous total nav-console crashes as their longitude shifted from 180° west to 180° east. Tanker planes accompanying the raptors had somewhat older navigation kits, so the tanker planes did not experience the same nav-console crashes. The pilots of these tanker planes were able to guide the Raptor pilots back to Hickman Air Force Base in Hawaii. The glitch was fixed later that month and the planes flew to Kadena (see Figure 4.5).[a,b]

Because of the accompanying tanker planes, the software bug wasted time and money, but did not cost lives.

[a] Page, Lewis. 2007. "US Superfighter Software Glitch Fixed." *The Register*, February 28 (accessed at http://www.theregister.com/2007/02/28/f22s_working_again).
[b] Johnson, Maj. Dani. 2007. "Raptors Arrive at Kadena." *Air Force Link,* February 19 (accessed at http://www.af.mil/news/story.asp?id=123041567).

The harms caused by bugs, glitches, and errors vary considerably with the larger system in which the software functions. Thus, threats to safety may be more difficult to predict than the results of mistakes in mechanical or chemical engineering. The software engineer's central responsibility is therefore best phrased as a responsibility to avoid errors that produce bugs/glitches, rather than to foresee which errors might cause the bugs that will present safety hazards and take special care to prevent those specific errors.[10]

Bill Wulf has pointed out some of the reasons for the enormous number of bugs typically found in software and the public's surprising tolerance of them. He gives the example that the number of bugs that are in the Microsoft Office suite at any given time is estimated to be between one-half million and a million bugs.[11] Not all of these bugs are due to human errors, and so not all are the responsibility of software professionals. As we saw in Chapter 1, Wulf argues that the problem of how to responsibly engineer software when one knows in advance that there will be some behaviors of the resulting system that one cannot predict is a problem for the engineering profession, rather than the individual engineer. (Here we will consider only those bugs due to mistakes and errors, not those that are emergent properties of digital systems.)

Some engineering codes of ethics suggest that engineers bear some individual responsibility not only for educating other members of the profession about their responsibilities but also for educating the public about engineering. For example, the 2006 revision of the NSPE Code of Ethics for Engineers states under the second professional obligation (to serve the public interest at all times):

> c. Engineers shall endeavor to extend public knowledge and appreciation of engineering and its achievements.

More obliquely, the Software Engineering Code of Ethics and Professional Practice (Version 5.2) issued by the IEEE-CS/ACM Joint Task Force on Software Engineering Ethics and Professional Practices lists as one of the areas of responsibility for furthering the public good that software engineers shall

> Be encouraged to volunteer professional skills to good causes and contribute to public education concerning the discipline.

This obligation of individual engineers to educate the public about engineering does not go as far as educating the public about the possible societal implications of new technology, as Wulf believes the engineering profession also ought to do.

*Why are bugs, glitches, and other errors more commonly the focus of attention for software and computer professionals, rather than safety problems per se? What does this mean about software engineers' or computer professionals' responsibility for safety?*

## Knowledge, Foresight, and Changing Criteria for Responsible Practice

Why is the ability to foresee dangers relevant to an individual engineer's responsibility for safety? How and why do specific criteria for responsible engineering practice change over time?

Experience with the consequences of engineering design decisions has broadened the range of consequences that responsible engineers are expected to foresee and the range of factors that they are expected to consider in controlling those consequences.

---

[10] I am indebted to Peter Elias for this insight.

[11] Wulf, William A. 2004. Keynote Address, *Emerging Technologies and Ethical Issues in the Practice of Engineering*. Washington, DC: The National Academies Press, 5.

There has been an increase not only in the number of factors but also in the *kinds* of factors engineers must consider. The engineers' responsibility to ensure that a device or construction is safe in its intended use is now only the beginning of what they must consider to ensure safety. For example, automobiles are not intended to have collisions, but inevitably, many do. Reducing injury and damage resulting from automobile accidents is therefore recognized as part of the responsibility of automotive designers.

The responsibility for making technology safe under extreme conditions (e.g., severe storms), under foreseeable misuses, or as affected by foreseeable mistakes, compounds the responsibility for considering unintended but frequently occurring circumstances.

As discussed previously, experience with the consequences of engineering design decisions tends to broaden the responsibility for safety. The engineer's responsibility to ensure that a device or construction is safe in its intended use and under normal conditions is only the beginning. The responsibility for making technology safe under foreseeable misuses is also related to the engineer's responsibility for safety in extreme but foreseeable circumstances. As we saw in the example of limiting students' access to first aid supplies, attempting to block every possible harmful misuse is paternalistic. It may even be harmful, because blocking every possible harmful misuse may block important beneficial uses. As an extreme example, society could ban the sale of knives because people often accidentally cut themselves.

A good illustration of a foreseeable misuse is the misuse of a carpenter's hammer by a farmer who lost an eye as a result.

## Injury from Misuse of a Tool

A farmer used the carpenter's hammer for a job in which a ball hammer was the appropriate tool. The forged head of the carpenter's hammer had become work-hardened with use, making it brittle and hence more likely to shatter when striking an object harder than itself. When the farmer used the hammer to drive a pin into a clevis[a] to connect a manure spreader to his tractor, a chip broke off the hammer and injured his eye. The farmer brought legal action against the hammer's manufacturer for the injury. The manufacturer's lawyer argued that his client was not at fault because a carpenter's hammer is not designed for the job that the farmer was doing. However, it was well known that carpenter's hammers were used for a variety of tasks, and work-hardening is a well-understood metallurgic phenomenon. Moreover, the manufacturer had received several chipped hammers that customers returned for replacement.

When this case was taken to court, the court found against the manufacturer holding that he *should have foreseen* the kind of use to which the farmer had put the hammer and done more to prevent such injuries.[b]

What more might the manufacturer have done to prevent such injuries?

---

[a] A clevis is a U-shaped metal piece with holes at the ends, through which a pin is run to attach a drawbar to a plow.

[b] Thorpe, James F. and William H. Middendorf. 1979. *What Every Engineer Should Know About Products Liability*. New York and Basal: Marcel Dekker, Inc., 34. Martin Curd and Larry May, in their booklet *Professional Responsibility for Harmful Actions* (Dubuque, IA: Kendall/Hunt, 1984), discuss retrospective responsibility (what, following the discussion in Chapter 1, can be called "judge problems") and criteria for ethical and legal fault, including the hammer case.

### Is "Idiot Proof" a Reasonable Goal for Engineers?

In "Is Idiot Proof Safe Enough?" Louis L. Bucciarelli analyzes idiot-proofing as an ideal in engineering design. He argues that by treating the public as idiots rather than educating them, such zealous protection ultimately makes members of the public more vulnerable to injury. Because idiot-proofing excludes the users from knowledge of the workings of the technology, they are less prepared to deal with the unknown and unexpected.[a]

Other considerations about legal liability are important for understanding the relation or lack of relation between ethical accountability and legal liability. The function of liability judgments in the U.S. legal system is not primarily to fix moral blame. Society's interest in obtaining care for injured parties, which in other countries is handled by social safety net measures, leads the U.S. legal system to look for sources of support – "deep pockets" – as often as for guilt. Furthermore, society's interest in having safer products is sometimes manifest in court decisions intended to drive the standard toward greater safety rather than punish an agent who violated the existing standard.

[a] Bucciarelli, Louis L. 1985. "Is Idiot Proof Safe Enough?" *Applied Philosophy*, 2(4): 49–57; reprinted in *Ethics and Risk Management in Engineering*, edited by Albert Flores, Lanham, New York, and London: University Press of America, 201–209.

We saw in Chapter 2 that having a responsibility to ensure some future state or condition, such as safety, requires more than simply carrying out some required acts (fulfilling predetermined obligations). To behave responsibly, an agent must decide what acts are required to attain the desired state of affairs. The special body of knowledge that characterizes a profession gives the practitioners of that profession an enhanced ability to foresee what combination of actions will produce the desired end. Engineering knowledge, both theoretical and practical, enables engineers to design devices, processes, and constructions that perform as required and are safe in foreseeable modes of operation and under foreseeable conditions. Increased foresight raises the attainable level of safety and adds to the complexity of moral responsibility experienced by the professionals involved.

The concern with safety has led to the development of many structured techniques for safety review, such as the review of safety in a chemical plant. One such technique is the hazard and operability (HazOp) analysis used by industrial chemists and chemical engineers. Although structured, this method is not a simple recipe or algorithm. Like methods to aid conceptual design, this one begins with brainstorming by a team leader and process experts (such as design engineers, process engineers, toxicologists, or instrument engineers) to identify potential hazards. The team then carries out a unit-by-unit, stream-by-stream analysis of possible hazards in the plant process.

Other structured approaches to the identification and control of hazards are event-tree analysis and fault-tree analysis. Event-tree analysis begins with an initiating event (a mistake) and explores the states to which that event may lead. Fault-tree analysis begins with a malfunction or accident and reasons diagnostically to the circumstances that might have caused the malfunction or accident in order to estimate the likelihood of such accidents in the future. The types of possible consequences considered include fires and explosions, toxic chemical effects, harm to an ecosystem, and negative economic consequences. Because such techniques are now part of the technical subject matter in science and engineering, I will not describe them further. Their development, however, further illustrates the intimate and constantly developing relationship between ethics and competence in professional engineering practice.

Criteria for responsible practice vary not only with profession but also develop with a profession's experience of accidents and failure. Henry Petroski's thesis that engineering commonly advances by learning from failures was mentioned earlier. Most of the failures Petroski discusses and discussed in the present chapter threatened human health and safety, so learning from failure occurs at a significant cost in terms of injury and loss of life.

Roland Schinzinger and Michael Martin, noting the extensive and often unpredictable character of the influence of technology on human life, including threats to health and safety, have argued that technological innovation amounts to social experimentation. In recent decades, informed consent has emerged as the primary criterion for the ethically acceptable use of experimental treatments on human subjects. Schinzinger and Martin have suggested adapting a similar standard for the adoption of new technology. Because of the human cost of learning through accidents, Schinzinger and Martin's proposal that new technology be regarded on the model of experimental medical treatment has some attractions.

### Why Compare Engineering to Health Professions?

We have frequently considered comparisons between the responsibilities of engineers and computer scientists to those of health care providers for several reasons: Health care professions, like engineering, draw heavily on scientific knowledge. The comparison provides useful information for the engineers and computer scientists who will work on biomedical technologies or in medical research. Understanding professional responsibility requires an understanding of it from both the client and practitioner perspective. Health care is a particularly useful profession to pair with engineering because everyone has had experience as a client of health care. The subject of the engineer's responsibility for health and safety, to which we now turn, highlights one of the striking differences between engineering practice and practice in most of the health care professions: Engineers/computer scientists generally have little or no face-to-face encounters with many of those whose safety and welfare depend on their actions.

Schinzinger and Martin propose using "proxy groups" composed of people similar to those who will be greatly affected by new technology.[12] Their mechanism for obtaining consent would not be the same as that used with human subjects in experimental studies. This difference is not surprising because Schinzinger and Martin develop their analogy not between engineering innovation and clinical *experimentation*, but between engineering innovation and use of experimental medical *treatment/care*.[13]

The use of an experimental treatment is governed by standards of competent care and informed consent for *care* rather than the more formal procedures required for human participation in *clinical experiments*. (Consent to participate in a clinical experiment will be discussed in Chapter 8, which deals with issues in research ethics.) A clinician who proposes to do a clinical study or experiment needs to have her plan for the study reviewed by the institutional review board of her institution. Suppose that clinician wished to give a patient an experimental treatment (perhaps because there were no effective treatments for the patient's condition, or because none of the customary treatments had worked with the patient in question). The clinician would only need some reason for thinking the treatment might be effective to proceed. The

---

[12] Martin, Mike W. and Roland Schinzinger. 1989. *Ethics in Engineering*, second edition. New York: McGraw Hill Publishing Company, especially pp. 63–78.
[13] *Ibid.*, 68.

clinical data from the use of the experimental drug in this "last resort" type case could not then be considered as part of a clinical study. The consent that the clinician would be required to get from the patient would be a willingness to proceed with the treatment after hearing the known facts. Consent for treatment (whether established or experimental) is generally *quite* loose, amounting only to a patient refraining from objecting to the treatment procedure. Although patients often sign forms to consent to surgical procedures, formal informed consent procedures are rarely used for nonsurgical care unless that care carries a great risk. You may have noticed that when you have had a blood test or X-ray, no one asked for your consent. Patients who cooperate with the testing are presumed to consent. More formal consent procedures are required to obtain the consent of experimental subjects, including prior review and approval of the study and the procedures for protecting human subjects by the facility's institutional review board (IRB).

enlrg The Martin and Schinzinger proposal speaks to the ambiguous character of many moral problems and, therefore, the importance of a moral agent's ability to devise courses of action that will be robust in the face of surprises (which we discussed in Chapter 3, Ethics as Design). These have special relevance to the professional responsibilities of those in the science-based professions, such as engineering and health care. These professionals, because of their special knowledge and experience, are likely to be the first to recognize threats to health and safety.

*Why is the ability to foresee dangers relevant to an individual engineer's responsibility for safety?*

*How and why do specific criteria for responsible engineering practice change over time?*

*Evaluate the Martin and Schinzinger proposal to treat the introduction of new technology on the model of an experimental clinical treatment.*

## Hazards and Risks

Is tolerating any threats to safety ever consistent with an engineer's responsibility for safety?

Thus far, we have considered mainly those safety hazards that threaten to cause accidents and malfunctions. This was true not only of the hazards that underlay major accidents but also of those that structured techniques like hazard and operability study and fault-tree analysis are designed to identify. The emphasis has been on identifying hazards and bringing them to the attention of those with the authority to reduce or eliminate them.

It is not feasible to eliminate certain hazards. Some can be eliminated or mitigated only by producing other adverse consequences. For example, an effective ingredient in repelling insects, commonly known as DEET, is known to be toxic to humans. Nonetheless, people continue to use repellents containing DEET because they have not found less dangerous substances that are equally effective in repelling insects. Besides discomfort, insect bites may carry serious diseases like dengue fever and encephalitis. The use of DEET in some circumstances is justified by health considerations. The decision to use DEET involves a tradeoff between risks and benefits. It might make use of one of the structured techniques of risk analysis, such as the risk-benefit analysis introduced earlier in Section 4 of the introduction on concepts.

## Concepts of Hazard and Risk

We saw in the introduction that the technical concept of risk used in risk analysis and in risk-benefit calculations differs from the ordinary notion of risk. We use "risk" colloquially when we speak of the risk of a traffic accident, or of running out of water, or of tripping and falling, focusing on the negative *event or situation* that we would seek to prevent or eliminate. This colloquial notion of risk is very similar to that of a hazard, except that a **hazard** is an *externally caused* threat: We may take a risk, for example, by carrying only enough drinking water for the number of days we expect to hike, but we do not "take a hazard" even when we expect to encounter one.

The hazards that are tolerated in one area may not be tolerated in another, however. We have just seen a risk tradeoff that is accepted for insect repellent. Tradeoffs that are accepted for lotions and cosmetics are not tolerated for food additives. The standard of safety currently required for food additives is that their addition must not cause harm to humans nor even cause cancer in other animals.[14] Furthermore, risks posed by additives are less tolerated than risks associated with substances that occur naturally. Many naturally occurring contaminants can produce significant harm but cannot be entirely eliminated from food. The presence of these contaminants has not led to banning such potentially affected food crops, notably peanuts and peanut butter, from the market.

Some notorious uses of cost-benefit analysis to trade off safety against money have been generally condemned as irresponsible. Perhaps the best-known example is Ford Motor Company's decision in the late 1970s about the Ford Pinto, which we discussed earlier. Recall from the technical definition of risk given in Section 4 of the introduction that the probability that a given course of action will produce some harm multiplied by the *degree* of harm defines risk in the technical sense. Therefore, for extreme harms, like death and injury due to a gas tank explosion, the risk will be extremely high unless the probability of it occurring approaches zero. Only if the likelihood of it happening were extremely small would it be comparable to the risk of other harms and costs and so only then would tradeoffs among them be reasonable.

Trading off safety against cost considerations is not necessarily morally objectionable, however. If the likelihood of some danger is at least partially under the control of those at risk, it is generally seen as less of a priority to further reduce their risk. Consider the design of a new highway. The terrain may necessitate curves in the road. How much banking should those curves have? The greater the banking (the smaller the radius of curvature of the highway surface), the less likely it is that cars will spin off the road, but the more expensive it will be to build. It seems reasonable to stop short of constructing the road as a speedway, but how big a safety factor should one build in? Is it enough to design for fair weather speeds of up to thirty miles over the legal speed limit? Forty? What about foul weather? What are the weather extremes in that locale in a normal

---

[14]This is specified in the "Delaney Clause" in section 409 of the Federal Food, Drug and Cosmetic Act (FFDCA) of 1958. See Senate Report No. 85–2422, 85th Congress, 2nd Session (1958) and 21 USC, section 348 (c) (A) (1976). For a discussion of proposals to change the standard for food additives, see the Congressional Research Service Report for Congress, "The Delaney Clause Effects on Pesticide Policy," by Donna U. Vogt, Analyst in Life Sciences, Science Policy Research Division, updated July 13, 1995, 95–514 SPR.

year? In a typical hundred-year period? In a five-hundred-year period? How safe is safe enough, or to put the matter another way, what is an acceptable level of risk? A judgment about the acceptable level of risk is implicit in any tradeoff of safety against other considerations, or even of one sort of safety risk against another.

One way of lowering risks is to increase the safety margins, or "overdesign" in one's work. The ability to overdesign varies with the type of engineering work. Civil engineers normally build in a safety factor of two, three, or more. That is, they often construct buildings to withstand stresses two, three, or more times what they expect them to experience. If aeronautical engineers tried to use such safety margins, their planes would be too heavy to fly. Using much smaller safety margins, however, makes it all the more important to avoid mistakes and identify all potential causes of failure.

Industrialized societies have accepted the risk of designing airplanes with relatively small safety factors, because the actual rate of airplane accidents is acceptable to the flying public. In the United States, the mortality risk in automobile travel is mile for mile greater than the risk of death in travel on a *regularly scheduled* airline trip.

After an accident, the adequacy of the safety factor that was used in designing the failed item is often compared with that used in the design of comparable items. The Department of Transportation Report on the 2006 collapse of part of the Ted Williams Tunnel in Boston found the tunnel ceiling to have a lower safety factor than comparable tunnels, for example.

In a society marked by rapid innovation, some adverse consequences can be expected. This is especially true about the manufacture and use of new chemicals.

---

### How Dangerous Is This Chemical?

This weekend is the first time in a month that Pat has had some time to relax. Pat's team at Colortex had just come up with a new dye, a sulfated alpha-napthol, that promises to give Colortex a larger share of the commercial dyes market. Pat recalls hearing that alpha- and beta-napthols are associated with high rates of bladder cancer and resolves to look into the matter on Monday.

On Monday, Pat checks the data on the carcinogenicity of the napthols. At least the alphas do not seem to be as potent carcinogens as the betas, and a sulfate radical might make the chemical even more benign. However, Pat resolves to take up the matter with the team leader, E.D. Able. E.D. is not an easy person to talk to, so Pat does not look forward to raising a "nonstandard" issue with E.D. but remains concerned about the dangers of a potential carcinogen.

When Pat finally gets a minute with E.D., E.D. dismisses the issue, saying, "If we were going to worry about every ring structure, we'd have to test cholesterol for carcinogenicity. Life can't be made risk free. Besides, assessment of carcinogenicity is the EPA's job. They will be notified 90 days before we market this dye. They can raise any objection then."

Of course, the EPA will eventually do its own assessment, but first many Colortex employees would be exposed to the chemical as Colortex moves from development to production of the dye. E.D. does have more experience than Pat, but Pat is not comfortable with the "innocent until proven guilty" stance that E.D. is taking toward this chemical.

What can and should Pat do?

*Is tolerating any threats to safety ever consistent with an engineer's responsibility for safety? What does your answer illustrate about the difference between saying that an engineer has a responsibility for safety and the assertion that an engineer is obliged to take whatever measures are necessary to eliminate a threat to safety?*

## The Scope and Limits of Engineering Foresight

What threats to human well-being are engineers generally qualified to recognize?

Engineers and scientists *are* likely to be the first to recognize many potential threats to health and safety. However, whether they are particularly well qualified to either predict or control *other* consequences of their work depends on the character of those consequences. For example, adoption of the automobile as a mode of transportation had many societal effects, including the growth of the suburbs, a reduction in the risk of tetanus (carried in carriage horses' feces), and a contribution to the rise in lead levels in the environment (from previous use of lead as a gasoline additive). It is unreasonable to fault those engineers who designed and developed automobiles for not foreseeing and mitigating all the negative consequences of the automobile, because we make no similar demands of clairvoyance on members of other professions. The experience with the automobile teaches many lessons about control of technology, only some of which pertain to engineering. In a democracy, all citizens, not only engineers and other technical experts, are expected to consider and have a voice in major decisions about such areas of technology as transportation policy. As was pointed out earlier, this gives all citizens of a democracy some responsibility for the uses of technology within that society. Because of the responsibility of citizens for technological choices, many engineering societies in democratic countries state that their members have a special ethical responsibility for education of the public. Such education is necessary if the citizenry is to understand engineering and technology. For example, under the heading of its responsibilities to promote the public interest the Software Engineering Code of Ethics and Professional Practice states that software engineers shall:

> 1.08. Be encouraged to volunteer professional skills to good causes and contribute to public education concerning the discipline.

As we have seen, engineers/computer professionals are expected to learn from past mistakes. If a negative effect *is* something that existing knowledge allows engineers to foresee, does that fact alone make the effect something that is the sole responsibility of engineers? An example will help to focus consideration of this question.

In an editorial, "A Sense of Sin," which appeared in the February 1988 issue of the *Biomedical Engineering Society Bulletin*, Steven M. Lewis argued that biomedical engineers are responsible for all the consequences of the devices they design and develop. Although the engineers' responsibility is clear enough in certain cases of safety problems under reasonably expectable use, operating conditions, and maintenance, Lewis claimed more. He contended that the engineers bore a special responsibility for the fact that life-support devices – respirators or ventilators, in particular – are often used inappropriately, such that their use merely prolongs patient suffering and wastes resources.

Now that society has experienced the harm resulting from inappropriate use of life-support technology, it is foreseeable that new life-support technology could be misused in the same way. Does this mean that it is irresponsible for engineers to develop any more life-support technologies? Notice that the knowledge on which the prediction of negative consequences would be based is not specific to engineering, but rather to North American society's reluctance to squarely face death and dying.

Lewis apparently believes that because engineers can now see that their work on medical technology may make it possible for this harm to be done, they therefore bear some responsibility for this misuse. Should devices such as respirators be taken off the market? Some types of devices have been removed from sale when hazards from certain uses are great. An example is the three-wheeled "dune buggy." These were intended to be recreation vehicles, but were removed from the market when it was found that children and adolescents driving these vehicles often had serious accidents.

Life-support technology, such as the respirator, provides incontrovertible benefits to a great number of people, which is why no one has seriously suggested a ban on life-support technology. Furthermore, respirators and other life-support technologies are not difficult or arduous to maintain properly. If they were one would expect engineers to modify their designs to remove or lessen the danger. Recall the earlier discussion of how engineers learn to consider new factors through learning from previous failure. When clothes dryers first became popular consumer items, they would resume their operation when the door was re-shut after opening in mid-cycle. As a result, some toddlers died or were severely injured after crawling into dryers that had stopped in mid-cycle. Now safeguards to protect toddlers and young children are an essential consideration in the design of household appliances. Safeguards are readily added in to the design of dryers, thus creating an "engineering solution."

In the case of hazardous waste, engineers now seek to avoid using or creating hazardous substances in processes of manufacture and maintenance as well as consider how to control the release of hazardous substances and means for safe disposal of such wastes. The creation of such processes is still a technical engineering task. The control of the clinical use of respirators is not.

Overuse of respirators and other life-support technologies differs fundamentally from using a dryer as a playhouse, or a dune buggy as a toy for children. What differentiates the appropriate use of a respirator from a misuse is not always obvious; in contrast, playing house in a dryer is very different from using it for drying clothes. In the dryer case, engineers created a solution to the hazard posed to toddlers by making operation of the dryer conditional on throwing a switch that adults can reach but toddlers cannot. What differentiates relevant use and misuse of the respirator, however, is the prognosis of the patient with whom it is used. Because the physical actions taken with the respirators do not differ between use and misuse, misuse cannot be prevented by engineering means.

Lewis takes the phrase "a sense of sin" from J. Robert Oppenheimer's famous description of science as having "known sin" in producing the atomic bomb. Lewis's use of the phrase suggests that producing devices like respirators is akin to producing a weapon of mass destruction. It draws attention to the value implications of work in science and engineering. The design and development of life-support technology differ from the creation of weapons of mass destruction in morally significant respects, however. Weapons of mass destruction are devices that, if they function *as intended*, have overwhelming negative effects for some people and arguably for humankind, although it has also been argued that possession of such weapons may deter others from using them. In any case used *appropriately*, life-support technology need not have any negative effects.

Lewis is right about the suffering caused by the overuse of life-support technology, but in this case, the responsibility for preventing or remedying the situation lies with parties other than biomedical engineers. As Stephen Lammers (1998), Professor Emeritus of Lafayette College, has pointed out, our society often looks to technology to eliminate the necessity of coming to grips with perennial human problems like death and vulnerability to disease and injury.

The example of the respirator illustrates the general point that although the possibility of foresight is required for engineers to be responsible for some harmful effect of technology, it also makes a moral difference if engineers are *able* to do something about the problem. The total configuration of consequences also makes a difference: If an engineer has good reason to believe that society is not able to appropriately control the use of some new technology or device and its effects will be wholly or predominantly harmful, that engineer has a good reason to refuse to work on that technology. If, however, the technology or device produces mixed results and only some uses are harmful, the responsibility for social control of that technology may lie with others or with citizens generally.

What examples do we have of such social control? Soon after Lewis wrote his editorial, there was some progress on the evident problem of the overuse of life-support technology.[15] This progress exemplifies the diverse means societies have for controlling the uses of technology.

## The Patient Self-Determination Act

The Patient Self-Determination Act is U.S. legislation that took effect in December 1991. It requires hospitals and other health care facilities (nursing homes, hospices, home health agencies, HMOs, and other facilities that receive Medicare and Medicaid payments) to maintain written procedures on advance directives. Advance directives for health care are usually in the form of a living will or a health care proxy statement. (A living will specifies in general terms the care

---

[15] In 1988, 31 percent of all Medicare expenses occurred in the last year of elders' lives. Because only 5 percent of those entitled to Medicare die in any one year, this is a disproportionately high expenditure. Forty percent of the final year total was expended for care in the last 30 days of a person's life. Although some of this care was expected to enable the patient to resume a meaningful life, as those of us who teach in hospitals know, some is futile and given primarily because the family, staff, or patient cannot accept the reality of the situation. Although the final year total is a large proportion of the total health care budget, that proportion did not rise markedly from 1976 to 1988, when the figure was 27 percent. The absolute dollar amount quadrupled in that period, however, from $3,488 to $13,316 per Medicare recipient. See Knox, Richard, 1993, "Care at the End Not as Costly as Assumed," *Boston Globe* 243(105): A1, April 14.

that a patient would or would not want to receive if no longer able to give or withhold consent to care.) A health care proxy directive (also known as a "durable power of attorney" statement) specifies whom the person would wish to decide her medical care were she unable to do so.

When patients are admitted to a facility, that facility is required to notify patients that they have a right to refuse treatment and ask if they have made out advance care directives. Health care facilities must also comply with state law regarding such directives.[a]

---

[a] *Medical Ethics Advisor,* January 1991, pp. 1–4.

Most states have now given legal recognition to either living wills or health care proxy directives, or both. These state laws, together with the Patient Self-Determination Act, help address the problem Lewis raises, although none of them requires engineering knowledge. Where, as in this case, engineering knowledge does not help either to foresee or to remedy some misuse of technology, engineers have only the same responsibilities as other citizens to prevent the misuse.

A different issue is raised when objections are made to the *intended* use of a technology. At one extreme are semiautomatic assault weapons and semiautomatic rifles, pistols, and shotguns. The argument that their intended use – namely, to kill and maim people – is ethically unacceptable led to a ban on the manufacture of many classes of these weapons in the 1994 Crime Bill.[16] (This is an example of a so-called end-use problem, which will be discussed in Chapter 8.)

These considerations about the scope of an engineer's responsibility for safety suggest the following questions:

- Which of the possible societal consequences of technology are currently practicing engineers able to foresee?
- Under what conditions or to what extent can such engineers judge which influences and outcomes are desirable, and when are others better able to make that determination? The engineering profession? The government? Commissions composed of experts and laypeople?
- Under what circumstances would such engineers bear some responsibility to ensure those outcomes through their engineering work (design, development, manufacture, repair)?

After answering those questions, we still need to clarify how one goes about fulfilling such a responsibility when one does have it. As we saw in the first chapter, deliberation about how best to fulfill a responsibility is complex, like a design problem. Even when the safety issue itself is relatively straightforward, circumstances may make it relatively difficult for an engineer, especially one who is new to the job, to devise a good response. Consider the following scenarios based on experiences of recent engineering graduates:

---

[16] The provisions of the Violent Crime Control and Law Enforcement Act of 1994 (the "1994 Crime Bill") are summarized at http://gopher.usdoj.gov/crime/crime.html. Among the newspaper articles summarizing provisions of the act is John Aloysius Farrell's "US House OK's $30.2b Crime Bill," *Boston Globe,* August 2, 1994, pp. 1, 8.

---

### Impaired Coworkers[a]

You are a student intern working second shift in testing of a million-dollar space-craft. Tasks on your shift are simple (regulating pressures, turning on or off valves, reading numbers from flow meters, etc.). However, keeping attentive and alert is very important, because turning a knob the wrong direction could cause major damage to the spacecraft.

You take your dinner break with two of the full-time employees. They pick out a bar/restaurant, and order large steins of beer. You pass when they offer you one. During dinner, they each have another and wind up using more than the allotted hour break.

You are certain that they must have been affected by the amount of alcohol they drank, although you do not see anything amiss in the way they are operating the equipment. The company does not allow drinking on the job, especially by those working on the spacecraft.

What, if anything, can and should you do about the situation?

Where might you go for advice?

#### Getting Started

As we shall see in Chapter 7, some companies have extensive services that you can consult anonymously or confidentially. How do you find out what your company offers?

---

[a]Based on a scenario by Peter Kassakian (MIT '94).

---

*What threats to human well-being are engineers generally qualified to recognize?*

## Matching an Engineer's Foresight with Opportunities for Influence

What, beyond having the knowledge of a likely safety hazard, does an engineer need to get the hazard reduced or eliminated?

The safety of the Chernobyl nuclear power plant had been criticized in Soviet technical journals before the 1986 "accident." Before the 1974 DC-10 crash, engineers had warned that the design of the airplane was faulty. Preparing engineers to recognize safety hazards, although vitally important, is clearly not enough to prevent many accidents. Because engineers often recognize a hazard but do not have the authority to remedy it and may be unable to get decision makers in their organization to attend to it, engineering ethics has widely discussed whistleblowing by engineers – that is, an engineer taking a concern outside her organization. However, whistleblowing in this sense always marks organizational failure.[17] A better way of matching an engineer's influence with her insights and foresight, one that is less costly for all concerned, is for organizations to become more responsive to their engineers. Although this alternative has been the subject of

---

[17]Rowe, Mary P. and Baker, Michael. 1984. "Are You Hearing Enough Employee Concerns?" *Harvard Business Review* 62(3): 127–135; Michael Davis. 1988. "Avoiding the Tragedy of Whistleblowing," *Business & Professional Ethics Journal* 8(4): 3–19; Peter Block. 1993. *Stewardship: Choosing Service over Self-Interest.* San Francisco: Berrett-Koehler Publishers; Keenan, Paul, Lockhart, Paula, Elliston, Frederick, and van Schaick, Jane. 1985. *Whistleblowing: Managing Dissent in the Workplace.* New York: Praeger Scientific.

much attention in industry, it has not been discussed to the same extent for other contexts in which engineers may work.

There is a growing consensus among engineering organizations on the subject of raising safety concerns, both through lodging complaints (within an organization) and through whistleblowing (outside the organization). First, engineers have a right to force attention to many types of error and misconduct – such as waste and misrepresentation in work done under government contract – even by going outside the organization. Second, engineers have not only a right, but a moral obligation, to bring the matter to light when human life or health is threatened.

For *some* negative consequences, such as threats to public safety, an engineer is morally *obliged* to go to special lengths – including contacting outside agencies – to draw attention to the problem if that engineer is unable to influence the situation by the means offered within the organization. Of course, an engineer must exercise judgment about appropriate places to take a concern. (NSPE Case 88–6 recounts a case in which the NSPE Board of Ethical Review judges the whistleblower to have handled the situation badly.[18]) A government agency charged with regulatory oversight is usually an appropriate place to go. Engineers' experience in going to the press is highly varied. At a minimum, it is difficult to describe technical matters in terms that a newspaper audience can understand. Furthermore, journalists are at least as subject to the temptation to sensationalize a story as managers are to downplay safety when they need to meet a deadline. Therefore, it is a good idea to know journalists and their motives before giving them sensitive information. (For a perspective from within the engineering profession on the appropriate use of the media in raising ethical concerns, consider the following case.)

## Publicly Criticizing a Project as Unsafe

Garcia, a renowned structural engineer, is hired for a nominal sum by a large city newspaper to visit the site of a state bridge-construction project. This project has been plagued by construction delays, cost increases, and litigation, primarily because of several well-publicized on-site accidents.

Garcia visits the bridge and performs a one-day visual inspection. In very general terms, her report identifies potential problems and proposes additional testing and other solutions. In a series of feature articles based on Garcia's report, that city newspaper alleges that the bridge has major safety problems that will jeopardize its completion date. Allegations of misconduct and incompetence are made against the project engineers, the contractors, and the state highway department. The state holds an investigation, in which Garcia states that her report only identified potential problems with the safety of the bridge and was not intended to be conclusive.

What is your ethical evaluation of Garcia's agreement with the newspaper?

In light of this experience, what safeguards might an engineer seek as a condition of accepting an assignment like Garcia's?

*Source:* Adapted from NSPE Case No. 88–7[a]

[a]This case, "Public Criticism of Bridge Safety," and the board's discussion of it and three related cases appear in *Opinions of the Board of Ethical Review* Volume VI, Alexandria, VA: National Society of Professional Engineers, 1989, pp. 117–119.

[18]This case, "Whistleblowing City Engineer," and the board's discussion of it and two related cases appear in *Opinions of the Board of Ethical Review* Volume VI, Alexandria, VA: National Society of Professional Engineers, 1989, pp. 115–116.

*What, beyond having the knowledge of a likely safety hazard, does an engineer need to get the hazard reduced or eliminated?*

## Summary

This chapter has discussed the central professional responsibilities of engineers (including those who are IT professionals). The standards of responsible professional practice are not static but change as new knowledge is acquired. However, at least among engineering groups there is a high degree of consensus that the responsibility for safety is foremost among the professional responsibilities of engineers. The scope of factors that can affect the safety of technology is unlimited. Some of these factors are discovered only through accidents, but the domain of facts that an engineer is expected to consider continually expands. The standards of responsible practice at any given time depend on what engineering knowledge of the time permits one to foresee and influence. It is in everyone's interest that engineers be heeded when they foresee risks and threats to the public welfare.

# 5 Computers, Software, and Digital Information

## What Is Different about Digital Systems and Digital Information?

What characteristics of digital systems and digital information set them apart from other technology and influence the morally significant problems faced by the engineers who work with them?

Digital systems and digital information have some special characteristics that influence the morally significant problems faced by engineers (IT professionals) who work with them, the engineering profession, and society in general. In Chapter 1 (the discussion of the engineering profession's criteria for responsible practice), we noted NAE past president Bill Wulf's observation that because digital systems are not continuous, a small change in a digital system (such as one bit in the memory of a computer) can produce a radical change in the behavior of the system.[1] As a result, some devastating effects of computer "bugs" are due not to human error or negligence but to *unpredictable* new characteristics of the system ("emergent properties"). As Wulf also argues, the lack of continuity in a digital system also creates insurmountable problems for testing such systems, for example, computers.

> Because digital systems are not continuous, a small change in a digital system (such as one bit in the memory of a computer) can produce a radical change in the behavior of the system. As a result, some devastating effects of computer "bugs" are due to *unpredictable* new characteristics of the system ("emergent properties") rather than human error or negligence.

These special characteristics of digital systems give rise to the question of criteria for responsible engineering of software *when one knows in advance that some behaviors of the resulting system will be unpredictable*. Wulf sees this question as one for the engineering profession, rather than the individual engineer, to answer. (We will examine such "macro problems" further in Chapter 10.)

Some special features of digital information create the possibility not only of new technologies but also of new sorts of crime. This fact has given rise to the expression "computer crime," although a more precise, if ungainly, name might be "digital information crime." We have no expression "automobile crime" or "telephone crime" to correspond to the term "computer crime," although automobiles and telephones may also be

[1] Wulf, William A. 2004. Keynote Address, *Emerging Technologies and Ethical Issues in the Practice of Engineering*. Washington, DC: The National Academies Press, 1–6.

used in the commission of crimes. However, the manipulation of digital information creates the possibility not only of new crimes but also of new *sorts* of crime and markedly changes the ease of committing, and therefore the prevalence of, some familiar sorts of crime. For example, the ability to steal information at a distance has challenged the legal system with new criteria for theft without "entering." The *plasticity* of digital information not only makes possible the transmission to and storage of information on a variety of devices, including PDAs and telephones, but also the desktop forgery of checks and even currency.

Digital information also possesses distinctive properties that markedly affect its status as intellectual property. (The policy issue of what, if any, intellectual property protection is appropriate for software will be considered in the next section.) Here we will consider the characteristics of digital information that make it easy to create unlicensed copies of licensed software, digitalized music, or digitalized video. The transformation of artistic works, including music, to digital form itself has been argued to threaten artistic integrity. Recognition of the *rights of authors and other creators to preserve the integrity of their works* as well as the property right known as "copyright" provides the basis of such an argument. We have seen that U.S. law is not based on a "natural right of ownership" on the part of a work's creator, but the United States has joined the Berne Convention, which does recognize a natural right of authors and other creators that transcends ownership. For example, French law recognizes rights of a painter to a share of appreciation in value of her painting, even if she has long since sold the painting to another.

The first characteristic of digital information that sets it apart from physical property is that when stolen or "pirated," it is not gone. This fact has immediate implications for the nature of the harm done to others by the theft of digital information. (The next chapter will address the subject of rights to intellectual property. Here we are concerned with novel features of software and other digital information, whether or not they have implications for property rights.) If Bob steals Carol's computer, the result is harm to Carol who no longer has the computer, becomes aware of her loss, and perhaps even knows that it was stolen. The manufacturer, distributor, and any identifiable inventor are not harmed. If Bob makes unauthorized copies of Carol's computer program or music file, Carol still has it (in the sense that she still possesses her copy and her original license to use it). She has not lost any property (although she may be morally corrupted if she is an accomplice to *unjustified* copying of the software). Others, such as the author of the software or the distributor of the software disk, may be harmed by being deprived of the profits from any purchase of the software that Bob would otherwise have made, although even they are unlikely to *know* that they have been so deprived in a particular instance. Consequently, an act of illegitimate copying of software is less likely to be known to those who are harmed by it. Because some people do not object to wrongdoing unless a wrong is done to them personally, condemnation of individual acts of copyright violation is less common than condemnation of individual acts of ordinary theft. This fact does not decide the moral evaluation of such copying, but only explains why some people may be especially tempted to do it.

An act of illegitimate copying of software is less likely to be known to those who are harmed by it. Because some people do not object to wrongdoing unless a wrong is done to them personally, condemnation of individual acts of copyright violation is less common than condemnation of individual acts of ordinary theft. This fact does not decide the moral evaluation of such copying, but only explains why some people may be especially tempted to do it.

There remains the question of whether copying, even illegal copying, might in some cases be morally justified. (The copying would *require* justification because, as we saw in Section 6 of the introduction, the burden of proof is on someone who claims that it is right to break the law in certain circumstances to show that is so.) Helen Nissenbaum has examined the question of whether making copies of software is always wrong when it violates licensing agreements.[2] She argues that licensing is not the only ethically relevant consideration and that sometimes copying that violates licensing agreements *is* morally justified. (She does not argue that one can justifiably ignore all restrictions on copying software protections, however.) Nissenbaum organizes her arguments in terms of the type of moral consideration in question: consequences, rights and duties, arguments about the property rights with respect to software, and the relationship between ownership and the prohibition of copying. Nissenbaum offers an example as representative of a situation in which the copying of software is justified, even though it violates licensing agreements. In her example, a person, "Millie," copies her home bookkeeping application for a friend who has trouble organizing his finances and cannot afford to purchase the software. From the way the case is framed it would seem that the copying has no implications for the income that the developer or publisher of the software would otherwise receive, which, as we shall see in Chapter 6, is one of the four criteria for "fair use" of copyrighted material. Fair use is a legal copying of copyrighted material, but one that stems from some moral considerations of what is *fair*. Here Millie's act fulfills *one* of the four considerations thought relevant for deciding if some copying is a *fair* use. Even if Nissenbaum is correct in arguing that the copying is morally justified in this case, the copying is illegal. As we saw in the introduction, if an act is illegal, that is *a* reason, although not necessarily a *conclusive* reason, for thinking the act is morally wrong. (I do not make a judgment on the Millie case, because that would require a more detailed consideration of Nissenbaum's argument.)

Millie will find it relatively easy to copy the software, and the copy may be expected to be as good as Millie's purchased copy (unlike, for example, a photocopy that degrades the image of which it is a copy). The ease of copying digital information and the fact that copying need not degrade the quality, in addition to the fact that Millie still has her purchased copy, make copying easy and effective and, therefore, *more tempting* than the theft of physical goods. Three other factors that contribute to the ease and efficiency of copying, but which do not enter into Millie's story, are that digital information can be stored compactly,

[2]Nissenbaum, Helen. 1995. "Should I Copy My Neighbor's Software?" In *Computers, Ethics, and Social Responsibility*, edited by D. Johnson and H. Nissenbaum. Englewood Cliffs, NJ: Prentice Hall, 201–213.

can be readily distributed to many recipients, even at great distances, and can be stolen at a distance.

The last feature is especially important for theft of personal information, rather than theft of software. The ease with which digital information can be searched and linked makes possible many new compilations and uses of personal information. Thus, it raises important issues of privacy, as we shall see later in this chapter.

*What characteristics of digital systems and digital information set them apart from other technology and influence the morally significant problems faced by the engineers who work with them?*

## Software as Intellectual Property

What characteristics of software require new ways of thinking about them as intellectual property?

If we take the legal protection of intellectual property as a given (an assumption we will examine in the next chapter), what intellectual property protections ought software have? That software code may be either patentable or covered by copyright shows how novel software is as a technology. Previous technologies were subject to patent only, with copyright being reserved for written works and works of art. Software is written work and therefore subject to copyright, but unlike most written work it has **functionality**; it does stuff. For example, browser software makes a computer function as a Web browser. It has functionality because, unlike other writing, it reconfigures logic gates and creates an electronic network that functions as a hardwired network would. If the hardwired network to which it is functionally equivalent would warrant a patent, then the software deserves one. (The actual history of granting patents for software is more complicated. Gregory J. Maier argues that the Patent and Trademark Office's concerns that it could not handle the workload produced by a flood of patent applications for software played a significant role in temporarily discouraging the patenting of software.[3])

Because one can use (source) code without reading it, intellectual property protections for digital information such as software work differently from the way they do for other technologies. When one "buys software," one buys only a license to use the software in certain ways. Certain other uses of the software, such as reverse engineering of it to obtain the source code, are generally forbidden by the license. (The exceptional case of open source software covered by a GNU license will be considered presently.) Source code is secret and object code (machine language) revealed. If source code is revealed to the buyer, then it is generally copyrighted.

[3] His historical overview of the legal history of intellectual property protection for software, "Software Protection–Integrating Patent, Copyright, and Trade Secret Law," is available at http://www.oblon.com/Pub/maier-3.html.

Because one can use (source) code without reading it, intellectual property protections for digital information such as software work differently from the way they do for other technologies. When one "buys software," one buys only a license to use the software in certain ways. Certain other uses of the software, such as reverse engineering of it to obtain the source code, are generally forbidden by the license.

Charges of copyright infringement often involve elements other than the code itself, the user interface, for example. Thus, Lotus won its 1990 suit against two software companies, Paperback Software and Mosaic Software,[4] that produced spreadsheet programs with the same user interface, more specifically the same "menu structure" or the arrangement of commands in the menu hierarchy as Lotus 1-2-3. The copied features are sometimes described as the "look and feel" of a program, although the lawyer who made the "look and feel" argument on behalf of Lotus did *not* intend that the "look and feel" become criteria for judging copyright infringement.[5] The question of what software features are copyrightable is better expressed in terms of specific attributes of the program, such as "menu structure." Subsequently Lotus sued Borland over the Quattro spreadsheet program, which had a different user interface from that of 1-2-3 but was able to interpret Lotus macros (which followed the Lotus menu hierarchy). Borland acknowledged that it copied the Lotus menu command hierarchy in its programs. In the initial case before the district court, one of Borland's arguments was that it was "fair use" (see Chapter 6), but when it lost in district court and the case went to the appeals court, its argument was simply that the menu structure was not copyrightable. The appeals court observed that Borland had copied the menu structure

> so that spreadsheet users who were already familiar with Lotus 1-2-3 would be able to switch to the Borland programs without having to learn new commands or rewrite their Lotus macros . . . . In effect, Borland allowed users to choose how they wanted to communicate with Borland's spreadsheet programs: either by using menu commands designed by Borland, or by using the commands and command structure used in Lotus 1-2-3 augmented by Borland-added commands.[6]

The case was ultimately appealed all the way to the Supreme Court, which split 4-to-4 on the decision. (Justice Stevens had recused himself.) This split decision let stand the appeals court decision that the menu structure was not copyrightable. (The appeals court had based its decision on somewhat different arguments from Borland's own.) The appeals court decision weakened subsequent attempts to claim user interface as a copyrightable feature.[7]

---

[4] *Lotus Development Corporation v. Paperback Software International and Stephenson Software, Ltd.*

[5] Thom Franklin, personal communication, January 1989.

[6] That appeals court decision, *Lotus Development Corporation v. Borland Intl., Inc.*, No. 93-2214, U.S. Court of Appeals for the First Circuit, March 9, 1995, is available at several places on the Web, including http://www.kuesterlaw.com/borlan2.html.

[7] Many of the judgments and *amicus curiae* ("friend of the court") briefs filed by various parties pro and con are available at http://www.swiss.ai.mit.edu/6805/articles/int-prop/software-copyright.html#Lotus.

One aspect of the court decision in *Lotus v. Borland* that deserves special attention is its use of a test for violation of a software copyright previously set out by the Second Circuit in *Computer Assoc. Int'l, Inc. v. Altai, Inc.,* 1992. As we saw in the introduction, copyright covers expression but not ideas, so the test is to decide whether what was copied was expression or idea. The Second Circuit Court's test has three steps: abstraction, filtration, and comparison. The abstraction step is to "dissect the allegedly copied program's structure and isolate each level of abstraction contained within it." This step establishes a framework within which to separate expression (which is copyrightable) from ideas (which are not). The filtration step requires examination of "the structural components at each level of abstraction to determine whether their particular inclusion at that level was an 'idea,' or was dictated by considerations of efficiency so as to be necessarily incidental to that idea; required by factors external to the program itself; or taken from the public domain." The comparison step compares the elements of the infringed work that survived the filtration to the corresponding elements of the allegedly infringing work to determine whether there was sufficient copying of protected material to constitute infringement.[8]

*What characteristics of software require new ways of thinking about them as intellectual property?*

## GNU/Free Software/Open Source Movement

What values does the open source movement seek to foster?

Even before Lotus filed against Borland, the GNU (a "recursive acronym" for "GNU's Not Unix") project was gaining ground as a movement to prevent copyright from hemming in programmers and limiting opportunities to create innovative software. It began in 1984 as a project to develop a complete Unix-like operating system, which is "free" in the sense of allowing users to copy, distribute, study, change, and improve the software, without worrying about copyright. (It is "free" as in "free speech," not "free" as in "free beer" as Richard Stallman expressed it.) The rebellion against the constriction that copyright places on software developers was evident in one of the alternative early names for this movement, the "copyleft" movement. This movement does not advocate the violation of copyright that is sometimes called "piracy," but rather the creation of an operating system and programs to run on that operating system that are not under copyright but are subject to certain conditions set out in the "GNU license."[9]

The GNU project and its organizational sponsor, the Free Software Foundation, seek to ensure freedom of users of *such* software (now commonly called "open source" software) to

[8] *Computer Assoc. Int'l, Inc. v. Altai, Inc.,* 1992, 982 F.2d 693 (2d Cir.) at 707–710.
[9] The GNU General Public License may be accessed at http://www.gnu.org/licenses/gpl.txt.

- Run the program, for any purpose
- Study how the program works, and adapt it to their needs (and therefore have access to the source code for the program)
- Redistribute copies of programs
- Improve the program, and release those improvements (which will then be in the same special class of open source software) to the public[10]

It may be clear that the user who would care to exercise all these liberties is one who has some interest in creating and improving software, not simply running it. The GNU project/open source movement is not directed solely at more *creative freedom* for software developers for its own sake, although it is clear that the project regards such creative freedom as intrinsically valuable. The project also aims to further the *social good* that it sees as threatened by having software staked out as private property with source code kept secret so that it cannot easily be improved upon. The four freedoms state this goal somewhat obliquely in the item about improving the program. It gives as the reason for improving the program and releasing the improvements "*so that the whole community benefits* [emphasis added]."

> The GNU project/open source movement is not directed solely at more *creative freedom* for software developers for its own sake, although it is clear that the project regards such creative freedom as intrinsically valuable. The project also aims to further the *social good* that it sees threatened by having software staked out as private property with source code kept secret so that it cannot easily be improved upon.

The specific proposals of the GNU project are directed toward fostering an environment in which computer professionals and "hackers" can freely create. The project does not say what sort of software would further the public welfare, however. Such things as the dangers and rewards of virtual reality programs or what software privacy protections are adequate – parallel to the question, "How safe is safe enough?" – are left unaddressed. The GNU project does discuss software *options for implementing* such things as selected levels of privacy protection, although it does not advise about when specific levels of privacy protection are required to promote the public welfare.

As one might expect, now that patenting as well as copyrighting software has become more common, the GNU project seeks to mobilize resistance to the patenting of software. For example, it urges U.S. residents to sign a petition against software patents.

[10] See http://www.gnu.org (last updated November 4, 2010; last retrieved November 7, 2010). These freedoms are numbered zero to three on the Web site, so to avoid confusion, I have dropped any references to numbers of the freedoms. The zero to three numbering, the choice of a recursive acronym GNU, and the term "copyleft" are bits of quirkiness that tend to be ignored in current discussions of open source software. For a 2004 overview of books discussing the many issues raised by the GNU/open source movement, see Michael Jensen, "Selected Readings on Open Source," *The Chronicle of Higher Education*, (September 24), 51(5): B23, retrieved from http://chronicle.com/article/Selected-Readings-on-Open/6060 on November 7, 2010.

## Is It Wrong to Copy a Vendor ID?

SCSI, an industry standard system for connecting devices (like disks) to computers, provides a vendor ID protocol by which the computer can identify the supplier (and model) of every attached disk.

First Company makes file servers consisting of a processor and disks. Disks sold by First identify First in their vendor ID. Disks from other manufacturers can connect to First's file servers; however, the file server software performs certain maintenance functions – notably pre-failure warnings based on performance monitoring – only on disks made by First Company.

Competitor Inc. decides to compete with First by supplying cheaper disks for First's file server. It quickly discovers that while its disks work on First's file servers, its disks are at a disadvantage because they lack the pre-failure warning feature of First's disks. The CEO of Competitor, therefore, directs the engineer in charge of the disk product to "find a solution to this problem." The engineer uses reverse engineering and discovers that by making the vendor ID on its disks match that on First's disks, the First file servers will treat Competitor disks as First disks. Competitor incorporates this change into its product and advertises the disks as "100% First-compatible."

Representatives of First charge Competitor with forgery; they maintain that, whether or not Competitor's practice is illegal, it clearly violates industry-wide ethics.

Competitor justifies its action on the grounds that the favored treatment of First's disks by First's servers is unfair and monopolistic. Moreover, it argues that using First's vendor ID is not forgery, because it does not mislead *people*: Competitor's disks are clearly labeled as coming from Competitor. Competitor's action at most "misleads" First's software.

If this action is not forgery, what is it? What, if any, ethical rights are infringed by copying the vendor ID of the Competitor disks?

### Getting Started

First and Competitor have given the arguments pro and con. If, in making your judgment, you have no further arguments to offer, say which arguments had the greater weight and why.

This case illustrates the novel legal questions that software and other digital information raise. You may be able to construct several lines of argument for different conclusions.

*Source*: Adapted from a scenario by Stephen A. Ward and loosely based on a legal case

*What values are promoted by making software open source? What values are promoted by encouraging everyone to create only open source software?*

## The Faces of "Hacking"

What is hacking? Is it prima facie wrong?

*Hacking*, in the morally neutral sense of making something function in a way it was not designed to, has long had an appeal for student engineers. Hacking in this sense can include taking some inexpensive device or components and making something valuable out of them.

There is a long culture of "**hacking**," in a second and clearly benign sense that is celebrated at many engineering schools. There, *a hack* is celebrated as "a clever, *benign, and ethical* prank or practical joke, which is both challenging for the perpetrators and amusing"[11] (emphasis added). The hacks perpetrated by MIT students at Harvard-Yale football games, such as the rocket that erupted from the sod at the zero-yard line during the 1990 game and shot a banner over the goal post with the letters "MIT" on both sides, may not have amused everyone from Harvard and Yale, of course. The MIT Hack Gallery at hacks.mit.edu elucidates the term "ethical" in the definition of hacking and gives a "code of ethics" for hacking. That code states that a hack must:

- Be safe
- Not damage anything [that is not later fixed by the hackers]*
- Not damage anyone, either physically, mentally, or emotionally
- Be funny, at least to most of the people who experience it.

> **Hacking**, in the morally neutral sense of making something function in a way it was not designed to, has long had an appeal for student engineers. Hacking in this sense can include taking some inexpensive device or components and making something valuable out of them.
>
> There is also a long culture of "**hacking**," in a second and clearly benign sense that is celebrated at many engineering schools. *A hack* in that sense is celebrated as "a clever, *benign, and ethical* prank or practical joke, which is both challenging for the perpetrators and amusing."

> "Hacking" is also used in a third sense to refer to gaining unauthorized access to computers, phone systems, etc., which is illegal.

The Grumpy Fuzzball Hack of 1989 is shown in Figure 5.1.

Some hacks in the MIT Web gallery predated this code of ethics, although most were consistent with that code.

"Hacking" is also used in a third sense to refer to gaining unauthorized access to computers, phone systems, and so on, which is illegal. The prevalence and destructive potential of identity theft have made many of the present generation of students less tolerant of this third sort of hacking.

Furthermore, the events of September 11, 2001, have brought home the point that threats to security may be serious. Some of today's computer students plan to work in computer security. Their interest in hacking in this third sense is directed to anticipating and preventing unauthorized access. In any event, it is generally recognized that, other things equal, hacking in the sense of gaining unauthorized access is both illegal and prima facie wrong. Of course, countervailing considerations might justify it in a given case. For example, if there were a reasonable suspicion that some significant harm was being planned and if hacking were the only feasible way to detect it to prevent it, hacking might be justified on those grounds. A further question would be *who* can legitimately do the hacking when there are reasons that justify

---

[11] See hacks.mit.edu.

* The 1962 "Great Pumpkin Hack" in which the Great Dome at MIT was decorated with eyes, nose, and mouth and flooded with orange-filtered light to make it look like a huge pumpkin rising over the campus is on the Web at http://alumweb.mit.edu/classes/1966/hack.html. The paint, especially that used for the eyes, was difficult to remove from the dome, so the perpetrators paid for its removal. See recollections of one of the perpetrators at http://hej3.as.utexas.edu/~www/writings/great_pumpkin.text.

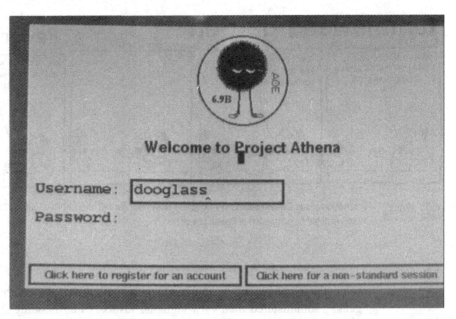

Figure 5.1 The 1989 Grumpy Fuzzball Hack to MIT Workstations (Photo: Douglas D. Keller). In 1989, the workstations at MIT ("Athena" workstations) showed students the image of the Athena owl when they signed on. At the end of the Fall Term, on the Monday of the last week of classes, a rather different character, the grumpy fuzzball, greeted them. Many students thought it fitting that the fuzzball resembled a burned-out owl.[12]

doing it. Do ordinary citizens with computing knowledge have a right to hack when they have such a reasonable suspicion? Do investigative reporters? Do the police? Does the National Security Agency? Do any of these parties need a search warrant or some other type of legal review before hacking?

*What is hacking? Is it prima facie wrong? Why or why not?*

## The Changing Culture of Computing

What would be a fair way to compensate authors and composers of music, computer games, and other digitalized entertainment? What, if any, DRM system that prevents copying of software and digitalized entertainment would be ethically acceptable?

Because of the special features of software/digital information, computing and software development were doubly new as an engineering field in the late twentieth century when it first became a popular undergraduate major. It had different experts and different journals, and some unquestioned assumptions of older engineering fields (such as that new creations would be candidates for patents and not copyrights, and stealing things required being physically adjacent to them) no longer applied. Furthermore, the online environment was so novel that norms for behavior, mostly rules about what counted as considerate online behavior or "netiquette," were constantly being proposed and reformulated. When some

[12] See http://hacks.mit.edu/Hacks/by_year/1989/grumpy_fuzzball/.

# tumbleweed-garden

## Pawan Sinha

DAVID LAMACCHIA:
A MARTYR IN THE CAUSE
OF SHAREWARE, OR ...

... A DESPICABLE
PERPETRATOR OF
SOFTWARE PIRACY?...

...OR JUST AN AVERAGE
20 YR.OLD WHO'S VERY
SCARED?

**Figure 5.2**    Contemporaneous Cartoon about the LaMacchia Case from the MIT Student Newspaper (*The Tech*, April 15, 1994; reprinted by permission of Pawan Sinha)

online source behaved obnoxiously and persistently violated the norms, computer "geeks" administered their own vigilante justice, often flooding the offender's email or spamming the offender's Web site, effectively shutting it down.

Many young people entered the computing field in the 1980s and 1990s, people too young to have experienced other work cultures that might have established their expectations. If any cultures influenced their work, it was school and university culture (always more freewheeling, not to say anarchic, than the work world). Even universities were scrambling to develop rules. For example, some universities initially argued that they had a right to read anything sent from a university computer because they owned the computer. Universities soon added specification of a burden of proof that would need to be satisfied before they would read documents and said they would read email messages etc. only if they had *reason to believe* a wrong was being done.

In 1994, David LaMacchia, an MIT junior, was accused of modifying an MIT workstation to allow people on the network to download copyrighted software without paying. He was indicted for conspiracy to commit wire fraud. The legal case raised major issues about liability of system operators and about the scope of computer crime and copyright laws,[13] but what is significant for the present discussion is that David LaMacchia thought that, if not okay, his actions were inconsequential.* A contemporaneous cartoon in the MIT student newspaper dealing with the LaMacchia case is shown in Figure 5.2.

One of my students at the time, Amy "Ringo" Gorin, wrote her term paper on the LaMacchia case. For many years, she made that term paper available as Web pages both in the Online Ethics Center and on Stanford's Web site (where

---

[13] Details are available in the OEC at http://www.onlineethics.org/Resources/19049/lamindex.aspx.
* The documents associated with this case can be accessed at http://www.onlineethics.org/CMS/computers/compcases/lamindex.aspx. The federal case was ultimately dismissed and the U.S. attorney on the case decided not to appeal the dismissal.

she attended graduate school). She shared important insights based on her own experience as an observer and participant in the changing computing culture. She concluded that LaMacchia probably thought his actions were no more than a harmless prank for which he would at most lose his Internet privileges at MIT but was surprised by a change in ethical norms for behavior on the Internet. Drawing on her own experience "as a long-time inhabitant of cyberspace" she witnessed the first federal prosecution of electronic copyright infringement and postulated a culture clash between the "'old-timers,' who thought little of sharing copyrighted material (ranging from source code to Playboy centerfolds), and the 'newbies,' who take their code of behavior from the real world, often without a full appreciation of the subtleties and implications of electronic communication."

The *statement* of a code of ethics for hacking that we read earlier reflects some melding of the two cultures that Gorin delineated.

---

## The Pirate Bay File-Sharing Case

A recent file-sharing case saw the four men who founded the Pirate Bay, a (torrent tracking) Web site found guilty of "assisting making available copyrighted material."[a] Three of them, Gottfrid Svartholm Warg, Peter Sunde, and Fredrik Neij, ran the site. The fourth, Carl Lundstrom, helped finance it. Charges were brought in 2008 and the trial was held in winter and spring 2009. The four are charged with accessory and conspiracy to break the copyright law.[b] Although their Web site did not *host* the copyrighted material, it allowed users to download copyrighted software (music, movies, and computer games) without paying for it. The trial has received unprecedented coverage in Sweden. Pirate Bay supporters have come out in force. The Pirate Bay set up a blog about the trial[c] in which the accused made light of the claims for compensation and damages amounting to 120 million kroner (about $14.3 million), saying they have no money to pay any damages. In addition to fines, they were also each sentenced to one year in jail, however.

The verdict is on appeal and scheduled for a hearing in fall 2010[d] while this book is in press. For the present, the Pirate Bay continues to operate.

---

[a] In 2005 the U.S. Supreme Court had unanimously found in the case of *MGM Studios, Inc. v. Grokster, Ltd.* that "We hold that one who distributes a device with the object of promoting its use to infringe copyright, as shown by clear expression or other affirmative steps taken to foster infringement, is liable for the resulting acts of infringement." It was not clear that a file-sharing site might not avoid liability by simply stating that it was only for legal sharing of digital information, however. The decision is available as a pdf at http://www.hrrc.org/File/GroksterDecision.pdf.

[b] Nordstrom, Louise. 2009. "Swedish Online Pirates Face Copyright Charges." *Christian Science Monitor*, February 19, 16. "Enigmax" News from The Pirate Bay Press Conference, February 15, accessed at http://torrentfreak.com/news-from-the-pirate-bay-press-conference-090215.

[c] http://torrentfreak.com/news-from-the-pirate-bay-press-conference-090215.

[d] "Pirate Bay Court Appeal Set after General Election," *Crenk*, March 12, 2010, available at http://crenk.com/pirate-bay-court-appeal-set-after-general-election/.

As evidenced by Steve Jobs' 2007 suggestion to get rid of digital rights management (DRM) measures on music files,[14] and the outcry after the recent discovery of notoriously intrusive DRM measures by the SONY Corporation, public opinion has not yet decided on the appropriateness of copy protection measures as a way of ensuring fair return for artists and the music industry.

---

### Sony's Installation of a DRM System in Its Music CDs Containing a Rootkit

In 2005, Sony-BMG released music CDs with copy protection systems XCP and MediaMax. Mark Russinovich discovered these systems while testing rootkit detection software that he had written. (A **rootkit** is software that is commonly classified as spyware.) Russinovich found that the rootkit was part of the CD DRM system, XCP, that had been installed on his computer when he had inserted a Sony-BMG music disk. Russinovich posted his finding in a blog, and news of XCP rapidly spread around the Internet. Further investigation revealed that XCP made computers more vulnerable to attacks, "that both CD DRM schemes install risky software components without obtaining informed consent from users, that both systems covertly transmit usage information back to the vendor or the music label, and that none of the protected discs include tools for uninstalling the software."[a] The Electronic Frontier Foundation brought a class action lawsuit against Sony-BMG. Sony-BMG recalled millions of CDs.[b]

---

[a] Halderman, J. Alex and Edward W. Felten. 2006. "Lessons from the Sony CD DRM Episode" extended version (February 14), accessed at http://itpolicy.princeton.edu/pub/sonydrm-ext.pdf. I thank Justin Rich, Sean Kelly, and Christian Miller, all CWRU 2006, for bringing this case to my and our class's attention.

[b] For an in-depth discussion of these two DRM systems and their weaknesses see J. Alex Halderman and Edward W. Felten. 2006. "Lessons from the Sony CD DRM Episode" available as a pdf at http://itpolicy.princeton.edu/pub/sonydrm-ext.pdf.

---

It would be a mistake to think that the culture of computer science and computer engineering has entirely abandoned its early freewheeling character, however. Many in the computer community continue to view free speech as an overriding value, that is, they regard the right to free speech as, if not an absolute right, at least one that outweighs many other ethical considerations. They typically regard all censorship as at least prima facie bad. (U.S. culture generally puts greater emphasis on the protection of free speech than do other technologically developed democracies, although Iceland is now making a bid to surpass the United States in this regard. See U.S. case law on pornography, for example – the emphasis on free speech is especially marked in U.S. computer culture.) The distinctive practices for raising concerns in cyberspace, which we will consider in the next section, reflect the emphasis on freedom of speech in the United States as well as the distinctive computer culture. These practices are without parallel in other areas of U.S. engineering.

*What would be a fair way to compensate authors and composers of music, computer games, and other digitalized entertainment? What, if any, DRM system that prevented copying of software and digitalized entertainment would be ethically acceptable?*

[14] Wingfield, Nick and Ethan Smith. 2007. "Jobs's New Tune Raises Pressure on Music Firms." *Wall Street Journal* CCXLIX, No. 31 (February 7), A1.

## Raising Concerns in Cyberspace

*What considerations would be morally relevant to deciding whether to post notice of a software defect on Slashdot or the Register?*

The freewheeling character of computer culture is reflected in the willingness of computer professionals to go public with concerns, and the special means for doing so. Notable among the latter are the Web sites *Slashdot*, http://slashdot.org/, and *The Register*, http://www.theregister.com/. *Slashdot* further characterizes itself as supplying "news for nerds, stuff that matters." *The Register* characterizes itself as "biting the hand that feeds IT." Thus, it makes explicit its willingness to publicize complaints and concerns of IT professionals when doing so will at least annoy and possibly hurt IT companies. Because the site solicits news stories from its readers, it readily provides a way of publicizing concerns. Therefore, IT professionals who have despaired of having their concerns heard within their companies and lacking protection for raising issues outside their organization often can publicize them on *Slashdot* or *the Register*. A post to such a forum gives the complainant a chance to air a concern under a pseudonym. (The nature of the complainant's knowledge may identify her, however.) In contrast a Web site that was set up to allow users (especially nurses) to report dangerous conditions that they witnessed at the hospitals where they worked quickly disappeared.

In Chapter 2 we mentioned the 2005 case in which a security researcher, Mike Lynn, discovered a serious security hole due to a router flaw (a flaw in the Cisco operating system that powered its routers). He publicly announced the security flaw in a conference paper. Subsequently, stories on the Wired News Web site, http://www.wired.com/, brought the case to wider attention, although the news stories did not include details of the vulnerability or of his attack that had penetrated the system's security.

### Publicizing a Router Flaw, the "Black Hat Bug"

In 2005, Mike Lynn resigned his job at Internet Security Systems in order to be free to deliver a talk at that year's Black Hat conference about a serious security flaw in Cisco IOS, the operating system used in Internet Security Systems' routers. Internet Security Systems tried to prevent him from speaking.

Lynn had announced the topic of his paper well beforehand, but attempts to block his presentation came only two days before it was scheduled. After he gave his talk (and was praised by the computer security community) the FBI began investigating whether Lynn had released trade secrets. The FBI closed its investigation the following November at about the time that Cisco systems released a patch for the "Black Hat Bug" that Lynn had discovered.

A February 2007 interview with Lynn detailing events is available at http://wired.com/news/privacy/0,1848,68365,00.html?tw=wn_story_page_prev2.[a] A second *Wired* article from November 2007, which contains an interview with Black Hat founder Jeff Moss about the case, gives a perspective on how the Digital Millennium Copyright Act (DMCA) may have influenced events. (The DMCA is discussed in Chapter 6, where we will consider property rights generally.) Cisco had claimed that Lynn's talk contained Cisco's proprietary source code, which under the DMCA would have been illegal to reveal. Moss and Lynn were willing to remove slides containing the code in question. It then became clear, at least to Moss and Lynn, that Cisco objected to Lynn's disclosure of the flaw. Moss suggests that the DMCA was inappropriately used to stifle full disclosure of security flaws. The interview brings out another impediment to computer professionals acting for the greater good by disclosing security flaws, namely that companies may ask such professionals to sign nondisclosure agreements when they sign contracts with those professionals to investigate the company's software for security flaws.[b]

---

[a] Kim Zetter. August 1, 2005. "Router Flaw Is a Ticking Bomb."
[b] Kim Zetter. November 7, 2005. "Black Hat Organizer Unbowed," http://wired.com/news/privacy/ 0,1848,69488,00.html?tw=wn_story_page_prev2. *Wired* has also reprinted information from the blog of Jennifer Granick, the lawyer representing Mike Lynn, that details her recollection of events and the *legal* ground for everyone's position at http://wired.com/news/technology/ 0,1282,68466,00.html.

---

Another twenty-first century case that highlights the question of what *limits* ought to be observed when disclosing security holes is the 2007 case of two self-styled computer security experts publicly identifying security flaws in the Mac OS X operating system throughout January 2007. One of the pair was in Ohio and the other, identified only as "LMH," was apparently in Spain. Identifying security flaws is important work, but unlike Mike Lynn, they did not first disclose the flaws to the maker of the flawed operating system, and for this they have been criticized. When LMH was asked if his action was the *responsible* thing to do, LMH was quoted as replying "The irresponsible thing is making someone pay more than 2k U.S. dollars for a nifty machine with broken software." Thus, he was arguing that Apple's actions, which he perceived as wrong, justified his public disclosure of information that might make Mac OS X users more vulnerable. The argument that "two wrongs make a right" is a rather notorious ethical fallacy.

Some Mac owners have stepped forward to help patch the security holes. Landon Fuller, a 24-year-old programmer who once worked at Apple but now heads operations for a maker of computer games, offered to create a fix for each flaw as soon as it was revealed. Others soon joined his effort. Fuller is critical of publicizing a flaw before revealing it to the software maker. The pair who announced the "Month of Apple Bugs" subsequently agreed to send to Fuller information about flaws in advance of making them public on the condition that Fuller not reveal the bug before they did.[15]

*What considerations would be morally relevant to deciding whether to post notice of a software defect on Slashdot or the Register?*

## Privacy in the Information Age

What, if any, new threats to privacy or means for reducing threats have emerged with the advent of computers, the Internet, and the "information age"?

---

[15] Gomes, Lee. 2007. "As Duo Publicize Bugs in OS X, Mac Owners Rush to the Rescue." *Wall Street Journal*, January 24, B1.

In the seventh section of the introduction we examined the notion of privacy and the right of privacy. There we distinguished physical, informational, decisional, and a new category that may be especially important for the dossiers of digitalized information that exist for most people in developed societies, *dispositional privacy*.

Along with sacrifices of both informational and physical privacy to increase security in the face of terrorist threats, we considered an example involving digitalized information that illustrated new threats to privacy; specifically, the Lotus Corporation's proposal to offer copies of its MarketPlace Data Base for sale and the furor that greeted it despite the fact that *the individual items of information aggregated were not the sort of information that is considered private.*[16] Had

> As Daniel J. Solove argues, dossiers of digital information currently compiled on individuals pose a threat to privacy that is more than the loss of control over personal information. He argues that an essential element of the threat is the bureaucratic processes for dealing with information – processes that are routinized and sometimes careless – and the lack of accountability for the handling of information. "This makes people vulnerable to identity theft, stalking and other harms," he concludes.

Lotus Corporation sold names, addresses, income brackets, and consumer choices of 120 million citizens,[17] that data would have enabled "targeted marketing," or spamming, or as Erik Larson calls it, "consumer espionage"[18] of those in the database. A similar but more frightening proposal was LexisNexis's 1996 proposal to sell P-TRAK Personal Locator. Although also retracted after widespread objection, P-TRAK would have provided the addresses, maiden names, and *Social Security numbers* of millions of Internet users. The information in dossiers such as those that Lotus and LexisNexis came to possess may be compiled with no particular *intent* to invade privacy, but the existence of these compilations threatens privacy. Other compilations of data, many of which have been amassed automatically in tracking credit card and Internet purchases, *have* been sold for marketing purposes.

Although some people like to receive marketing targeted to their tastes and many will readily give personal information in return for discounts and other incentives, Daniel J. Solove argues that such dossiers pose a threat to privacy that is more than the loss of control over personal information. He urges that we reconceptualize privacy, or at least privacy law, to take account of harms

---

[16] "Lotus – New Program Spurs Fears Privacy Could be Undermined." *Wall Street Journal,* November 13, 1990, p. B1 and "Lotus is Likely to Abandon Consumer-Data Project." *Wall Street Journal,* January 24, 1991, p. B1. It is important to distinguish information that is private (rather than public) from information that is personal in the sense that it is information that would be intrusive for others to demand, obtain, or discuss. For example, in the United States in the early part of this century, some people (especially women) rarely disclosed their age. They considered this information highly personal even though their birth dates were matters of public record. The judgment of what matters are personal is highly cultural. For example, the Dutch consider it intrusive to look over the books in a person's bookshelf without first asking permission. In some cultures, it is considered impolite to speak of a woman's pregnant condition even when it is evident.

[17] Gurak, Laura J. 1997. *Persuasion and Privacy in Cyberspace: The Online Protests over Lotus MarketPlace and the Clipper Chip.* New Haven, CT: Yale University Press.

[18] Larson, Erik. 1994. *The Naked Consumer: How Our Private Lives Become Public Commodities.* New York: Penguin.

to individuals and to society arising from the existence of large databases that aggregate nonprivate facts. He argues that an essential element of the threat is the bureaucratic processes for dealing with information – processes that are routinized and sometimes careless and lack accountability. "This makes people vulnerable to identity theft, stalking and other harms."[19] Solove does not see the compilation of the dossiers as motivated by any malevolent plan or attempt to use the information to dominate people (as in *1984's* story of a totalitarian government, "Big Brother," that watched every move its citizens made). Because the threats to privacy arising with these dossiers do not fit either the model of intent to invade privacy or intent to dominate (the models that he thinks have been predominant in thinking about previous threats to privacy), Solove thinks they will not be properly understood. He argues that these are threats not only to the individual but to society, because the determination of identity of its members is a crucial feature of any cultural group. Solove gives most of his attention to the actions of businesses, although he discusses public records and government access to personal information as well. His argument is more plausible if one frames it, as he does, in terms of legal conceptions of privacy and recent laws and legal decisions that attempt to address these threats.[20]

> *Evaluate the threats to privacy or means for reducing threats that have emerged with the advent of computers and the "information age."*

## Challenges of the Information Age

> What threats, other than to privacy and to rights to intellectual property, have appeared with the advent of the information age? Which of these are amenable to control or reduction by technical innovation? What ethical issues do those technical innovations raise?

We saw in the previous sections how new kinds of threats to privacy have emerged in the information age. In the case of rights to intellectual property, we have seen that some countermeasures, specifically some digital rights management (DRM) measures, have raised additional privacy concerns. What threats, other than to privacy and rights to intellectual property, have appeared with the advent of the information age?

It may be that some distinctive features of the age in which we live will become apparent only in the future, but at least some new challenges are apparent now. One class of new threats, including the previously mentioned identity theft, arises from the distinctive features of an online encounter. The early days of the Internet gave rise to the frequent, often gleeful observation, "On the Internet, no one knows you are a dog." That possibility seemed less charming as it became clearer that on the Internet it is harder to prove that *you* are you and that other people are *not you,* and **authentication,** in the sense of proving that someone or something is who or what it seems or claims to be, has become a bigger concern.[21]

---

[19] Solove, Daniel J. 2004. *The Digital Person: Technology and Privacy in the Information Age.* New York: New York University Press, p. 9.

[20] *Ibid.*

[21] Kent, Stephen T. and Millett, Lynette I. (Eds.). 2003. *Who Goes There? Authentication through the Lens of Privacy.* Washington, DC: National Academies Press.

Authentication is a matter of authenticating not only people but also servers and Web sites. Flaws in the Domain Name System (DNS) protocol would make it possible to "poison" the cached information about which server was associated with a given domain name and send people to **phishing** sites, sites that deceive people into giving financially important information, such as passwords that go with their accounts.

---

### A Flaw in the Domain Name System (DNS) Protocol

When looking for a better way to stream videos, Dan Kaminsky stumbled on a flaw in the DNS protocol in early 2008. He developed an attack that worked with disquieting ease. He did not reveal it publicly or sell it to the highest bidder. Instead, he organized an industry-wide response that culminated in a gathering of researchers in March 2008 to address the problem. Because reverse engineering the patches issued to fix security holes can identify those holes, all the researchers decided to issue patches for major DNS software simultaneously.

A flaw in the DNS protocol had made it possible to "hijack" the lookup procedure that identifies (by numerical address) the server associated with each domain name. As a result, users trying to reach, say, their bank, could be sent to another Web site masquerading as their bank. The bogus server identification was stored in the user's server; thus the attack is often called "cache poisoning." This flaw had been known. Some "cache poisoning" was carried out as early as 1989 and new security features were added in the 1990s to thwart such attacks. Kaminsky had found a way to defeat those security features, however.

Kaminsky waited and asked the convened researchers to wait for a month after the DNS patches were released (and, presumably, most of the DNS patching had been done) to discuss the nature of the flaw publicly.

Some found his behavior laudable but others thought it fed skepticism about the danger of such a flaw. IT professionals who needed to evaluate the patch and decide whether installing it was worth the disruption the installation would cause were frustrated by the lack of information about what it was supposed to fix.

As many agree, the Internet is filled with flaws, and more may become exploitable as technology changes.

The controversy about the way this flaw was patched draws attention to the absence of an agreed upon process for identifying and fixing flaws in the Internet.[a]

---

[a] Messmer, Ellen. 2008. "Major DNS Flaw Could Disrupt the Internet." *Network World*, accessed at http://www.networkworld.com/news/2008/070808-dns-flaw-disrupts-internet.html. Naone, Erica. 2008. "The Flaw at the Heart of the Internet." *Technology Review* (November/ December): 63–67.

---

Viruses, worms, and Trojan horses are also widely recognized threats.[22] Although in the early days of the Internet much malicious software was traced to youthful hackers, such software has become a tool in phishing schemes and, since 9/11, the possibility that terrorists will use those means to shut down commercial and financial systems has become a concern.

**Spam**, unwanted messages (usually advertisements) sent to large numbers of email addresses, is another side effect of information age communication. Outcry

---

[22] "Cyber Insecurity, Viruses, Spam, Spyware," *Consumer Reports* (September 2006): 20–24.

about one particularly offensive form of spam, unsolicited pornography, gave rise to the CAN-SPAM Act (Controlling the Assault of Non-Solicited Pornography and Marketing Act) of 2003, which took effect January 1, 2004, and sets out requirements for email marketers (including those advertising content on a Web site). It establishes penalties for those who violate the law, and specifies that consumers have a right (and requires advertisers to give consumers the means) to opt out of future advertisements. "Transactional or relationship messages" – email sent within an existing business relationship – are exempt from most provisions of the CAN-SPAM Act, except that it may not contain false or misleading routing information.

The Federal Trade Commission (FTC), the nation's consumer protection agency, is charged to enforce the CAN-SPAM Act, along with the Department of Justice, and companies that provide Internet access may sue violators. The FTC maintains a consumer complaint database of violations of the laws that the FTC enforces. Consumers can submit complaints online at www.ftc.gov and forward unwanted commercial email to the FTC at spam@uce.gov. See the FTC Web site at www.ftc.gov/spam for updates on implementation of the CAN-SPAM Act.

The CAN-SPAM law is often violated, however, especially by marketers outside of the United States, and use of the "opt-out" measures that the law requires may just inform the marketer that one's email account is current and so encourage spammers to send more spam. To limit spam to their clients, Internet service providers (ISPs) often give those clients the option of putting filters on their account to catch likely spam or even to filter out messages that are not from the client's address book or "white list." This example illustrates how technical measures may be more effective than legal prohibitions in curtailing negative behavior when enforcement of the laws is difficult or prohibitively expensive.

*What threats, other than to privacy and intellectual property rights, have appeared with the advent of the information age? (Your answer need not be limited to the examples discussed in this section.) What, if any, means do you know of to effectively control any of them?*

# 6 Rights and Responsibilities Regarding Intellectual Property

## Individual Credit and the Ownership of Innovation

> How broadly should one share ideas? How readily should one copy the ideas of others? Does it matter what the ideas are or the human wants and needs that those ideas help meet?

The best-known philosophical argument for the existence of property rights is that of John Locke,[1] mentioned in Section 4 of the introduction. Locke argued that people have some rights that are "natural" in the sense that they exist prior to any contracts or agreements; among these are certain property rights. The basic right for which Locke argues is the right to the fruits of one's labor. Locke assumes the right to one's own body and argues that if one performs work or mixes one's labor with some freely available material, one owns the product. Locke gives the example of gathering acorns leading to one's ownership of the resulting accumulation of acorns. (Acorns are nourishing although bitter tasting. They were plentiful in England, and sometimes people had subsisted on them.) Locke recognized that people might make trades and other agreements that lead to the acquisition of property rights other than those that are the direct fruits of one's labor.

By extension (and assuming one has a right to one's own mind or intellect parallel to one's right to one's body), one may argue that *intellectual* labor involved in the creation of research, artistic, and technological works provides the basis of property rights. If the creators of the product in question are paid for producing the product, then arguably the product and any resulting trademarks, patents, copyrights, or other *property* rights belong to the employer or client who paid them (although the creators still deserve *credit* as authors or inventors of those patented or copyrighted creations). Saying that patents and copyrights are "property rights" and therefore alienable allows that they may not reside with the creators of the items patented or copyrighted.

The framework of laws and conventions covering intellectual property may be questioned, however. The view that people should freely share their good ideas has found some strong advocates even in modern times (i.e., from the seventeenth century onward). For instance, the Shakers, a celibate religious sect

---

[1] Locke, John. (originally published: 1690). *Second Treatise on Government*. Indianapolis, IN: Hackett Publishing Company, 1980.

that lived in separate communities and flourished in the United States at the end of the eighteenth and through the nineteenth century, regarded work as a form of worship and believed that the ways they found to reduce drudgery should be freely offered to the world. The names of the individuals who invented the remarkable number and range of Shaker designs (for farm implements, household tools, furniture, and clothing) remain largely unknown. For a while, Shakers refused to take out patents. They felt they profited enough by using, making, and selling their inventions and that the inspiration behind their inventions was God's gift to humankind. Thus, the clothespin and the flat broom have become part of Americana. Only when outsiders began to patent Shaker inventions, such as the screw propeller, did this community also apply for patents. Whether or not one finds this Shaker outlook on inventions compelling, it is an alternative to the assumption that inventors should receive (recognition and) special financial rewards via a period of competitive advantage in the market. How broadly should one share ideas? How readily should one copy the ideas of others? Does it matter what the ideas are?

Consider a surgeon who develops a technique that can save lives but keeps the technique a "trade secret" to enhance her prestige. Is withholding the technique morally wrong? Would it be wrong for another surgeon to try to learn that technique, by, say, electronic eavesdropping or asking an operating room assistant? Are the ethical limits on publicizing another's medical technique the same as that for other sorts of innovation? What about techniques for sanitizing water supplies for populations whose health is suffering from the lack of clean water? What about techniques for growing better crops? Is the difference between legal and moral obligation (or between moral *obligation* and more generally what one *ought to do* or what would be *good to do*) relevant here? What boundaries should be established by public policy, and which ones should be left to individuals or communities to draw?

Recall that the rationale for *creating* special property rights for authors and inventors in the U.S. Constitution was *not* an argument like Locke's, that authors and inventors *have such a moral right* because of the intellectual work they have done. The framers of the Constitution argued that *it would promote the public good* to encourage technological innovation and artistic creation by *giving* the creators of such works certain potentially lucrative *legal* rights. Such rights would reward the creators and give them a personal incentive to create and to make their creations known. When we examined the GNU project and open source movement in Chapter 5, we found contemporary counterarguments to the effect that the public good (and advances in the computer fields) is best served by having *fewer copyright restrictions on software*. For the present, it will be sufficient to recognize the *legal* rights to intellectual property, especially patents and copyrights, and recall the arguments in Section 5 of the introduction for a moral obligation to respect the law in general (and, therefore, legal rights to intellectual property in particular).

Ethics codes and guidelines of engineering professional societies also provide some guidance. The section of the NSPE's code of ethics on intellectual property addresses more than proprietary (i.e., ownership) interests and gives standards for fairly crediting others as well. It outlines the following as professional obligations:

10. Engineers shall give credit for engineering work to those to whom credit is due, and will recognize the proprietary interests of others.

    a. Engineers shall, whenever possible, name the person or persons who may be individually responsible for designs, inventions, writings, or other accomplishments.

    b. Engineers using designs supplied by a client recognize that the designs remain the property of the client and may not be duplicated by the Engineer for others without express permission.

    c. Engineers, before undertaking work for others in connection with which the Engineer may make improvements, plans, designs, inventions, or other records that may justify copyrights or patents, should enter into a positive agreement regarding ownership.

    d. Engineers' designs, data, records, and notes referring exclusively to an employer's work are the employer's property.

In Chapter 8, Ethics in the Changing Domain of Research, we examine the stringent standards for fairly crediting intellectual contributions in academic and research contexts. These standards implement the obligation expressed in 10a above. In academic and research settings, stringent rules of crediting apply because these settings have as their mission the *advancement and transmission of knowledge*. In those contexts, it is *never wrong to give credit to others when one knows their contributions*. The advice to give credit where credit is due immediately raises the questions of what is due when and why. The same level of crediting is not due in all contexts. In informal and noncommercial settings, such as when telling a joke to friends or passing on a recipe that is not of one's own creation, one is ordinarily not expected to say where one heard the joke or found the recipe. Attempting to acknowledge others for everything we learn from everyone and adhering to the scholarly standards of citation and acknowledgment in *every* aspect of life would be an enormous burden.

*How broadly should one share ideas? How readily should one copy the ideas of others? Does it matter what the ideas are or the human wants and needs that those ideas serve?*

## Copyrights, "Fair Use," and the DMCA

Why are some creations accorded copyright protection? Under what, if any, circumstances is it fair to copy a copyrighted work without explicit permission?

As we saw in Section 7 of the introduction, a copyright is a legal right to exclusive publication, production, sale, or distribution of some work. A copyright is most commonly held by the author, the composer, or the publisher of a work. It may be assigned to others or inherited, however, so the copyright holder need not be the party who deserves credit for authoring the work. The intellectual property that is protected by the copyright is the "expression," not the idea. Ideas cannot be copyrighted.

Most of U.S. copyright law functions in the context of literary works such as novels, plays, and films. When copyright protection was applied to such works, erring a bit on the side of strong protection was thought to cause no great harm:

It would only require later authors who treated the same themes to do a bit more to differentiate their work from the original expression. Today the term *literary works* includes computer databases and computer programs, however. As we shall see in the next chapter, there are arguments to the effect that too much protection for software could have more serious negative consequences.

The 1998 Digital Millennium Copyright Act (DMCA) tightened copyright protection. Part of the reason this bill was created was to bring U.S. law into agreement with two international copyright treaties of the World Intellectual Property Organization. The act also sparked much debate because of two of its provisions: The first is its limitations on the "fair use" of copyrighted material in educational settings. The second is its restrictions on devices to defeat anticopying protections.

The idea behind **fair use** of copyrighted material is that some copying of copyrighted material may be justified if it does not undermine a copyright holder's property interest or is in the public interest (e.g., because it facilitates education). The fair use doctrine allows such uses as the copying of a recent relevant news article and distribution to the members of a class. The four criteria for deciding whether some copying of copyrighted material is a **fair use** (as stated in Section 107 of title 17 of the United States Legal Code) are:

- The purpose and character of the use, including whether such use is of a commercial nature or is for nonprofit educational purposes;
- The nature of the copyrighted work;
- The amount and substantiality of the portion used in relation to the copyrighted work as a whole; and
- The effect of the use upon the potential market for or value of the copyrighted work.
- The fact that a work is unpublished shall not itself bar a finding of fair use if such finding is made upon consideration of all the factors.

(Notice that the law specifies that all four of these factors must be considered in deciding whether some particular use of copyrighted material *is* a fair use. It does not give a formula or algorithm for *weighing* these four criteria, however. Judgments about fair use, therefore, exemplify the characteristics of a professional judgment discussed in Chapter 4.)

The DMCA, like prior copyright law, allowed some copying of copyrighted material in educational settings in which students meet with the instructor face-to-face, but did not extend those fair use exemptions to online courses. It also did away with long-standing rights of libraries and university educators to archive and lend out copyrighted material and to use reverse engineering for certain purposes – for example, to use reverse engineering of software programs to detect viruses.[2]

*Why are some creations accorded copyright protection? Under what, if any, circumstances is it fair to copy a copyrighted work without explicit permission?*

*Consider a case in which an instructor makes copies of a newspaper article that appeared within the last week and distributes it to her class without obtaining permission of the copyright holder. How does such a use measure up on each of the criteria for fair use?*

[2] Blumenstyk, Goldie. 1998. "House Approves Copyright Bill, but Amendments Do Little to Ease Scholars' Worries," *Chronicle of Higher Education,* Thursday, August 6.

# Patents and Trade Secrets

> How does one know what knowledge or information is proprietary? What considerations are relevant in deciding how a computer professional or engineer can best keep confidential the proprietary knowledge of a client or employer?

In the last chapter, we examined some unique characteristics of digital information that lead to novel moral questions for computer professionals and for society as a whole. Here we will address some moral problems about credit and intellectual property that are common to *all* areas of engineering. One of these is the question of what factors determine the extent to which knowledge acquired in working for one client or employer can be used when working for another. A frequent complication is that one's clients or employers may be competitors of one another. Computer professionals, like other engineers, need to know how to distinguish between standard design elements (which one may learn about in courses or on a job and use in another job) and customized (and perhaps patented) knowledge belonging to one employer or client. This problem situation is illustrated in the scenario, "One Client Teaches You Something that Would Help Another Client."

---

### One Client Teaches You Something That Would Help Another Client[a]

You are the lead software developer working for a small software developing company. You develop a specific type of software for several companies that are maneuvering for market share in a competitive industry. In your job as lead developer you work with clients to assess their specific needs and implement patches and updates to the software that you have developed for them. It is a big job to update the software in response to complex requests. Therefore, if a solution can be found without needing to update the software, resources are saved for all parties.

A few weeks ago Company A came to you about a major difficulty with your software. You were busy resolving another issue with Company B at the time and the Company B project had priority. Before you got back to Company A, its IT person called to inform you that he solved the issue by using a very specific configuration of Company A's network. He described the configuration to you in detail and you were satisfied that the issue was solved without the need to update your software.

Now, Company B contacts you with the same issue. What should you do?

What facts in this situation are morally relevant to deciding whether (or how much of) Company A's configuration solution is confidential information? For example, does it matter whether Company A volunteered its configuration solution or you asked for it? Does it matter if part of the configuration solution is something you learned about elsewhere but had not thought about in this application? Do you know how much, if any, of what you learn from Company A in the course of delivering services to it is proprietary and therefore confidential? If you do not know, how could you find out?

---

[a] Adapted from a scenario by Kyle Kaliebe (CWRU '05).

Certainly, it matters if the information you have received is proprietary. It could be so by being part of a confidential business plan or other sort of trade secret, a patented device or process, or copyrighted code. Your responsibility for being aware that some knowledge is proprietary knowledge is very different in the case of trade secrets. Suppose you invented a synthetic polymer to fill the flaws in emeralds together with the method for inserting it into the emeralds, and it turned out to be the same as a trade secret polymer and method for filling the flaws in emeralds. If you then sold the emeralds that had been enhanced by this method, you would not be subject to any legal penalties.[3] However, had the polymer formula and method been patented and you independently discovered it, used it, and sold the resulting emeralds, you would be subject to penalties for patent violation. If someone discloses to you something that is not obviously or not ordinarily confidential information, but they wish you to treat it confidentially, they are obliged to *tell* you that. Of course, you are obliged to know that certain items, such as engineering designs, *are* proprietary.

---

### Distinguishing Generic Software Code from Proprietary Code[a]

You are a programmer for Big-Time Software Corp., a large company that develops a wide range of products. Having come to Big-Time two years ago for a better location and better pay, you are beginning to be bored with projects that you are assigned. When Big-Time announced it was starting a small Game Development division, you volunteered to be transferred to this new group.

Before you came to Big-Time, you worked at Small-Time Games, which was located not far from your old college and was made up almost entirely of graduates from the school. You had enjoyed that work and were looking forward to similar satisfaction working in the new Game Development division. Your enthusiasm cooled when, at the first meeting, it was made clear that the new division and positions within it were provisional. Your boss, Willie, informed the team that the company was just testing out the profitability of a Game Development division. If the division fails to be profitable, then it will be dismantled and some of you might be laid off.

After a few weeks, everything seemed to be coming together. Willie then assigned you to work on a system for keeping track of player data on the user's machine and on a central server. This would allow players who use this service to continue playing on different machines.

The next day Willie asked you about progress on the system. When you said you needed more time, Willie said that the design from your previous experience with Small-Time Games could be easily adapted to the current assignment. Willie claims that the prior work was a standard design, so there were no copyright issues.

What criteria would distinguish generic standard design from intellectual property belonging to Small-Time Games? How might you find out if these are met?

What are the ethical criteria for evaluating the behavior of the supervisor, Willie?

What other factors should you consider in deciding what to do and how to go about it?

#### Getting Started

A company that accepts unethical behavior from its employees is at risk for unethical treatment *from* those employees, so a company has prudential as well as ethical reasons for supporting ethical behavior in its employees.

---

[a] Adapted from a scenario by James Post and Bryan Drake (CWRU 2006).

---

Two sorts of problems about credit and compensation commonly face engineers in private practice. The first, which is prudential, is how much to trust that parties who solicit bids from them and competitors will properly credit and compensate them for their work. The second, which is ethical, is how they in turn should handle information from other engineers doing work for the same agencies, or credit or compensate those engineers for their work. Both are raised in the following scenario.

[3] Aeppel, Timothy. 2007. "How Arthur Groom Riled the Emerald Business," *Wall Street Journal* CCXLIX (Wednesday, February 7): A1 and A12.

## Intellectual Property of Engineers Submitting Bids

Randy, an engineer, submits a proposal for a project to the county council. The proposal includes technical information and data that the council requests. A staff member to the council makes Randy's proposal available to another engineer, Thornton. Thornton uses Randy's proposal to develop another proposal for a different project and submits it to the council. The parties dispute the amount of Randy's information that Thornton used.

What are the county council's responsibilities in handling Randy's proposal?

What factors are morally relevant in this situation? For example, does the amount of information used make an ethical difference? How does the nature of the information bear on the ethical question?

*Source*: Adapted from NSPE Board of Ethical Review Case 83-3[a]

---

[a]The NSPE Board of Ethical Review (BER) cases and opinions from 1976 through 2001 (only) are available at http://www.niee.org/pdd.cfm?pt=NIEE&doc=EthicsCases.

Two related questions are: What, if any, ethical issues are raised by performing unpaid work in hopes of receiving a later paid assignment? How ought an engineer treat the work of other, potentially competing engineers in this context?

## The Use of Work from an Unpaid Consultation

A state agency considers designing a facility that requires special expertise in the field of solar energy. It learns from a federal agency that the Moreau firm previously developed a plan for a similar facility for that agency, and so it contacts the Moreau firm. The Moreau firm submits preliminary data to the state agency, which in turn includes that information in a proposal to a private foundation to secure additional funds for the project. The state agency holds many informal discussions with Moreau's firm and so that firm comes to believe that, if the project is approved, it will be awarded the contract.

Several months later, the state agency tells Moreau's firm that the public and private funding it received will not be sufficient to fund the full scope of the facility. It asks the Moreau firm to evaluate the possibility of a more limited facility. Believing that it will be awarded the design contract, the Moreau firm (at its own expense of several thousand dollars) investigates the possibility of a more limited project, and submits a revised proposal to the state agency.

Subsequently, the chief state engineer informs Moreau's firm that he had turned over all of its data to the Barron firm, and is conducting initial negotiations with it. The chief says that if these negotiations fall through, it will contact Moreau's firm to negotiate the project. All the while, the Barron firm had been aware of the involvement of the Moreau firm in the project but it has not contacted the firm to discuss the project or obtain Moreau's earlier submissions to the state agency. The Moreau firm protests to the state agency and accuses the Barron firm of violating the NSPE Code of Ethics.

*Source*: Adapted from NSPE Board of Ethical Review Case 77-5[a]

---

[a]The NSPE Board of Ethical Review (BER) cases for 1976–2007 with judgments offered by the BER based on application of the then current NSPE Code of Ethics are available in hard copy in the volumes 5–9 of *Opinions of the Board of Ethical Review*, from the National Society of Professional Engineers. See the reference guide with an index of cases through 2009 at http://www.nspe.org/resources/pdfs/Ethics/EthicsReferenceGuide.pdf. Cases and opinions from 1976 through 2001 (only) are available at http://www.niee.org/pdd.cfm?pt=NIEE&doc=EthicsCases.

Does Moreau's firm have an agreement with the state agency?

How should one interpret its provision of unpaid services to the state agency? As a bribe? As foolish? As an appropriately cooperative act? Something else? In what ways was it like or unlike the unpaid work that goes into submitting any proposal?

How would you evaluate the conduct of each of the firms? Of the state agency?

Is there other information that would be morally relevant in evaluating the situation? If so, what is it and what ethical considerations does it influence?

> *How does one know what knowledge or information is proprietary? What considerations are relevant in deciding how a computer professional or engineer can best keep confidential the proprietary knowledge of a client or employer?*

## Property Rights Contrasted with Credit for Invention or Authorship

> What is the difference between having a *property right*, such as a patent or copyright for something one has invented or written, and *credit* for having written or invented it?

Recognition for design work and other innovative technical contributions is manifest in a variety of ways and settings. Naming the device for an individual (e.g., the Jarvik heart) or for a group or corporation – for example, an "NCS knee" or "Microsoft's new operating system" – may reflect credit for an engineering design. Even when a device is named for a person, that individual need not be the designer(s) or inventor(s). Many medical devices that are named for individuals carry the name of the physician who stated a need for such a device, or collaborated on designing it or even who was the first clinician to use it. An inventor's name goes on the patent (which may be owned by some other party), but unlike an author's name, which is usually included in a copyrighted work (whether or not the author retains the copyright), the inventor's name may not appear anywhere except on the patent. In that case, it may be less likely that most users of an invention will ever know the name of the inventor. (As we shall see in Chapter 9, the order of the listing of inventors on a patent is not significant as it often is for the ordering of authors.)

The research that goes into developing a new product and the testing undertaken to identify the causes of any defects that appear in production are rarely published in journals or books. Therefore, those who conduct research in industry do not regularly receive publication credit as authors of articles on their research, although, as we just saw, if some device is so novel as to warrant a patent, the name(s) of the inventor(s) will go on the patent. Research and testing in industry are commonly credited in the same way as design and manufacturing work in that setting, namely, by promotions, raises, and the like, rather than by ownership or association of one's name with a design or device. The results of work done for an industrial employer are generally recognized as the employer's property. As we saw, the NSPE stipulates that an "engineer's designs, data, records, and notes referring exclusively to an employer's work are the employer's property." In contrast, when investigators working under a grant or a contract conduct research at a university or research facility, they may publish the results. If research is funded by an industrial sponsor, university investigators may delay publication for an agreed upon period (usually not more than a year) to give an industrial sponsor a head start in using the results. Classified (confidential, secret, or top secret) research cannot be published, of course. That is one reason why many universities will not accept any grant or contracts to do classified research on campus. (Another is that it would create on-campus research that would be closed to students who were not eligible for security clearances and thus curtail the open exchange of ideas that is a central value of universities.)

The proprietary rights embodied in patents and copyrights work differently from crediting mechanisms that have no property implications. The patent arrangements that attend industrial sponsorship of university research are independent of criteria for fairly crediting authors and other contributors to a research article. Consider the following situation:

---

### Failure to Credit the Source of Research Data

Ramos is the head of a chemical company. As a part of a research and development effort, Ramos offers to provide funding to the chemical department of a major university for research on the removal of poisonous heavy metals (chromium, copper, lead, nickel, zinc) from waste streams. In return, the university agrees to give Ramos's company the exclusive rights to any technology developed in the field of water treatment or waste stream management. As compensation, the university will also receive a royalty from the company from the profits resulting from the use of the technology.

At the university, a group of professors, led by Polinski, decides to form a company to exploit the technology obtained except for water treatment and water waste management that Ramos's company will develop.

Meanwhile, while the university is conducting this research, Ramos's company is conducting its own parallel research. Both teams obtain data and performance figures, and Ramos's company freely shares its results with the professors in Polinski's company.

Later, Depasquale, a professor of civil engineering at the same university, decides to conduct research and publish a paper on sewage treatment technology. He contacts the professors in the chemistry department, who furnish him with data from their tests, as well as with data from Ramos's company. Depasquale is unaware that some of the results come from Ramos's company.

Depasquale is successful in her research, and her article is published in a major journal. The data obtained by Ramos's company are displayed prominently in the paper, and make up a major portion of the article. The paper credits the members of the chemistry department, but nowhere mentions the contributions of Ramos's company, even though its funds supported both projects. Depasquale later learns that Ramos's company was the major contributor to the data in her paper.

Is it plagiarism for Depasquale to publish the data without crediting all of the sources? Why or why not? Is it Depasquale's obligation to give full credit to Ramos's company for its data?

What, if any, action should Ramos take after discovering the article? What, if any, additional information would you want before deciding what to do, if you were in Ramos's position?

*Source*: Adapted from NSPE Case 92-7

---

Under the agreement with Ramos's company, the university agreed to give the company exclusive use of any resulting technology developed for water and wastewater treatment. The company left the faculty members free to exploit applications of the technology other than the treatment of water and wastewater. Presumably, any patent on the technology would belong to the university or its faculty members, with Ramos's company having exclusive license to develop or apply the technology for water treatment.

Under the terms of the 1980 Baye–Dole Act, universities are encouraged to patent inventions made with federal funds, but the government retains the rights to use those inventions without paying royalties. One of the intentions of the act was to encourage universities to disseminate the results of government-sponsored research. Giving universities patents that they can license to others gives them a financial incentive to disseminate those research results. After passage of the Baye–Dole Act, what had been offices of "patenting" or "patents" in universities with strong engineering and science departments often were renamed offices of "technology licensing" or "technology transfer." Critics of these changes charged that universities would, and indeed did, compromise their identities and change their cultures for the worse as they became more entrepreneurial and directed more of their effort to selling licenses to use the technology that their faculty members invented.

Certainly new conflicts did arise as universities and their faculties had more financial incentives that potentially competed with the fulfillment of other responsibilities. However, because engineering schools were long accustomed to having their faculty members start companies and make money based on some of their academic research and development, many of those schools had rules to prevent abuses brought on by greed. For example, some schools had rules against faculty members hiring their own thesis students to help with their consulting (which included their own businesses).[4] In this way, those schools sought to prevent a situation in which a faculty member might seek to delay the graduation of a talented student because that professor needed the student's help on consulting work.

## Patenting of Inventions Contrasted with Publication of Research

As we saw in the introduction, a patent is a legal right granted by the government to use, or at least to bar others from using, one's invention. This right, like a copyright, may be assigned to others, so the owner or "holder" of the patent or copyright is not necessarily the inventor or author. Property rights are alienable in the sense discussed in the introduction, whereas the status of being an author or inventor is inalienable. Obtaining patents and defending them are costly. Therefore, inventors sometimes prefer to assign patents to others and receive alternative forms of compensation for their work.

Once one has decided to file a patent, questions of priority enter in much the same way that they do in scientific discovery. As we shall see in Chapter 9, *submission for publication* establishes the date and hence the priority claim for published work. In filing for a patent, however the crucial date for claim of priority is not the date of the filing – although in the United States one must file within a year of any "public disclosure" of the design or plans – but the *date of conception of the idea*. Documentation of the date (and even time) when the invention was

---

[4]Many private universities allow their full-time faculty members to work "one day a week" at consulting, which is broadly interpreted to mean a potentially income-producing activity. Some public universities put a much tighter limit on the amount of time their faculty members can engage in such outside activities.

conceived is needed when competing applicants file for the same claims. Two ways to establish when inventions are conceived are:

- Keep a permanent design notebook (bound, with sequentially numbered pages) documenting your work and ideas (dated and in ink), and have it periodically (or for a particular idea) dated and signed by your instructor.
- Document your idea with annotated sketches, explain it to a fellow student and/or instructor, have that person sign the document and indicate that she has understood the idea, and then send the document to yourself by registered letter, which you save, unopened. The postage date and time stamp on the letter document the time you conceived of the idea.[5]

The rules for keeping a design notebook are the same as those for keeping a laboratory notebook on a research project. Your *design* notebook is your own possession. Unlike a laboratory notebook, you can take the design notebook when you graduate. Students' rights and responsibilities regarding their *laboratory* notebooks will be discussed in Chapter 8, Ethics in the Changing Domain of Research.

U.S. patent law requires that for an invention to be patentable, the patent application for it must be initiated within one calendar year of the "public disclosure" of the invention. European and Japanese patent law requires patent application *before* public disclosure, but most countries honor patents taken out in other countries. Presentation in a class or elsewhere in a university community does not constitute public disclosure. If someone in the audience for, say, a class presentation, were to discuss the ideas outside the university, however, that discussion might constitute public disclosure. Because students may make such a disclosure, they need some understanding of the laws and conventions regarding intellectual property in order to avoid undermining their own rights or those of others in the university community.[6]

Neither public disclosure nor filing for a patent are precise analogs of publication of research. Public disclosure makes work public and *may* enhance an inventor's reputation, but it is not a mechanism for establishing priority unless publication attends public presentation. Filing for a patent establishes a *claim* to ownership, but, unlike the date of a publication, the date of filing does not *establish* priority. After public disclosure, designs and devices that have not been patented are "in the public domain," that is, open to free use by anyone.

Disclosure of research results works differently. No strict time limits apply to the interval between reporting on one's research at scientific meetings and publishing a detailed report of the work, although as the scenarios in Chapter 8 exemplify, announcing a research result that one does not describe in detail can cause problems for others and diminish one's standing in a field.

These questions, like the rationale for intellectual property rights in the U.S. Constitution, focus on issues of public's welfare. Does the present system of intellectual property ownership and social control ensure that members of the

---

[5] I thank MIT Professor Igor Paul for these two suggestions.

[6] For an extensive discussion of intellectual property rights, see Vivian Weil and John W. Snapper, 1989, *Owning Scientific and Ttechnical Information: Value and Ethical Issues*, New Brunswick, NJ: Rutgers University Press.

professions serve *the public interest*? In the case of patents, the temporary exclusive right gives inventors an incentive to make public the knowledge of their innovations. Trade secrets do not increase public knowledge and so have fewer protections. The protections for trade secrets are primarily protections against others learning those secrets by dishonest means, such as industrial espionage or breaking of confidentiality agreements.

> *What is the difference between having a property right, such as a patent or copyright for something one has invented or authored/composed, and credit for having written or invented it?*

## Benchmarking and Reverse Engineering

> On the one hand the "not invented here" attitude, which disregards advances made outside of one's own organization, is widely blamed for slowing advances in quality and safety. On the other hand, legal specifications of copyright and patents and other intellectual property protections are intended to limit the use that others can make of one's designs. What are fair and prudent means of learning from others? What other ethical issues arise in learning from the innovations of others?

A commonly accepted first step in the design process is benchmarking. The common meaning of "benchmark" is a standard by which a thing can be measured. In engineering "**benchmarking**" refers to obtaining a competitor's devices or publicly available information before one designs and manufactures a new product. If the product is not prohibitively expensive – as a nuclear power plant would be, for example – samples of the competitor's product are commonly purchased, examined, and analyzed. Benchmarking may or may not involve copying anything from the competitor. A company might wish to benchmark for reasons other than to copy the competitor's design, say to examine its competitors' products or pricing structure to learn about the competitors' cost of manufacture and to judge whether, with some new manufacturing process, the benchmarking company can enter the market and produce a competitive product at a much lower price.

One means of obtaining information about a competitor's product is to reverse engineer it. **Reverse engineering** is the examination of a product to understand the technology and process used in its design, manufacture, or operation. It commonly involves disassembling the product and testing ways to destroy it. Often, reverse engineering is used to learn what a competitor has done in order to copy or improve on the competitor's work. For example, engineers might photograph and enlarge pictures of silicon chips to learn about the architectural features of the chips, such as whether it uses one function twice or two different functions once. (It would put a company at a competitive disadvantage to *simply copy* chips or other rapidly evolving technology, because then the copying company would lag behind its competitors in bringing out the same product.) Figure 6.1 shows the latest generation of the multiple-channel silicon-based sensing chip. (It consists of sixty-four nanosensors, and is less than one square centimeter. Each side has sixteen nanosensors – all that is required for cell phone use.)

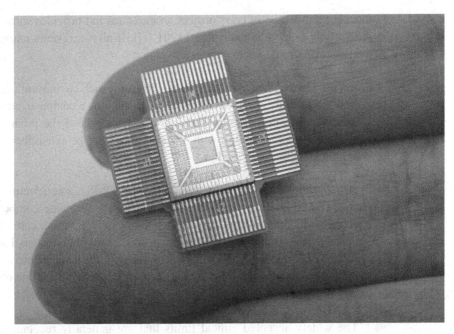

Figure 6.1 An Enlarged Photograph of a Silicon Chip (Photo: Dominic Hart/NASA)

Is copying based on reverse engineering ethically justified? In reverse engineering, no credit is given to the individuals who originate the innovations or the company that employs them. Should copying some features of a competitor's product be understood on the model of plagiarism, the copying of another's words or ideas? As we saw in the discussion of intellectual property earlier in this chapter, the ethical violation called "plagiarism" is *not* the same as the legal concept of copyright or patent violation. Plagiarism is taking credit that one does not deserve, rather than violation of a property right. For a patented device, the patent holder has the property right to the innovation, but the patent also gives lists (and, therefore, credits the inventor(s)). There is no established practice for crediting those who create an *un*patented technological innovation that is comparable to listing the author of some literary work that was never under copyright. (Recall that items in the public domain, as contrasted with private intellectual property, are available to be freely copied by anyone.) Shakespeare's plays, the King James authorized translation of the Bible, and government reports such as the *Report of the Presidential Commission on the Space Shuttle* Challenger *Accident* are all in the public domain, although the individual or committee that authored them are identifiable and credited.

Unlike the Shakers, technologically developed societies do recognize intellectual property and have legal systems that accord property rights in the form of copyrights and patents. Those legal rights, and the absence of them for innovations that are not sufficiently novel to be patentable, color the ethical norms for copying of unpatented innovations. One issue that has been recently disputed is whether copying that aims to create compatibility/interoperability (rather than device improvement) is acceptable. The previously mentioned DMCA forbids most circumvention of copy protection measures but provides some exceptions. Particularly notable is exception (1201 [f]). It allows reverse engineering of a computer program to achieve circumvention

> and the development of technological means for such circumvention, by a person who has lawfully obtained a right to use a copy of a computer program for the sole purpose of identifying and analyzing elements of the program necessary to achieve interoperability with other programs, to the extent that such acts are permitted under copyright law.[7]

Some printer manufacturers argued that the DMCA's anticircumvention provisions should be interpreted in a way that would eliminate competition in the aftermarket toner industry. Companies that manufacture toner cartridges, but not the printers themselves, reverse engineer toner cartridge designs for printers in order to make toner cartridges that will work with the same printers. As we shall see in the next chapter, a printer manufacturer, Lexmark, sued one of these companies arguing that the DMCA prohibited such copying, but the courts found against Lexmark in a far-reaching decision.[8]

The widely accepted ethical limits that are generally recognized in benchmarking and reverse engineering (other than respecting legal property rights) are constraints on the *means* one can use to obtain information rather than on the nature of the information or the use one makes of it. An argument for placing the ethical limits in this place is that patents and copyrights already prohibit the copying of patented and copyrighted materials, so that the remaining proprietary information that may be learned through benchmarking and reverse engineering is analogous to a trade secret. Like a trade secret, learning and using the information are not prohibited as long as one has not used unfair means to learn it. (In Chapter 7, Workplace Rights and Responsibilities, we will examine some guidance that the Texas Instruments (TI) Ethics Office has given its employees on many matters.) Here we will consider advice that pertains to benchmarking and reverse engineering for what it shows about where an ethically concerned company draws the line between ethical and unethical behavior. The TI Ethics Office lists the following as acceptable benchmarking practices:

- Asking customers about equipment and prices of TI competitors
- Asking employees of well-run businesses that do not compete with TI about their practices
- Searching for information through public resources
- Reading books and publications describing other companies
- Encouraging other TIers who come in contact with customers to be observant of practices that might be useful to TI

---

[7]The Digital Millennium Copyright Act of 1998, U.S. Copyright Office Summary, December, p. 5.
[8]Fisher, Ken "Caesar." 2004. "DMCA Dealt Serious Blow by Sixth Circuit Appeals Court," *Arstechnica*, October 26, accessed at http://arstechnica.com/news.ars/post/20041026-4352.html.

Practices that Texas Instruments considers unethical include:

- Misrepresenting oneself as working for another employer
- Colluding in fixing prices or allocating markets or customers
- Disparaging a competitor's business to customers or to others
- Attempting to gain confidential information about other businesses[9]

Because misrepresentation is deception, TI finds it as morally objectionable as lying.

**Collusion** is a secret agreement for a deceitful or a fraudulent purpose. Finding out about the competition's pricing structure *might* lead to price fixing, but it need not. Collusion to fix prices, which is illegal, is more of a temptation for *companies* than for *individual engineers*. Avoidance of even the appearance of fixing prices is one reason that TI tells its employees never to attempt to gain competitive information directly from a competitor. Indeed, companies occasionally purchase benchmarking from third parties to avoid direct communication with competitors.

> Like learning a trade secret, learning and using the information gained through benchmarking and reverse engineering are not wrong as long as one has not used unfair means to learn it.

Disparaging a competitor to customers may be a poor policy rather than unethical, assuming one is truthful in one's assessment of the competitor's failings. Like negative campaigning by a politician, disparagement is likely to lead people to believe that the speaker focuses on the competitor's failing because the speaker has little positive to offer, or that one's whole industry is corrupt or incompetent.

The first three objectionable practices are relatively easily recognized, but the line between the fourth, attempting to gain confidential information, and the *acceptable* practice of obtaining information through public resources, is sometimes harder to discern. What about going into the showroom of a competitor, as though one were a customer, and asking questions of the salesperson? TI (and some other reputable companies) holds that in that circumstance it is not necessary to tell the salesperson the name of one's employer or the reason for one's interest. From the previous prohibition of misrepresentation of one's employer, this only means one need not volunteer the information, not that one can lie if asked for it.

The standard of disclosure required of an engineer engaged in benchmarking is rather different from the standard of disclosure required of, say, a journalist getting a story. It is wrong for either the engineer giving a professional opinion or the journalist writing a story to tell lies and misleading half-truths, of course, but now we are considering how much each must disclose to third parties from whom they seek to get the information necessary to do their jobs. A journalist who *does not disclose* that she is a journalist will be judged for behaving deceptively, tricking others into speaking more candidly than they would if they knew their remarks were to be reported. An engineer engaged in benchmarking who does

---

[9] Article Number 72 from the TI Ethics Office, available in the "Ethics in a Corporate Setting" section of the Online Ethics Center at http://www.onlineethics.org.

not disclose to a competitor's salesperson what she is doing is seen as acting deceptively only if she *does something to deceive* the salesperson. As we saw in Chapter 1, what is entrusted to engineers is different from what is entrusted to members of other professions. Understanding what is central to fulfilling the public's trust in each profession is a complex matter. The provisions in ethical codes and guidelines are justified insofar as they express what is necessary or important to fulfilling that trust.

Consider whether it is ethically permissible for a company to send a product obtained from one supplier to a second supplier so that the second supplier can reverse engineer it. If one has a confidentiality agreement with the first supplier, clearly the action would be illegal as well as unethical. If there were no confidentiality agreement, would the action be ethically permissible? The currently accepted practice is for the second supplier to purchase the product from the first supplier (or distributor) and then to use reverse engineering to find out about the competitor's product. It may be unfair to the first supplier for the customer to play an active role in the second supplier's attempt to reverse engineer the product.[10] If the customer had some special advantage in obtaining the product or information about it, the action would certainly be unfair.[11]

*Under what circumstances are benchmarking and reverse engineering fair and prudent means of learning from others?*

## Conclusion

In this chapter, we have examined fair credit primarily in terms of prevailing laws and customs in technologically developed democracies with market economies. We have also reflected briefly on the legitimacy of those laws and customs by comparing them with the practices of a community that once produced many technological innovations but measured success in terms other than market competition.

The laws and standards now covering technological innovation show that learning from others, even in the absence of any explicit means of according them credit, is the norm in technological innovation so long as one recognizes ethical boundaries in acquiring information and does not violate copyrights, patents, or trademarks.

[10]This is the TI Ethics Office's assessment of the situation. See TI Ethics Office Article 142, "Reverse engineering and patent infringement," in the Online Ethics Center at http://www.onlineethics.org.
[11]*Ibid.*

# 7 Workplace Rights and Responsibilities

As we saw in Chapter 3, it is in *everyone's* interest that engineers be heeded when they recognize risks and threats to the public welfare. It is in a client or employer's interest to see that engineers' concerns are heard within their organization, so that no dangers or defects will be overlooked. Those organizations that disregard their engineers or even try to silence them, leaving them no alternative to "**blowing the whistle**" (i.e., going outside their organization to get attention to their concerns), lose the benefit of their engineers' expertise and the respect of the public.[1]

In prior chapters, we have focused on the moral skills that enable engineers to fulfill their responsibilities both in responsive and unresponsive organizations. In countries like the United States where employee engineers usually have no written employment contracts that protect engineers against retaliatory discharge, less reputable companies may retaliate against engineers for pursuing ethical concerns that clash with the company's short-term business objectives. Therefore, creating a workplace that is relatively free of the risk of such retaliation is a much larger ethical issue for engineering in a country like the United States than in countries where employee contracts are the norm.[2]

The case of Roger Boisjoly in Chapter 2 shows that sometimes even concerns arising out of engineers' most fundamental responsibilities are ignored. In this chapter, we examine organizations – corporations, government agencies, universities, and research facilities – to see what makes them more able to listen. In Chapters 2 and 3, we considered many problems of engineers in private practice, but most engineers work as employees. They are immersed in organizational cultures that significantly influence their moral lives. Especially important are the practices an organization has developed to respond to "bad news" and to promote fair treatment of employees.

Organizational cultures vary greatly. The best companies focus on the goals of consumer trust and public goodwill, quality products and service, and high morale and continued professional growth of employees. These factors generally contribute to *long-term* profitability. Other companies have management practices that aim to maximize short-term profits. Economist Milton Friedman's view that

---

[1] The point that companies lose when their engineers must "blow the whistle" is argued in a 1988 essay by Michael Davis, "Avoiding the Tragedy of Whistleblowing," *Business & Professional Ethics Journal* 8(4): 3–19.

[2] For a brief comparison of the work situation for engineers in Germany and in the United States, see "Engineering Ethics in the U.S. and Germany" by Adolf J. Schwab in *IEEE Institute*, June 1996, which is available at http://www.institute.ieee.org/INST/jun96/ethics.html.

the sole responsibility of managers is to "make as much money as possible" (consistent with obeying the law and conforming to ethical custom) would be a caricature of sound management philosophy, if read as an injunction to maximize *short-term* profits.[3] Companies devoted to *long-term goals* are often explicitly concerned to meet ethical standards even *in advance* of custom. A recent example of this was the decision of some energy companies to take the initiative to act more proactively to reduce greenhouse emissions.

One widely held management view is that in a rapidly changing world, organizations must continually *learn* if they are to succeed. This view emphasizes the importance of not only hearing the bad news and anticipating problems *before* they arise but also of fostering *personal and* professional growth of employees and responding to employees' desire to build something important, as well as pursuing their own self-interest.[4] Such theories do not always guide practice, however. An organization's actual practice may reward managers for maximizing short-term profits or for failing to report any expensive problems if the consequences will be attributed only to others. If you wish your engineering knowledge to be respected, it is wise to avoid working for such companies or clients.

## Engineers and Managers

How do managers relate to engineers in good companies and how does this contrast with the relations between managers and engineers at companies that are not as good?

A recent study of communications between engineers and managers by researchers at the Center for the Study of Ethics in the Professions at the Illinois Institute of Technology[5] (IIT) reveals how managers respond to unwelcome news from engineers in well-run high-tech companies. The study identified three value orientations of companies depending on whether the company gave first priority to

Customer satisfaction
The quality of its work/products
The financial bottom line

---

[3] Friedman, Milton. 1970. "The Social Responsibility of Business Is to Increase Its Profits," *The New York Times Magazine* (September 13) reprinted in *Ethical Issues in Engineering*, edited by Deborah Johnson (Englewood Cliffs, NJ: Prentice-Hall, Inc., 1991), 78–83.

[4] Senge, Peter M. 1990. *The Fifth Discipline: The Art and Practice of the Learning Organization.* New York: Doubleday.

[5] Davis, Michael. 1997. "Better Communications between Engineers and Managers: Some Ways to Prevent Many Ethically Hard Choices," *Science and Engineering Ethics,* 3(2). (The report is sometimes called the "Hitachi Report," for its sponsor.) The picture of communications at companies that put profits above all else derived largely from the reports of engineers who had *formerly* worked at such companies and were now working at a quality or customer-centered company, and so is less detailed.

The difference between the picture of communications in the Hitachi Report and the negative picture of communications in Robert Jackall's *Moral Mazes* (New York: Oxford University Press, 1988) may be due to Jackall's selection of bottom line companies or his focus on management at higher levels.

Although this is a rough typology and the priority given those factors is a matter of *degree*, for simplicity the report speaks of three *types* of companies. I shall call the first, "customer-oriented" companies, the second, "quality-oriented" companies, and the third, "finance-oriented" companies.[6] (The identification of both customer-oriented and quality-oriented companies comes as welcome news to some young engineers, who fear that concern with the bottom line always dominates other concerns as Friedman argued it should.)

The types of companies differ in several ways: In the quality-oriented companies, quality (and of course safety) takes priority over cost and the customer's desires. Cost is still considered, but as one engineer put it, "Cost comes in only after our quality standards are met."[7] Quality-oriented companies listen to their customers, but take pride in being willing to say "no" to them. In one manager's words, "If a customer wants us to take a chance, we won't go along." Such companies try to convince customers to keep their applications of a product within the specifications for the product's appropriate use, but if they fail to convince the customer, will forfeit the business rather than supply a part or a device that will not perform the customer's job well. Although this strategy does not maximize *short-term* profits, the quality-oriented companies in this study had secured a large and growing share of the markets in which they competed, so their reputation for quality seems to have contributed to their *long-range* success.

Even in the quality-oriented companies, managers and engineers had different concerns and priorities. The engineers were likely to see managers as more concerned about cost or more superficial in their judgment, and the managers to view the engineers as likely "to go into too much detail."

In the customer-oriented companies, customer satisfaction was the main objective. They replaced the internal standard of the quality-oriented companies with an external standard of satisfying the customer. Predictably, in such companies, engineers' quality concerns often conflicted with managers' desire to please the customer.

Davis and his colleagues found both engineers and managers to be critical of finance-oriented companies (perhaps because all of the interviewees in the study who had worked for finance companies had *left* them and now worked for customer- or quality-oriented companies.) In finance-oriented companies the desire to maximize the number of units shipped conflicted not only with the engineer's concern for quality, but in some cases even with other ethical standards, such as when engineers or managers were pressured to adjust test results to make it seem that the product met the customer's specifications.

An important result that ran contrary to the investigators' expectations was that managers expected engineers to "go to the mat" for both safety and quality concerns. This expectation held at the customer-oriented and quality-oriented

---

[6] Michael Davis and his coinvestigators call these companies the "engineer-oriented" companies, "customer-oriented companies," and the "finance-oriented" companies. Quality and safety are values central to engineering, in the sense that an engineer who does not uphold them is seen as a poor engineer, but because quality rather than the engineers (e.g., their happiness or career development) is the focus in these companies, I call them "quality oriented."

[7] *Ibid.*, 29.

companies that made up the total of the companies *directly* studied. (The companies in which financial outcome was the dominant concern were studied through the reports of engineers and managers who had *formerly* worked for such companies, and so evidence about their interest in having engineers fight for quality or safety is largely unexplored.) Even managers at customer-oriented companies, in which they expected to sometimes overrule the engineers on matters of quality, if not on safety, wanted to hear the strongest case for quality from their engineers. The engineers studied generally felt their safety judgments were accepted. The managers studied stressed the importance of appreciating the engineers' evaluations to doing their own (managerial) jobs well.

Many factors influence the relationships between engineers or scientists and their supervisors in corporations. The relationship of an engineer or a scientist to a supervising manager in a corporation is quite different from the relationship of a graduate student in engineering or science to a research supervisor, for example. Graduate students are somewhat like medical students in relation to their research supervisors, except that the relation to the thesis supervisor dwarfs all other supervisee relationships that graduate students have. Both graduate and professional students are in a vulnerable position in case of conflict with those supervisors. Despite the differences, issues of not only quality, product safety, productivity, and customer or sponsor satisfaction but also laboratory safety, harassment, prejudice, and the hostile work environment are common to university, agency, and corporate settings. University or departmental cultures, corporate cultures, and agency cultures all vary in the support they give for raising of ethical concerns, and in their willingness to monitor or control the activities of their members. Organizations vary significantly in their policies and procedures for resolving conflicts about reading employee or student email or computer files stored on the university, corporate, or agency computers, or subjecting their members to drug and other biological testing.

> Even good patterns of communication may fail on occasion. Recognizing the importance of heeding the warnings of engineers, many large U.S. companies have instituted complaint procedures or "hot lines" to ensure both that difficulties are recognized and appropriately addressed and that those who raise concerns in *good faith* are protected from retaliation.

The study by Davis and his colleagues of communications between engineers and managers reveals how managers respond to unwelcome news from engineers in high-tech companies. Even good patterns of communication of the sort revealed in the study by Davis and his colleagues may fail on occasion. Recognizing the importance of heeding the warnings of engineers, many large U.S. companies have instituted complaint procedures or "hot lines" to guard against such failures. Good complaint procedures ensure both that difficulties are recognized and appropriately addressed and that those who raise concerns *in good faith* are protected from retaliation.

*How do managers relate to engineers in good companies and how does this contrast with the relations between managers and engineers at less good companies? Is the difference important to the work situation you want? What are some ways of finding out about how managers and engineers in a company relate before you accept a position?*

## Organizational Complaint Procedures

Suppose you have an ethical concern, but the person or office to whom you are supposed to take your concern is unresponsive. Is there anything you can do other than keep quiet, quit, or "blow the whistle"?

"**Complaint procedures**" may sound vaguely repellent, because of the negative connotations of "complainer," but "complaint procedures" is the general term for the procedures by which organizations ensure the ability to hear inconvenient truths. (An organization may have several different sorts of complaint procedures for different sorts of complaints. For example, universities generally have some sort of research integrity office for dealing with allegations of research misconduct and only such concerns.) "**Complainant**" rather than "complainer" is the term for someone who uses such procedures.

Frequently the occasion for a complaint by an engineer is a difference in judgment rather than an accusation of malfeasance. (For that reason, some of the procedures in question are called "dispute resolution procedures.") Some disagreements stem from reasonable differences of opinion, some from innocent mistakes, others are due to someone's negligence or, more rarely, from evil intent. Often what is morally blameworthy is not an initial mistaken judgment, but *the failure to heed arguments and evidence brought forward to show that a judgment is mistaken*. Failing to heed arguments and evidence is a way in which an unresponsive organization often transforms simple mistakes into negligence.

> Often what is morally blameworthy is not an initial mistaken judgment, but *the failure to heed arguments and evidence brought forward to show that a judgment is mistaken*.

By having good complaint procedures, an organization can ensure that bad news is not repressed. Not all complaints are ethically significant or even well founded. The ethics officer of one large high-tech company said that the majority of complaints that came to her office were about food in the cafeteria. Engineers who have worked at that company assure me that the food at that company is not bad. Food is something that people readily complain about, however. Scattered among the food complaints are matters that really require the attention of the ethics office.

In 1986, in response to public outcry about the high cost of items obtained under defense contracts and about outright financial fraud, defense contractors formed the Defense Contractors' Initiative, which established standards for government contractors to handle employee complaints, including those of financial fraud. As part of this initiative, participating contractors established complaint procedures and an office for handling concerns. (This is often called the "ethics office" although it is sometimes called a "compliance office." The difference in name reflects a significant difference in thinking about its function: whether it is intended to foster ethical values and conduct or simply ensure that the organization complies with regulations.)

Safe and effective complaint procedures come in many forms, some formally instituted and others arising de facto. Large organizations may provide separate routes for raising concerns about product safety, laboratory or worker safety, a coworker's substance abuse, misuse of funds or fraud, and questions of fairness in promotion or work assignment. In small companies or start-ups, the procedures may be entirely informal. Large companies may announce an "open door policy" in which employees may bypass lower layers of management and take concerns directly to the top, or may employ an "ombuds" or "**ombudsman**" whose job it is to remain neutral in controversies and to inform complainants of their options or facilitate their exercise. A start-up company in which everyone routinely deals directly with everyone else and good advice is available from many sources may have no need to announce an open door policy or designate an ombuds. Some companies legitimize the delivery of bad news through "screwup boxes." These work somewhat like suggestion boxes, and people may use them anonymously. Complaints are posted on bulletin boards, along with management's responses. Anonymity is a two-edged sword, because if the complainant is not identified, the complainant cannot be protected against reprisal from someone who guesses the complainant's identity and is offended by the complaint. For this reason many larger companies provide both anonymous and identified means of raising concerns.

> Anonymity is a two-edged sword, because if the complainant is not identified, the complainant cannot be protected against reprisal from someone who guesses the complainant's identity and is offended by the complaint. For this reason many larger companies provide both anonymous and identified means of raising concerns.

Whatever their form, complaint procedures must have certain characteristics if they are to work. Privacy theorist Alan Westin lists characteristics of complaint procedures that make them effective and eliminate the need for whistleblowing and litigation.[8] Though originally proposed for employees within companies, they apply as well to universities, government agencies, and hospitals or other organizations and to their employees, students, and trainees.

1. The complaint and appeals mechanism must fit the organizational culture.
2. The means of dispute resolution must inspire general confidence.
3. Top management must display continuing commitment and involvement in the process.
4. The organization must reward merit.
5. Formal procedures must guarantee the process, without creating a legalistic atmosphere.
6. The organization must continually emphasize the availability of channels.
7. Employees must have assistance to bring forward their complaints.
8. Someone must be the advocate of fairness itself, rather than of any particular group or position.
9. All who raise issues or give evidence must be protected from reprisal.
10. Line managers must support the procedures.*
11. The organization must accept the responsibility to change in response to what the process reveals.
12. The organization must, without violating privacy, make public the general nature of the problem, the procedure used to examine it, and the outcome.

[8] Westin, Alan F. 1988. *Resolving Employment Disputes without Litigation.* Washington, DC: Bureau of National Affairs.
*  **Line managers** are managers who make decisions central to the work of the company.

13. Probing employee surveys that actively seek concerns must supplement the concerns individuals bring forward.
14. Employee representation must be part of the process.
15. A fair dispute procedure must be established as permanent.

The fit between the complaint and appeals mechanism and the organizational culture is important for the mechanism to be accepted and trusted. For example, the high value placed on academic freedom in a university environment contrasts with the high value placed on obedience in the military. Complaint procedures in a university and in a military organization would necessarily reflect differences in practices of the two types of organization. Even organizations of the same general type may have their own traditions and ways of working that must be considered when setting up procedures.

The specific characteristics Westin identifies as desirable may conflict with one another in a given circumstance. For example, in the mid-1980s many research universities realized that their procedures to deal with complaints of wrongdoing or misconduct in research were inadequate. Complaints of misconduct had been mishandled and some complainants had experienced retaliation. Universities revised their procedures. Those universities in which decision making centers in departments often choose to handle at least the initial "inquiry" stage at the department level. Handling the inquiry at the departmental level fits with existing culture and practice at those universities, in accord with the first of Westin's characteristics, but that practice is at variance with his fifth and fifteenth requirements. Because charges of misconduct are infrequent, those within so small a unit as a department will almost all be dealing with such charges for the first time. Even very intelligent and conscientious people make mistakes when they are new at their task. What they learn from one experience does not become institutional learning if each department handles things in its own way and has little opportunity to benefit from the experience of others. Handling the inquiry at the departmental level violates the requirements that there be formal guarantees of the process, and that a fair process be established as permanent. Westin's criteria are useful for evaluating complaint procedures; the experience of

1. The complaint and appeals mechanism must fit the organizational culture.
2. The means of dispute resolution must inspire general confidence.
3. Top management must display continuing commitment and involvement in the process.
4. The organization must reward merit.
5. Formal procedures must guarantee the process, without creating a legalistic atmosphere.
6. The organization must continually emphasize the availability of channels.
7. Employees must have assistance to bring forward their complaints.
8. Someone must be the advocate of fairness itself, rather than of any particular group or position.
9. All who raise issues or give evidence must be protected from reprisal.
10. Line managers must support the procedures.
11. The organization must accept the responsibility to change in response to what the process reveals.
12. The organization must, without violating privacy, make public the general nature of the problem, the procedure used to examine it, and the outcome.
13. Probing employee surveys that actively seek concerns must supplement the concerns individuals bring forward.
14. Employee representation must be part of the process.
15. A fair dispute procedure must be established as permanent.

many disparate organizations confirms their importance. They provide valuable design criteria but are not a recipe for creating a procedure.

Before accepting a job in any large organization, one would do well to inquire into its complaint procedures. If engineers accepted positions only at companies with good complaint procedures, not only would those individuals be better off, but companies that did not have good procedures would go out of business. In assessing such procedures one would assess not only how they measure up to Westin's criteria and how well they fit the organization's culture but also the level of awareness of the procedures by members of the organization. A further indication that such procedures work is that employees are willing to talk about the company's procedures. If they say the "open door comes around and hits you in the rear," that is important information, as is any reluctance they show about giving you their opinion of the procedures.

Some ethically objectionable situations lend themselves to organizational solutions, others to technical solutions, still others to legal or legislative solutions. Some objectionable situations are so objectionable that everyone should speak out against them. Others are less grave and best left to those who are familiar with those situations or specially prepared to address them. In an imperfect world, many things call for reform. The question of which imperfect situations you should work to change, like the question of what profession to choose, or whom, if anyone, to marry, is a question *larger* than one of professional responsibility. It is a question of what is important in life and what responsibilities or vocation a person should take up.

> *Suppose you are considering whether to take a job with a certain company. Although it seems like a generally reputable company and the people in the group you worked with there last summer seemed ethically aware and concerned, you know it will be a big hassle if you have to quit. What can you do to find out more about the strength of its commitment to ethics and truth telling?*

## Government Agencies

Agency culture, good or bad, is often slower to change than that industry, but agencies, like companies, are coming to recognize the importance of programs to recognize employee concerns. The Department of Energy (DoE) is among them. It has established an internal avenue to receive all types of employee concerns; to ensure that these concerns are reviewed, referred, or investigated; and to guarantee the person who originates the concern an appropriate response. The DoE program encourages employees to resolve disputes with their first-line supervisor unless that individual is a factor in their concern.

### Difference of Professional Judgment within the Nuclear Regulatory Commission (NRC)

> What does the recent history of the NRC tell you about how well the organization can respond to bad news?

The NRC is the federal agency charged with oversight for the United States' 110 commercial nuclear reactors, and is the agency to which employees of nuclear power plants may report safety problems that go unheeded in their home facilities. In the mid-1990s, the NRC was found to be ignoring serious safety violations so that nuclear power plants could avoid shutting down or could operate more cheaply. George Betancourt and George Gatalatis,[9] two senior engineers at Northeast Utilities, tried to bring to light dangerous violations at their company's plant, Millstone Unit 1 in Waterford, Connecticut. The resistance they encountered from supervisors at Northeast Utilities led them, after 18 months, to take their concerns to the NRC, only to find that the NRC had been winking at such violations for years. Other engineers at Northeast who had objected to safety violations had also found little support from the NRC.

In 1987, the Nuclear Regulatory Commission had revised its previously existing procedures – called the "Differing Professional Views (DPVs) and Differing Professional Opinions (DPOs)" programs – for dealing with employee concerns.

In 1994, the U.S. NRC reviewed the Differing Professional Opinion/View Program (DPO/DPV). (Transcripts of NRC meetings before 1996, and therefore the 1994 report on the DPO/DPV, are not available on the NRC Web site.) A leading engineer in engineering ethics, Stephen Unger, considered the review report in the second edition (1994) of his book, *Controlling Technology*. The evidence that most workers who have used the new procedures would not do so again led Unger to conclude that the NRC procedures worked poorly even after the 1987 reform.[10] Several of Unger's criticisms focus on the aspects of the procedure Westin finds crucial. Unger found no general confidence in the means of dispute resolution, and, specifically, no conviction that those who raise issues or give evidence will be protected from reprisal. Westin's twelfth requirement of making public the nature of the problem, the examination of it, and the outcome may be compromised as well, for in the name of national security, the NRC is required to make public only portions of a case. Unger was concerned that this loophole would allow the NRC to cloak blunders or manipulate public opinion.

In 2002, the U.S. NRC again reviewed the Differing Professional Opinion/View Program (DPO/DPV). The transcript of that review meeting is available at http://www.nrc.gov/reading-rm/doc-collections/commission/tr/2002/20020813.html.[11] According to that 2002 report, it drew heavily on a 2000 IG* audit of filings to the DPO/DPV. Key elements in the report given at that meeting are:

> The [DPO/DPV] panel chairman and staff members, members of the panel considered it a good means of evaluating differing views. Filers [i.e., complainants], on the other hand, exhibited a wide divergence of opinion. Many found or felt that it was a valuable experience, that it did adequately treat their concerns. Others, however, were very critical of the process. They were frustrated by the timeliness or

[9] Pooley, Eric. 1996. "Nuclear Warriors." *Time*, March 4, 46–54. Wald, Matthew L. 1996. "Two Northeast Utilities Plants Face Shutdown." *New York Times*, March 9.
[10] Unger, Stephen. 1994. *Controlling Technology*. New York: John Wiley, 226.
[11] Page dated August 13.
* I interpret "IG" to mean the inspector general.

lack thereof, the quality of the ad hoc panel, in their opinion, and lack of feedback that they experienced.

. . . .

With regard to the organizational climate, the senior managers felt that it was important to the Agency and they valued it and they respected it. There is, however, a perception among the filers that using the process is dangerous to your NRC career, (sic) that you might be subject to retaliation for indulging in it.

This is a persistent impression. This is an impression that was found as far back as the 1987 audit. And obviously, the perception, even if not founded in reality or actuality, taints the process. It certainly reflects poorly on the organizational climate.

With regard to understanding the process, we found that the Management Directive 10.159 is resorted to by the filers and that the managers use it as a helping hand in guiding them through the process and we found the process is generally well understood.

The Panel developed five findings with attendant recommendations. And the findings . . . are the current process lacks Agency level oversight. It is duplicative and could be made more effective. The time frames set forth in the Management Directive are not being met. No points for information exchange are identified in the process and open discussion of views is very important to NRC safety culture.

> Given the flaws in the NRC program, it is not surprising that the NRC persists in having a mixed record in supporting those who report safety violations at nuclear power plants.

The report's five findings are discussed in detail in the transcript at http://www.nrc. gov/reading-rm/doc-collections/commission/tr/ 2002/20020813.html.

Given the flaws in the NRC program, it is not surprising that the NRC persists in having a mixed record in supporting those who report safety violations at nuclear power plants.

*What does the recent history (through 2002) of the NRC tell you about the difficulties the organization has had in hearing bad news? We have seen that NASA did not learn from its mistakes that led to the* Challenger *explosion to avoid repeating many of them in the* Columbia *flight. How would you go about learning whether the NRC reformed its practices?*

### Professional Judgment in the American Forestry Service

As a practical matter, what incentives might an employee have for reporting bad news about something that will happen long after that employee has moved on to another position?

The handling of differences in professional judgment has been a sore point within the Forestry Service as well as the NRC. It lacks a culture that leads "line officers" (the counterpart of line managers in industry) to listen to their technical experts and to develop a complaint procedure with the characteristics that Westin describes. As Doug Heiken of the American Forestry Service Employees for

Environmental Ethics (AFSEEE) points out, communications and decision making are jeopardized at the Forestry Service because negative consequences are often far removed from the decisions that caused them. This circumstance combined with a policy of holding district rangers accountable only for the effects on a locale during their service at that locale undermines the incentive to make careful decisions. Below is an edited version of a scenario that Heiken gives to illustrate the problem.

---

### The Wrong Incentives

A Forestry Service hydrologist finds that her predecessor boosted timber targets by violating forest plan standards designed for the protection of watersheds, and now many of the watersheds in the district are in poor condition. The watersheds are healing, but could degenerate rapidly if there is greater than normal precipitation in the coming years. If bringing this bad news simply puts her in an unwelcome role, neither the hydrologist nor anyone else will want to pass it on. The hydrologist will not even want to recognize the danger herself. She has strong incentives to say nothing and simply hopes the rains will not be too heavy.[a]

[a] This scenario was distributed by Doug Heiken in a draft statement, "Making Whistleblowing Obsolete through Forest Service Reform," dated December 14, 1994 and distributed to the large AAAS Science and Engineering Ethics (AAASEST) email list, Thursday, March 2, 1995, 09:30:13-0800.

---

In the Forestry Service, the failure to heed evidence about risks does not lead to something like an explosion the following day, as it did in the *Challenger* disaster. Repressing bad news in the Forestry Service does not even lead to an event (like the collapse of a skyscraper) that even if delayed, will be readily traceable to the technical experts who made the errors that produced the catastrophe. As Heiken points out, judgments such as those to allow cattle to overgraze may have no dramatic effects for years and then a large rainstorm causes noticeable erosion and downstream flooding. Overcutting of old growth in the forest may be widespread before it is realized that certain wildlife species are rapidly declining. Heiken proposes assigning responsibility to managers for their decisions even when the consequences do not become apparent until later, even long after they have taken different positions. His general point is that accountability encourages people to deliver the bad news as soon as they learn of it.

> Accountability encourages people to deliver the bad news as soon as they learn of it.

Heiken's suggestions are in accord with Westin's criteria for good complaint mechanisms and Unger's observations about the weaknesses in NRC communications. However, as was shown by the engineer–manager communications that the IIT study found to work well, certain changes in routine procedures would lessen even the need to use complaint procedures, much less to "blow the whistle." Findings about changes in routine procedures are consistent with the finding that in organizations that are more authoritarian and less responsive, employees have little confidence that internal means of redress will prove effective. Therefore, in those less responsive organizations, employees are more likely to become whistleblowers.[12]

*As a practical matter, what incentives might an employee have for reporting bad news about something that will happen long after that employee has moved on to another position?*

## The Hanford Nuclear Reservation

If you know that engineers at some facility have been retaliated against in the past for raising important ethical issues, what would it take to restore your trust that you could raise issues of a similar nature at successor organizations (i.e., organizations that took over from the first), and why?

One famous whistleblower, Inez Austin, was an engineer employed by Westinghouse at the Hanford Nuclear Reservation, a nuclear weapons facility in Richland, Washington. (As we shall see in Chapter 10, the Hanford weapons facility is the site of the worst radioactive contamination in the United States.) In the summer of 1990, she refused to approve a plan that would have pumped radioactive waste from one underground tank to another, a transfer that risked explosion. She was subsequently harassed, sent for psychiatric evaluation, and had her home broken into. Her case brought attention to the abuse of complainants as well as to safety, environmental, and security lapses at the Hanford Reservation. In February 1992, Inez Austin, like Roger Boisjoly before her, was awarded the AAAS Award for Scientific Freedom and Responsibility for her exemplary efforts to protect the public health and safety. After many instances of abuse of complainants who reported threats of a nuclear accident or pollution of the environment with toxic chemicals and nuclear wastes, strong measures were needed at Westinghouse Hanford to begin to rebuild the trust of employees and of the public.[13]

A landmark study commissioned by Westinghouse Hanford Company and carried out by the University of Washington's Institute for Public Policy and Management in 1992 confirmed that severe retaliation had often followed the raising of a concern at the Hanford facility. This finding led to the formation of The Hanford Joint Council for Resolving Employee Concerns. Among the study's findings were that every complainant they interviewed was sincere and credible and that Westinghouse's practice of responding to whistleblowing incidents by commissioning security department investigations of the cases and sending whistleblowers for psychiatric evaluations was unwarranted. This retaliation

[12] Keenan, Paul, Lockhart, Paula, Elliston, Frederick, and van Schaick, Jane. 1985. *Whistleblowing: Managing Dissent in the Workplace*. New York: Praeger Scientific.
[13] I happened to witness one bit of continuing peculiarity when in the early 1990s a student working on the Online Ethics Center first attempted to contact Inez Austin about putting her story in the Online Ethics Center by writing to her *at the Hanford facility*. (It is now in the Online Ethics Center at http://onlineethics.org/cms/9090.aspx.) The letter was returned to us marked "Moved, not forwardable." I then used telephone information to obtain Inez Austin's home phone number in Richland, Washington, and found her living in the same house she had occupied for decades.

A few years later I took a photo of the welcome sign that proclaims Hanford's commitment to safety.

included multiple instances of illegal surveillance. In 1991 the inspector general of the Department of Energy, John C. Layton, had found an array of sophisticated eavesdropping equipment in the possession of Westinghouse and Battelle, another contractor at the Hanford site, and at other weapons sites in Idaho and South Carolina. Only state or congressional law enforcement agencies are allowed to use such surveillance equipment.[14]

The Hanford Joint Council was an innovative attempt to restore public trust and secure effective cooperation in accomplishing a difficult and dangerous cleanup, which received praise when it was formed.[15] It began considering cases in January 1995. The council was made up of three public interest representatives; two unaffiliated, neutral parties; an ex-whistleblower; and two managers from Westinghouse, the contractor for the Hanford Nuclear Reservation. This council was a chartered, nonprofit organization with no legal ties to Westinghouse Hanford or the Department of Energy. It had full endorsement from Westinghouse Hanford Company, the State of Washington Department of Ecology, and the U.S. DoE[16] and a commitment from the DoE to operate for a minimum of five years with the flexibility to expand its mandate and hear concerns from employees of other contractors from the Hanford site. (For years, it had a Web site at http://www.halcyon.com/tomcgap/www/hjc.html. The Government Accountability Project [GAP] maintained that Web site, *but that link is now dead*. The current GAP Web site, as of June 5, 2010, is at http://www.whistleblower.org.)

In 2005, the Hanford Concerns Council replaced the Hanford Joint Council[17] (see http://www.hanfordconcernscouncil.org/doc/council_history.htm). The Hanford Concerns Council proclaims its trustworthiness based on the diversity of the people who serve on it. The members are listed under various headings, including two company names (with eight members on the council), "Advocacy," "Neutral," (with three and four members, respectively), and two ex-officio positions for people from the U.S. Department of Energy. The chair, whose field is dispute resolution, chaired the Hanford Joint Council from 1994 to 2003. We have seen the intimate connection between professional responsibility and trustworthiness: To carry out the ethical responsibilities that go with being engineers,

---

[14] Schneider, Keith. 1991. "Inquiry Finds Illegal Surveillance of Workers in Nuclear Plants," *New York Times,* August 1, A18.

[15] See for example Walter Elden's "Resolving Ethical/Technical Dissent through Due Process," which was published in the August 1996 (vol. 6, No. 2) issue of *Engineering Ethics Update*, the newsletter of the National Institute for Engineering Ethics.

[16] See the Hanford Joint Council's Web site at http://www.halcyon.com/tomcgap/www/hjc.html. This Web site is maintained by the Government Accountability Project (GAP). The GAP is a public interest group that supports conscientious employees who seek to raise issues on waste, fraud, abuse, threats to public health, and worker safety and environmental hazards. The GAP has been in operation since 1977, and for a decade its attorneys have worked with whistleblowers from the Hanford Nuclear Reservation in Richland, Washington. The GAP reports that whistleblowers at Hanford had experienced "harassment and retaliation, ranging from management intimidation, security clearance revocation, professional blacklisting, dismissal, and even home break-ins."

[17] The Hanford Joint Council for Resolving Employee Concerns operated from 1994–2003, and was closed because one contractor/partner (Fluor) no longer wanted to participate. The Hanford Concerns Council was started in 2005 under an agreement between CH2M HILL Hanford Group, the Government Accountability Project (now Hanford Challenge), and the Department of Energy, Office of River Protection (email communication, June 28, 2009, in response to my query to the Hanford Council Web site).

Figure 7.1 Hanford Proclaims Its Commitment to Safety (Photo: Caroline Whitbeck)

engineers need not only to be trustworthy people but also must have trustworthy means of bringing forward concerns that arise out of their engineering knowledge and experience. Hanford has since proclaimed its commitment to safety – see Figure 7.1 – but restoring trust after it has been seriously betrayed is a difficult matter. A review body that represents all relevant perspectives provides one important mechanism for restoring trust.

> *If you know that engineers at some facility have been retaliated against in the past for raising important ethical issues, what would it take to restore your trust that you could raise issues of a similar nature at successor organizations (i.e., organizations that took over from the first), and why?*

## Disagreeing with Your Supervisor

### IEEE "Guidelines for Engineers Dissenting on Ethical Grounds"
Suppose you find yourself disagreeing with your immediate superior about whether some action on the part of the organization is ethically acceptable. How do you go about voicing your concern or otherwise acting on it?

In 1996, the IEEE Ethics Committee issued "Guidelines for Engineers Dissenting on Ethical Grounds."* Although its advice is broadly applicable to any situation

---

*Although these guidelines were revised in 2002 by the successor IEEE committee, the IEEE Ethics and Member Conduct Committee, that revision did not change the advice given, but only eliminated a statement to the effect that the IEEE might give further support to engineers with ethical concerns. (The NSPE currently has a "hot line" to give ethical advice to its members, but the IEEE has ceased offering such a service, and now has a clause in its charter that forbids it to offer such advice. For a period the IEEE "Guidelines for Engineers Dissenting on Ethical Grounds" were available at http://www.ieee.org/web/aboutus/ethics/dissent.xml, on the pages of the IEEE

in which someone in a science-based profession dissents from decision makers in an organization, it is specifically addressed to engineers acting on ethical concerns. Its advice is born of deep and broad experience with ethical problems in engineering, and many points in the guidelines mirror actions that Roger Boisjoly took in seeking to prevent the *Challenger* explosion.

Ten guidelines are given:

1. Establish a clear technical foundation
2. Keep your arguments on a high professional plane
3. Try to catch problems early, and work with the lowest managerial level possible
4. Make sure that the issue is sufficiently important
5. Use organizational dispute resolution mechanisms
6. Keep records and collect paper
7. Resigning [pros and cons of]
8. Anonymity [pros and cons of]
9. Outside resources
10. Conclusions

The stated goal of these guidelines is "to provide general advice to engineers, including engineering managers, who find themselves in conflicts with management over matters with ethical implications." As it accurately observes, "much of this advice is pertinent to more general conflicts within organizations." For this reason, the guidelines are doubly valuable for U.S. engineers, who unlike physicians and lawyers in their country, are expected to make the transition from high school to full professional status in about four years. The writers explicitly assume that those consulting the guidelines will have the dual objective of preventing some serious harm (an ethical value), while minimizing career damage (a prudential concern).

> The writers explicitly assume that those consulting the guidelines will have the dual objective of preventing some serious harm (an ethical value), while minimizing career damage (a prudential concern).

The IEEE Ethics Committee, which authored these guidelines, envisioned the sort of case that leads an engineer to consult the guidelines as one which two desirable goals could not be satisfied simultaneously and an elegant solution described in Chapter 1 is not fully achievable (although the guidelines do help the dissenter to satisfy many constraints simultaneously). The example it gives is of a situation in which adequate testing of a product will cause one to miss a deadline. Because the failure to meet the deadline will be a highly visible failure with clearly defined penalties, but failure to test adequately will not, it will be tempting to proceed with less than adequate testing. If one's immediate supervisor wants to give in to the temptation but the company has a deep commitment to safety and quality, then if carried out wisely, one's dissent is likely to succeed

Ethics and Member Conduct Committee, but those excellent guidelines no longer appear on the IEEE Web site (as of August 17, 2010). Copies of the original version (1996) supplied by one of the guidelines' authors are still available at http://www.onlineethics.org/Resources/ethcodes/EnglishCodes/IEEEguidelines.aspx http://temp.onlineethics.org/codes/guidelines.html.

within the company. If upper management is also easily tempted or self-deceived, then whistleblowing may become an issue.

The first of the guidelines' recommendations is to "establish a clear technical foundation" for one's dissent. It warns that this "does *not* mean that you must be able to validate your position with near mathematical certainty." It is enough that you have good reason for your concern.

In connection with this first guideline, the committee recommends several additional elements:

Get the advice of colleagues
Carefully consider counterarguments
Be willing to revise your position if arguments or evidence convinces you that
    you should

The committee is working from the experience of a great many engineers and it suggests how to implement their recommendations. For example, to help engineers carefully consider counterarguments, it suggests, "A good way to ensure that you understand someone else's position is to restate it to the satisfaction of that person." (Careful consideration of others' views also shows respect and makes them more likely to consider *your* position.) Seeking the advice of colleagues may also persuade them of your concern, but make sure not to demand that they become advocates of you or your concern. One factor that can make people leery of dissenters, especially in a situation that has become polarized, is the risk that dissenters who do not get total agreement and vigorous support for their position will turn on those advisors, see them as advocates of "the other side," and attack those would-be advisors.

> Seeking the advice of colleagues may also persuade them of your concern, but be sure not to demand that they become advocates of you or your concern. One factor that can make people leery of dissenters, especially in a situation that has become polarized, is the risk that dissenters who do not get total agreement and vigorous support for their position will turn on those advisors, see them as advocates of "the other side," and attack *those would-be advisors.*

The second guideline is about formulating your concern. It advises one to keep "your arguments on a high professional plane, as impersonal and objective as possible, avoiding extraneous issues and emotional outbursts. For example, do not mix personal grievances into an argument about whether further testing is necessary for some critical subsystem." It advises against *impugning the motives* of others. This is important even if you *are* suspicious of others' motives, because impugning those motives adds nothing to your technical case and makes it harder to achieve another objective that the committee emphasizes, namely *minimizing the embarrassment* to those who are being asked to change their position. On the subject of minimizing embarrassment to others, recall from Chapter 2 that Roger Boisjoly showed his memo (the memo to the Morton Thiokol vice president) to his direct supervisor when he went to a higher level of management. Informing your immediate supervisor of your actions is a wise policy (unless it is the character of the immediate supervisor, rather than some decision of the supervisor or others, that is the matter of concern), because creating an unpleasant surprise for the supervisor cannot help your cause and telling the supervisor that you are taking the issue higher may even cause her to rethink any part she had in the situation.

> Avoid extraneous issues and emotional outbursts. For example, do not impugn the motives of others even if you *are* suspicious of others' motives, because impugning those motives adds nothing to your technical case and makes it harder to achieve another objective, namely *minimizing the embarrassment* to those who are being asked to change their position.

The committee rightly emphasizes *not overstating your case*. Overstating or exaggerating seriously undermines one's credibility. The virtue of honesty (and a reputation for it) is best practiced consistently, apart from raising any particular concern. (In my experience, a favorite form of exaggeration among undergraduate students who want to be exempt from some class requirement is to assert something to be true of large numbers of other students in the class. With surprising frequency, what they assert about other students is something that, as the instructor, I am in a position to know to be false. The students who are prone to such exaggeration probably do not know that their assertion is false, but they have no basis for it and are just *hoping* that it is true. When their assertion does not "fly," they seem to think no more about it. Such reckless assertions do undermine their credibility, however.) Pre-law and pre-med students have seven years or more to make the transition from high school to professional life. Engineering students have to learn standards of professional conduct much faster. If a habit of reckless assertion is taken into the engineering workplace, it can have serious consequences, at least for the engineer's own career and perhaps for the public welfare.

> Avoid overstating your case. Overstating or exaggerating seriously undermines one's credibility.

The committee adds to the second guideline the warning that if "the matter turns into a serious conflict, efforts will be made to portray you as some sort of crackpot. Avoid behavior that could be used to support such an attack." In my experience, this advice is especially valuable for those in graduate school, because in *every* one of many instances that I have seen of conflict between a faculty member and a graduate student or post-doc, the graduate student or post-doc has been alleged to be unbalanced. Some may have actually been unbalanced, but it is unlikely that all were and so the accusation is too readily applied to trainees.

The third guideline advises engineers to seek to "catch problems early, and keep the argument at the lowest managerial level possible." We will see the advice about handling matters at the lowest managerial level echoed by corporate advice in the *Gray Matters* game. Dealing with a problem at an early stage usually makes it easier to solve, and at an early stage it is not usually appropriate to take one's concern very far up the management ladder. Raising the issue at an early stage, even if it means dealing with many unknowns, also may prevent others from taking positions from which they may be reluctant to later retreat for fear of losing face.

The fourth guideline advises engineers to "make sure that the issue is sufficiently important" before "going out on a limb." "Out on a limb" is an exposed and risky place to be. As we observed earlier, in an imperfect world, many things go wrong. If one asks for attention to every minor imperfection, others will stop listening. The committee does not counsel self-sacrifice, but rather that the engineer consider how important the matter is and whether it warrants taking great risks. It considers that if a matter involves only financial risks for the employer, dissenting from a manager's unreasonable decisions is not worth risks to your career.

Fifth, the guidelines advise engineers that if managers are unresponsive to engineers' concerns and there is no powerful figure who is able to mediate a discussion with their managers, the engineers in question should make use of any organizational dispute resolution mechanisms that are available. Using dispute resolution mechanisms, including grievance procedures, "will almost certainly damage relations with your manager;" therefore the committee urges one to take this step "only after a careful review along the lines discussed in guidelines 1 and 2." Using such dispute resolution procedures does show one's willingness to use any avenues that one's organization provides, and that may be very important.

Further, it advises that if there is no dispute resolution mechanism, you consider championing the creation of a good one, although it admits that doing so would be difficult while you are in the midst of pursuing a concern.

Sixth, the guidelines advise you to keep written records "as soon as you realize that you are getting into a situation that may become serious." The records it mentions include a log in which you record the "steps that you take (e.g., conversations, email messages, etc.)" with times and dates. It advises that to the extent permitted by law, "you keep copies of all pertinent documents or computer files at home, or in the office of a trusted friend – to guard against the possibility of a sudden discharge and sealing off of your office."

> Dealing with a problem at an early stage usually makes it easier to solve. It also may prevent others from taking positions from which they may be reluctant to later retreat for fear of losing face.

The seventh guideline considers the question of whether to take the steps of resigning or of "blowing the whistle," if you are unable to resolve the conflict with your organization. It advises that unless you have a job that is protected by civil service or the like, it is unlikely that you could stay at your organization once you are known to have taken your concern outside.

Resigning has pros and cons. The positives the IEEE committee identifies are that

- It adds credibility to your position – makes it obvious you are a serious person.
- It cannot be argued that you are a disloyal employee if you are no longer an employee.
- You may be fired; in which case, resigning may look better on your record.

The negatives the committee identifies are that

- Once you are gone, it may be easier for the organization to ignore the issues you raised, as others in the organization may be unwilling to carry on the fight.
- The right to dissent from within the organization may be one of the points you wish to make.
- You thereby lose pension rights, unemployment compensation, and the right to sue for improper discharge.

It advises that it would be wise to consult an attorney before making a decision about resigning.

The 2005 example of the resignations of Drs. Susan Wood and Frank Davidoff from their positions with the U.S. Food and Drug Administration over rejection of their advice for what they saw as political ends illustrates that if one has a position that is in the public eye, resignation itself may be a form of whistleblowing and conscientious refusal.[18] In this case, their resignations were over an issue that was already a matter of public record, so Davidoff and Wood did not need to disclose any information to the press in order to explain why they were resigning. In contrast, a scientist or engineer at a company would be less likely to have the matter over which they were resigning be a matter of public record. Disclosing the information to the press, as contrasted to a body charged with oversight, such as the Environmental Protection Agency (EPA) or the Occupational Safety and Health Administration (OSHA), would be whistleblowing of a sort likely to make future employers wary of one's judgment and doubt one's loyalty. In contrast, Wood and Davidoff appeared to be people who tried to work within their organization but who stood on principle and sacrificed a job rather than play along with what they saw as political intrusion on a matter that should be settled by the science. The resignation did add credibility to their position; it made it obvious that they were serious people.

The eighth guideline considers trying to blow the whistle anonymously to someone who may be able to take action: "a regulatory agency, a senator, or a reporter." Anonymity seems to have the advantage of protecting the engineer from retaliation, but to make the concern credible, the engineer may have to provide so much information that the source of the complaint can be identified. As the committee points out, if you are discovered to have made an anonymous report, that discovery may precipitate more damage to the engineer than openly making the report. It also warns that *a reporter* may seek to sensationalize the story and so distort it in such a way as to jeopardize satisfactory resolution of the situation. It advises that *if one decides to talk to a reporter one should be especially careful not to malign any individuals and to state how one's claims can be verified*, and, as it says in guideline ten, "When given a choice among media organizations, choose those with reputations for fairness and accuracy." The committee also advises that you "take special pains to be accurate and clear when dealing with journalists so as to minimize sensationalism and distortion." To this advice I would add that if one speaks to a reporter it is best to choose one who has a record of careful reporting, because my own experience is that even reporters within the same organization will vary in their interest in accuracy.

The ninth guideline discusses to whom (outside one's organization) one might take one's concerns, if internal complaint and conflict resolution procedures fail. Like the advice in the *Gray Matters* Mini-case 58, the IEEE guidelines look first to the appropriate regulatory agency (or law enforcement agency, if your

---

[18] *Reuters*. 2005. "Second Expert Resigns over FDA Delay." October 6 (retrieved from http://www.cnn.com/2005/HEALTH/10/06/contraceptive.resignation.reut/index.html on October 6, 2005).

organization's actions break the law). (Of course, if you know the regulatory agency is so seriously underfunded and understaffed that it is a decade behind in its investigations, you can conclude that that regulatory agency is not likely to help.) The IEEE guidelines list as the next appropriate places to take one's concern: "Members of Congress (from one's own district or state, or the head of a relevant committee), state or local government officials or legislators, or public interest organizations. Of course some combination of these might be chosen." When the media do pick up the story, the committee advises engineers to "take special pains to be accurate and clear when dealing with journalists so as to minimize sensationalism and distortion."

The committee also recommends that engineers seek guidance and support from their professional societies (although, in my experience, engineering and science societies vary considerably in their willingness to provide their members with advice and support, with the NSPE being the most proactive of the engineering societies).

Finally, the committee repeats its advice to consult an attorney, but cautions that in considering the advice of attorneys, "one must take into account the tendency of attorneys to discourage any acts accompanied by legal risks." It is best not to *threaten* to consult an attorney *before doing so*, because that is only likely to induce your organization to turn the matter over to their attorney and tell no one to talk with you.

Having served on many review bodies, including university grievance panels, I am impressed by the wisdom of the guidelines' advice and how often otherwise intelligent people get themselves in trouble by failing to follow one or another provision of them.

Rather than summarize the resources that the IEEE Ethics Committee assembled in 1996, I refer you to a more recently updated list at http://www. ethicscasediscussions.org/forum/resources.

*Suppose you disagree with your immediate superior (or perhaps the whole organization that employs you) about whether some action on the part of the organization is ethically acceptable. How do you go about voicing or otherwise acting on your concern?*

### Employment Guidelines from Engineering and Scientific Societies
What are your rights, obligations, and responsibilities vis-à-vis your employer?

For a new engineer or scientist the first chance to gain an impression of the organizational culture of a potential employer is usually the job interview. The "Guidelines to Professional Employment for Engineers and Scientists," was adopted by the IEEE and other signatory organizations in 1975. It provided a framework of expectations for both employees and employers. The guidelines have since been independently revised by some of the original signatory organizations, including the ACS. Many of these revisions are useful for assessing the ethical climate at a potential employer and for understanding one's own obligations in conducting a job search.

The original edition of these guidelines received wide endorsement by about thirty professional societies of engineers and scientists beginning with the NSPE, the American Institute of Chemical Engineers, American Society of Civil Engineers, American Society of Mechanical Engineers, and the Institute of Electrical and Electronics Engineers. (Some of those endorsers have now revised those guidelines and claim subsequent editions as the property of that society.*) Many employers as well as employees contributed to formulating these guidelines, so they do represent what employers and engineering societies agree is good professional conduct by engineers and employers. The guidelines make good reading in preparation for transitioning from college to the professional world and are an appropriate topic for discussion in an extended job interview.

The objective of the guidelines is to help professional employees and employers establish a climate that enables them to fulfill their responsibilities and obligations. The guidelines list a wide range of responsibilities of professional employees and their employers. These responsibilities are grouped under four headings: recruitment, employment, professional development, and termination and transfer. The topics they cover range from already familiar engineering codes of ethics to expectations about performance reviews.

The original guidelines and the IEEE second edition identified the following as prerequisites for an ethical climate that supports the fulfillment of responsibilities:

1. A sound relationship between the professional employee and the employer, based on mutual loyalty, cooperation, fair treatment, ethical practices, and respect
2. Recognition of the responsibility to safeguard the public health, safety, and welfare
3. Employee loyalty and creativity in support of the employer's objectives
4. Opportunity for professional growth of the employee, based on employee's initiation and the employer's support
5. Recognition that discrimination due to age, race, religion, political affiliation, or sex should not enter into the professional employee-employer relationship. There should be joint acceptance of the concepts that are reflected in the Equal Employment Opportunity regulations.
6. Recognition that local conditions may result in honest differences in interpretation of and deviations from the details of these guidelines. Such differences should be resolved by discussions leading to an understanding which meets the spirit of the guidelines.[19]

The employment guidelines are intended to draw as clear boundaries as possible between behavior that is ethically acceptable or desirable and that which is not and

---

\* The second edition, which was substantially unchanged from the first, is available at http://temp. onlineethics.org/codes/empintro.html. The 2003 IEEE revision of those guidelines is available at http://www.ieeeusa.org/volunteers/committees/ECSC/06MarAgenda/ProfessionalGuidelines. pdf). The ACS 2008 revision (*its* eighth edition) is at http://portal.acs.org:80/portal/ PublicWebSite/careers/ethics/CTP_004009.

[19] The foreword to the guidelines states, "Where differences in interpretation occur, they may be referred to the headquarters office of any of the endorsing societies."

to give general guidance in the many areas where discretion must be exercised. They recognize some clear abuses, such as the acceptance of one job and then refusing the job to take another offer, or summarily rescinding a job offer that an applicant has accepted. Such behavior breaks a commitment, violates trust, and may seriously harm the other party's career or staffing plans. (On this point the guidelines state, "Having accepted an offer of employment, the applicant is morally obligated to honor the commitment unless formally released after giving adequate notice of intent," and "Having accepted an applicant, an employer who finds it necessary to rescind offer of employment should make adequate reparation for any injury suffered.")

> The employment guidelines are intended to draw as clear boundaries as possible between behavior that is ethically acceptable or desirable and that which is not and to give general guidance in the many areas where discretion must be exercised.

The original version also disapproved of such practices as employers forcing employees to promise that if they leave their company they will not work for a named competitor, and the current IEEE version urges that noncompete restrictions "be introduced only with the minimum requirements for the relevant employee." The original and revised provisions are balanced with a strong statement of an employee's obligations not to divulge proprietary knowledge of a former employer. Although it is clear that one should not tell trade secrets of one employer to another, there are many subtle issues about proprietary information. We dealt with some of these in the previous chapter but others remain such as: Under what, if any, conditions should one refuse to work in an area similar to that in which one had worked at a previous employer? Under what, if any, conditions (other than legal requirement) should one refuse to work for a competing company?

*Consider your rights, obligations, and responsibilities vis-à-vis your employer as outlined in one of the major sections of the Guidelines for Professional Employment, such as Professional Employment or Employment Separation, and evaluate its advice.*

## Organizational Control and Individual Privacy: The Biological Testing of Workers

What privacy rights do you have and how are these weighed against your employer's right to know how well you are performing on the job?

Some matters of the employer-employee relationship remain controversial and are not covered in the employment guidelines just discussed. One of the most controversial areas is that of sacrificing the **privacy** of employees to an employer's need or desire to have information about the employee. As was remarked earlier, organizations vary greatly in policies on such matters as whether they may read employee email or the computer files that students or employees store on the university, corporate, or agency computers.[20] **Biological testing** is another area of conflict between an organization's interest in knowing and an employee's right to

---

[20] The Electronic Frontier Foundation monitors the issue of the privacy of employee and student computer files.

privacy.[21] On the horizon is the question of whether companies should be allowed to use DNA and other biological testing to exclude from the workplace workers who would be genetically predisposed to occupational diseases associated with contaminants in their workplace.

The refusal in 1995 by two young U.S. marines, John Mayfield and Joe Vlacovsky, to give DNA samples for a military DNA registry brought national attention to DNA data banks. (The two were court-martialed for disobeying a direct order. The final disposition of the case was that the two marines were given a reprimand and confined to the base for seven days, but they received honorable discharges and full veteran's benefits and kept their DNA.[22]) The U.S. military started this data bank in 1991 and planned for it to contain DNA samples from all enlisted personnel.

Because the infringement of privacy rights with which engineering students are most likely to be familiar is drug testing by employers or potential employers, that is the example we will consider in some detail. Drug testing raises issues of both the justification for acquiring information from the test and the demeaning circumstances of the testing procedure. The issues are separable even if a reliable test result requires testing conditions that are only *somewhat* demeaning. (The test is clearly not worth doing if results can be easily faked, but the measures taken to ensure that results are not faked, from shutting off the water in the restroom sinks, to the more stringent measure of having the giving of the sample witnessed, make the test more demeaning.) Why *do* organizations from the armed forces to high-tech corporations want to conduct drug testing, and when, if ever, is the infringement of privacy justified? If justified, how is the test best conducted to be fair and respectful of those affected?

"Screening" is testing of a large number of individuals designed to identify those with a particular characteristic or biological condition. Random or universal testing is screening. Testing a person because there is reason to believe he has some condition is *not* screening.

**Screening** for drugs is the drug testing of people whom the tester has no reason to think are using drugs. The first and most obvious justification given for screening for drugs is that subtle impairment of workers endangers the public safety. The argument of public safety is plausible only for occupations in which response time and coordination are critical to the safety of others and affected by small doses of psychoactive substances. Air traffic control, truck driving, surgery, and piloting aircraft are all

---

[21] See the following works for more on biological testing in the workplace:

Bird, Stephanie and Jerome Rothenberg. 1988. "To Screen or Not to Screen: Drugs, DNA, AIDS," unpublished manuscript.

Murray, Thomas. 1983. "Warning: Screening Workers for Genetic Risk," *Hastings Center Report* 13(1): 5–8.

Ashford, Nicholas A. 1986. "Medical Screening in the Workplace: Legal and Ethical Considerations." *Seminars in Occupational Medicine* 1(1): 67–79.

Rothstein, Mark A. 1987. "Drug Testing in the Workplace: The Challenge to Employment Relations and Employment Law," *Chicago-Kent Law Review* 63: 683–743.

Murray, Thomas H. 1992. "The Human Genome Project and Genetic Testing: Ethical Implications." In *The Genome, Ethics, and the Law*, AAAS-ABA National Conference of Lawyers and Scientists. Washington, DC: AAAS.

[22] Mnookin, Seth. 1996. "Department of Defense DNA Registry Raises Legal, Ethical Issues." *Gene Watch* 10(1) (August): 1, 3, 11.

such occupations. Although engineers and scientists sometimes perform work in which response time and coordination are critical, that is not usual. Let us look for possible justifications of the general testing of the engineering and science workforce.

Substance abuse by workers in safety-critical jobs poses a hazard to public safety. Apart from that, substance abuse by any workers reduces productivity. The Research Triangle Institute in North Carolina estimates that in 1983, productivity losses due to drug abuse were $33 billion and losses due to alcoholism were $65 billion. Health care costs resulting from substance abuse are estimated at another $9 billion. (These figures do not include the even greater costs to substance abusers and their families.) The loss to companies takes many forms – from absenteeism, which is four to eight times higher for alcohol abusers than for nonabusers, to fatal accidents, 40 percent of which are attributable to alcohol abuse.[23] Furthermore, companies can be held legally liable for what an affected employee does outside the workplace. In one case, an employer was found liable for sending home an intoxicated employee who while driving home hit two people and killed them.[24]

Clearly, employers have an interest in reducing productivity losses. Drug testing helps to identify substance abusers. Furthermore, testing deters drug use. Drug testing could be done on urine, blood, saliva, breath, hair, or brain waves, but urine tests are most common.

Tests are imperfect and each has a false negative and false positive rate associated with it. These are rates at which the test, when *correctly* done, will give the wrong indication. The **false negative rate** is the proportion of tests in which traces of a drug or its characteristic metabolites *are* in the sample tested but not detected by the test.[25] **False positive rate** is the proportion of tests in which the test result is positive but due to factors *other* than drug use. In addition to the question of the fairness of testing without cause, another important issue of fairness is that of appropriate protection for employees who might have false positive results. In one notorious case, two members of the armed forces were court-martialed and dishonorably discharged after testing positive for opiates. It was eventually established that their positive tests were due to the poppy seeds on bagels they had eaten the morning of the test.

To some extent, the percentage of false positives will vary inversely to the percentage of false negatives in a test depending upon the *concentration* of a detected substance that *counts as a positive* result. As one might expect, the tests that have both low false positive and low false negative rates – that is, the tests are best in identifying those and only those who are using drugs – are also the ones that are more costly to use.

---

[23] Lachman, Judith A. *Issues in Management, Law and Ethics,* chapter 22, unpublished manuscript. See also *Toward a National Policy on Drug and AIDS Testing: Report of Two Conferences on Drug and AIDS Testing,* Washington, DC, October 20–21, 1987, and Racine, Wisconsin, March 8–10, 1988, Washington, DC: Brookings Institution, 1989; and Walter E. Scanlon. 1980. *Alcoholism and Drug Abuse in the Workplace: Employee Assistance Programs.* New York: Prager.

[24] *Otis Engineering v. Clark,* 668 S.W. 2d 307 (Tex. 1983)

[25] The length of time that traces of a given drug or its characteristic metabolites can be found in the urine after use depends on a host of factors including drug metabolism and half-life, the user's physical condition, fluid intake, and method and frequency of drug use.

Most companies that conduct random or universal screening of their employees for drug use test all samples with an inexpensive test and then retest any positive samples with a more expensive test. This procedure reduces the likelihood of false positive results without raising the likelihood of false negative ones.

For established employees, the usual consequence of a first-time positive drug test is referral to an "employee assistance program." Employee assistance programs are intended to help with the many problems and traumas that can interfere with job performance. Substance abuse, personal problems, marital difficulties, worry about one's own or a family member's health, and bereavement at the death of a friend or family member are among the situations that can lead employees to go to their company's employee assistance program. Many are problems that people cannot entirely avoid. Employee assistance programs offer counseling and other services to help workers cope with such difficulties and suffering. These are difficulties that can affect work performance. Ought an employer also screen for these difficulties? Is there a justification for screening only for substance abuse?

> Employee assistance programs are intended to help with the many problems and traumas that can interfere with job performance. Substance abuse, personal problems, marital difficulties, worry about one's own or a family member's health, and bereavement at the death of a friend or family member are among the situations that can lead employees to go to their company's employee assistance program.

For good or ill, few companies take steps to monitor their work force for problems other than substance abuse. Most large companies (and all government contractors) provide employee assistance programs, however. Referring people to employee assistance usually occurs after their work performance falters, although it might occur simply because a person is visibly distressed. Complete confidentiality is required for employees to trust using such programs. If the content of any session were not protected by confidentiality, employees would have the experience of being under surveillance. Such a perception would only add to the employees' stress and make them reluctant to use the service. For this reason, many companies subcontract employee assistance services so that the counselors in the program are not company employees. When employees are referred to an employee assistance program, it is usually up to the employees to follow through, although in the case of referral for positive drug tests, employees might be required to demonstrate that they did go to the service.

Except for the case of drug testing, few employers bio-screen their employees for signs that they are in difficulty. Indeed, some companies subject employees to drug *testing* only if the employees gave reason to believe that they have been abusing drugs. Other companies do subject their employees to random or universal drug screening, and many subject job candidates to a pre-employment drug test. What, if any, justification is there for putting drug use in a special category and screening employees for it?

If the U.S. government were to do random drug testing of its citizens, that would be a violation of the Fourth Amendment to the Constitution (in the "Bill of Rights"), which states:

The right of the people to be secure in their persons, houses, papers, and effects, against unreasonable searches and seizures, shall not be violated, and no warrants shall issue, but upon probable cause, supported by oath or affirmation, and particularly describing the place to be searched, and the persons or things to be seized.

The Bill of Rights applies to the rights of people vis-à-vis the *government*, but those rights also influence thinking about the rights of individuals more generally. A requirement of probable cause for a warrant to search someone's person for drugs is met if that individual gives *signs* of being under the influence of drugs, or gives *evidence* of having used drugs on the job. Arguably, it is met by the policy, used by some companies, of testing those involved in accidents. The requirement of probable cause is not satisfied for random or universal screening of employees.

Drug testing as a condition of employment is different from government screening of citizens in that potential employees have some freedom not to seek employment with certain companies. (One offensive feature of Ford's treatment of a student who was subjected to drug testing in the case described at http://www.onlineethics.org/CMS/workplace/workcases/riggsford.aspx was its failure to inform him that drug testing was a part of the plant trip. Had the company informed him, he could have decided if he wanted to take the trip under those conditions.)

Companies commonly require a pre-employment physical exam. Is the requirement of a pre-employment physical and a pre-employment drug test similar? Is either justified?

To answer these questions, we must know about the substances tested for, the accuracy of the tests, and the conditions of testing. Illegal drugs are usually included, together with any of a variety of prescription medications.

Some substances are common to both illegal drugs and prescription medications. Some people argue that managers may have a right to know about *illegal* drug use, but not to monitor the *details of one's health care*, and find the testing for prescription drugs more objectionable. The Americans with Disabilities Act protects workers with disabilities. Some workers require medications to cope with their disabilities. Does testing for such medications put them at special disadvantage or invade their privacy in a way that is discriminatory?

Some question a company's right to test for *illegal* drugs. Some argue that it is intrusive for an employer to go looking for legal violations that occur outside work time, which is when most detectable drug use occurs. The length of time that substances remain detectable in a user's urine varies greatly. Alcohol lasts not much longer than the significant influence on performance – from 4 to 12 hours; cocaine is detectable for 2 to 4 days and marijuana for several weeks (depending on use). Drugs like barbiturates, Valium, and Darvon are somewhere between these extremes of alcohol and marijuana. Drug testing might pick up marijuana used weeks earlier, perhaps while on vacation, and miss the alcohol abuse that affected the job performance earlier the same day.

If illegality itself is considered as sufficient justification for a drug test then why focus only on that one type of infraction? Would it be acceptable for an employer to go looking for other illegal activities by employees, say by searching records of unpaid parking tickets or using surveillance to discover employees' illegal gambling? Are such violations closely analogous to screening for illegal drugs?

One justification given for drug testing is that parties who receive government grants and contracts, as do many employers of engineers, must provide a "drug-free" workplace. Universities also receive such grants and contracts and make such certifications, however, without random or universal testing of their employees for drugs.

As we have seen, alcohol causes the greatest losses to employers, yet its traces disappear quickly. Alcohol intoxication is also more readily observable than marijuana intoxication. Why do companies not train managers to be alert to the signs of alcohol and substance abuse and, when they find such signs, *then* require testing? Substance abuse, like anxiety, depression, and other "employee assistance" problems, proves a difficult topic for many managers to raise. Rather than do intensive management training on signs of substance abuse, many companies simply use a technological fix and test their employees randomly or universally. Perhaps the requirement that supervisors (as well as managers) deal appropriately with issues of sexual and other harassment will increase the competence of supervisors to handle sensitive issues and open the way for personal rather than technological responses to substance abuse.

**Employee Rights in Refusing Drug Tests**

Some landmark legal decisions in California held that employees had rights of recovery against their employers if fired *precipitously* after refusing to take a drug test. Therefore, the right of an employer to test is limited by many countervailing employee rights.

Some organizations have respected employees' or recruits' objections that drug *screening* is an invasion of their privacy. In some cases, companies have waived requirements for pre-employment drug tests for students and recent graduates who have argued that without probable cause such a body search is demeaning and unjustified. When one high-tech company instituted random drug testing in the 1980s, a respected employee announced that he found such testing an affront and would never submit to drug testing. Refusal to take the test is supposedly grounds for dismissal at this company, but in the eight years that this drug testing has been in place the employee's name "has never come up" despite the supposed randomness of the selection.

Some of the arguments offered for drug testing have not been supported with ethical justifications but only arguments that what might be ethically desirable is not practical, at least as things now stand. Such arguments raise the further question of whether the present situation (including organizational practices, the technical limits of existing tests, legal guarantees for employers and employees) should, as a moral matter, be reformed, and if so, in what way.

*What rights of privacy do you have and how are these weighed against your employer's right to know how well you are performing on the job?*

## Limits on Acceptable Behavior and Resources for Resolution of Problems in a Large Corporation

What standards of behavior are considered ethical at the company/ies where you are considering working? What resources for addressing employees' ethical concerns about their work exist within those companies?

Companies, even those that screen their employees for drugs, do not simply put their interest in increasing productivity ahead of such employee rights as privacy. In fact, ethics training at reputable companies encourages employees to consider their own rights and interests and to further them so long as furthering them is consistent with other important values. The responses to problems that such companies favor are *not* the most self-sacrificing, but those that implement the following goals:

1. Promptly getting to the root of any difficulty
2. Preparing/educating managers to prevent difficulty
3. Protecting the public's interest, especially in health and safety
4. Respecting the law
5. Keeping the company honest
6. Protecting the company's *reputation* for honesty and fairness
7. Promoting trustworthiness and good working relationships among people in the company
8. Making appropriate use of organizational channels
9. Minimizing the aggravation that attends measures taken to meet the other criteria

Pursuit of these practical goals helps a company flourish. Considerations such as protecting the company's reputation and making appropriate use of organizational channels have ethical implications: Fulfilling them helps maintain the trust both inside and outside the company that is necessary to meet other, more obviously ethical, criteria. Ethics training in reputable high-tech companies often emphasizes developing the knowledge and discretion to design responses to meet all these criteria if possible, so everyone "wins."

The ethics materials from two large high-tech companies, both government contractors, Lockheed Martin and Texas Instruments, are good examples of such training materials. *Gray Matters*, for instance, is a game that companies use to teach ethics. George Sammet originally authored it for Martin Marietta, now Lockheed Martin. Other high-tech companies such as Boeing, Honeywell, McDonnell Douglas, and General Electric have since used the *Gray Matters* game. Lockheed Martin now uses that same content in a board game called "The Dilbert Game."

### Lockheed Martin's Gray Matters *Ethics Game*

In what, if any, respects do the judgments reflected in the scoring in the *Gray Matters* ethics game differ from your own? For any points of difference, what ethical justifications can you offer for yours and for those of the *Gray Matters* game?

The game consists of more than one hundred mini-cases that very briefly present ethically significant situations that call for a response. These range from observing a coworker snorting cocaine, to being instructed to mischarge your time, to communicating with subcontractors. The point of the game is to make employees aware of ethical problems that can arise in their day-to-day responsibilities, enable them to think through the consequences of their decisions and actions, and teach them what resources and company channels are best used in a large company to resolve the problem.

The point of the *Gray Matters* game is to

- Raise employee awareness of ethical problems that can arise in their daily responsibilities
- Enable them to think through the consequences of their decisions and actions
- Teach them what resources and company channels are best used in a large company to resolve the problem

Each mini-case is accompanied by four potential answers. (Usually the game is played in groups. The group discusses the case and comes to a decision about which answer is the best course of action.) The answers are scored (from −20 to +15) and an explanation or a rationale for the score is provided. The potential answers and evaluations of those answers inform employees about the company's values and standards on business ethics, develop their skills in applying company standards, and help employees find the best procedures for addressing a variety of ethical concerns within their company.

Consider mini-case 68:

You have been assigned to work on a proposal to the government. The proposal manager tells you and several other non-exempt workers that he'd like you to stay home Thursday and Friday and then come in and work Saturday and Sunday, but report that you worked on Thursday and Friday. That way, you would work 40 hours for the week, but the company would not have to pay you overtime for the weekend. "After all," he says, "proposal money is short."

What do you do?

The answers offered are:

A. Grudgingly comply thinking these days a job is a job.
B. Check with Human Resources to see if company policy permits this.
C. Call the ethics officer and allege unfair treatment.
D. Speak up immediately and question the manager's right to impose such a condition.

Answer A is scored −5 points with the comment "To go grudgingly along with a company imposition is not conducive to good morale. Isn't there a better way?" The negative points indicate that this is a mildly bad answer and the comment discourages employees from allowing themselves to be exploited.

Answer D, directly challenging the supervisor, is scored +5 points, that is, a moderately good answer, with the comment, "Certainly you are within your rights to do this."

Answer B receives the highest number of points of this set, +10 points, with the comment, "This is a sensible approach. If company policy doesn't permit this, it will be corrected. If it does, you have the facts needed to make a decision. Most companies and the government would consider this falsifying your timecard, thus denying the manager's right to ask you to do it." From this comment employees learn both that the practice requested by the proposal manager counts as *falsifying one's timecard (and thus is a form of financial fraud)* and that some companies may nonetheless allow the practice. Presumably, if your company did permit this practice, you could complain to the government. (It is excessive to "blow the whistle" over a *single occurrence* if it is the action of a *single individual and the company corrects the situation*.) In any case, the company *does not consider* blowing the whistle to the government in this instance, although it does in other mini-case answers. Thus in this instance it directs employees to seek *only internal* means of redress. If your company does allow chiseling, that would tell you something important about its ethics. Response B is the course of action that will get you information most quickly and with a minimal risk to yourself. The remaining answer, C, "Call the ethics officer and allege unfair treatment," is scored +5 points with the comment: "This response will take longer, but will eventually arrive at the same answer as 'B.'" Note that the stated reason for ranking answer B higher than answer C is *not* alleging unfair treatment is more aggressive than necessary, but that going through the ethics office in this case would *delay* resolution. This ranking reflects a concern for a *prompt* resolution.

Mini-case 15 is:

A coworker is injured on the job. You are a witness and could testify that the company was at fault. What do you do?

The answers provided are:

A. Don't get involved.
B. Contact the injured coworker and offer to appear on her behalf.
C. Report to the company what you saw to ensure that the safety hazard is corrected.
D. Protect the company by refusing to appear as a witness for the injured.

Not surprisingly, both answers A and D are scored –10 points. Answer A is termed a "cop-out." Unfairness to the coworker is cited in the comment on D. B receives +5 points for showing compassion but faulted for not addressing the unsafe condition, and C is scored +10 with the comment, "Gets at the cause of the injury. Whatever happens after that, happens. If the injured wants you as a witness that is [within] both your rights." The company here clearly says that it wants hazards removed and employees treated fairly. Removing hazards and treating employees fairly encourage an atmosphere of trust, which in turn fosters cooperation necessary for a productive work environment. Fairness to other employees is put ahead of the company's short-term financial interest.

Case 4 addresses an issue in which coworkers are at fault; exploiting other coworkers. For several months now, one of your colleagues has been slacking off, and you are getting stuck doing the work. You think it is unfair. What do you do?

The candidate answers are:

A. Recognize this as an opportunity for you to demonstrate how capable you are.
B. Go to your supervisor and complain about this unfair workload.
C. Discuss the problem with your colleague in an attempt to solve the problem without involving others.
D. Discuss the problem with the human resources department.

Answer A is scored 0 with the comment that although this may solve the workload problem, *if* you hold up, it does not address the ethical issue (which is equitable distribution of work) and so receives no positive points. Answer C is presented as the best *initial* response although, as the comments make clear, it might not work in which case you would have to "take the next step." That next step presumably is B. That answer is given +5 points, because this strategy brings your colleague's behavior to your supervisor's attention and may give you expanded responsibility. Answer D is scored –5 with the comment, "Pushes the problem solving onto someone else. The problem is between you, your supervisor and your colleague. Solve it there." In its comments on answers C, D, and B the company makes clear that it wants conflicts with co-workers solved at the *lowest level* possible.

The scoring of answers to other mini-cases makes clear the limits on what you can do for yourself and your friends. For example, consider mini-case 14:

> A friend of yours wants to transfer to your division but may not be the best qualified for the job you have open. One other person, whom you do not know, has applied. What do you do?

Scoring and comments seek to guide one through this conflict of interest. Predictably, it discourages putting your friend's wishes, or your preference to work with someone you know and like, ahead of the company's interest in finding the most qualified person. (The answers to this mini-case secondarily teach employees how to make use of human resources departments to help with the selection of the appropriate person.)

Finally, on the topic of whistleblowing, several mini-cases make it clear that just as it makes a great difference whom *within* the company you tell about a particular difficulty, so it makes a great difference *where* you go when taking concerns *outside* the company.

Consider mini-case 58:

> You are working on a government contract and are convinced that a serious mis-charging incident has occurred. You also believe that it was deliberate since the program was running out of funds.

Here the stakes are much higher than in mini-case 68, considered earlier. The candidate answers are:

A. Call the Department of Defense hotline.
B. Inform the local newspaper of your suspicions.
C. Discuss it with your local audit office.
D. Send an anonymous note to your corporate ethics office.

Answer A is scored 0, neither positive nor negative, but *the company wants* you to report it *within the company first*. The office that is best equipped to deal with the complaint is the local audit office (so answer C gets +10). An anonymous note to the ethics office will get the situation investigated, but the company regards it as less desirable because anonymity will slow the investigation (+5). The scoring implies that although some wrongdoing *may* occur within the company, the company believes that it provides employees with appropriate means of reporting the misconduct without undue risk to them. (If fraud was a *regular* occurrence at one's company and the company did little to stop it, the situation would be different, ethically speaking. In that case, going straight to the DoD *would* be appropriate.)

Going to the press is scored –10. Telling the press (unlike going to the DoD) elicits a company comment about "breaching confidentiality." The underlying value is better described as one of loyalty in giving the company a chance to remedy the situation, rather than confidentiality, because misdeeds do *not* automatically become confidential matters. In addition to potentially embarrassing the company, going to the press is a *less reliable* means of getting the abuse addressed than going to the audit office or to the DoD, unless the DoD hotline as well as the company audit office is failing in *its* function.

Taken as a whole, the mini-cases give a picture of both the ethical values of the company that regularly uses them and of its understanding of the characteristics of a well-functioning work community. The *Gray Matters* game consistently encourages employees to both maintain their moral integrity and not allow themselves to be exploited, as well as to be honest and fair to others inside and outside the company. The recommendation is to go to specific persons or offices as appropriate to the situation, such as human resources, the legal department, and the local audit office, for expert advice rather than choose the most self-sacrificing course, *because* going to these offices will lead to resolution.

> The *Gray Matters* game encourages employees to both maintain their moral integrity and not allow themselves to be exploited, as well as to be honest and fair to others inside and outside the company. It counsels behavior that is in the company's *long-term* interest.

The *Gray Matters* material counsels behavior that is in a company's *long-term* interest, much as the NSPE code and the judgments of the NSPE Board of Ethical Review show distinctive concern about the collegial relations among engineers. Although *Gray Matters* generally recommends getting information when one is uncertain, it also recommends addressing some problems directly rather than referring the problem to someone else.

Readers may be surprised at how much information the material instructs employees to take to their supervisors – for example, reporting seeing a company quality manager snorting cocaine at a party (mini-case 23), or reporting that the reason one was ill and missed work was that he was hungover from partying (mini-case 3). Some may question *Gray Matters'* consistently negative view of warning one's coworkers about some perceived unfairness on the part of one's supervisor. However, the rules implicit in the *Gray Matters* judgments, including the rule against gossip, find some justification in fostering a work community able to achieve corporate objectives. The cases provide a benchmark set of expectations for an ethically concerned company. Like the guidelines on employment discussed earlier in this chapter, the *Gray Matters* mini-cases might be useful in discussion with potential employers to see how they would recommend handling the same issues.[26]

> *In what, if any, respects do the judgments reflected in the scoring in the* Gray Matters *ethics game differ from your own? For any points of difference, what ethical justifications can you offer for yours?*

### Advice from the Texas Instruments Ethics Office
How ought one deal with ethical problems in large organizations (given that those problems may not be what they first seem)?

---

[26] The complete game contained 105 mini-cases. It is now out of print and Lockheed Martin will soon replace it with new materials, also employing mini-cases.

The values and organizational arrangements assumed in the advice from the Texas Instruments (TI) Ethics Office* are very similar to the values and organizational arrangements implicit in the answers and scoring of Lockheed Martin's ethics game. In both cases, much of the ethical content concerns responsibility of employees for a well-functioning work *community*. Sometimes such advice is alleged to be about etiquette, but like netiquette (an early name for rules of behavior on the Internet), those bits of advice may have ethical significance. In this respect they are quite *unlike* the etiquette of using the correct fork. Many are ethically significant because if they are frequently violated, then *major moral responsibilities become difficult to fulfill*.

The advice format of the TI Ethics Office articles leaves room for detailed explanation of policies and reflection that find no place in an ethics game, and the advice recognizes that situations may not be what they first seem. One of the TI advice articles discusses situations (presumably similar to ones that had been reported to the Ethics Office) that *looked suspicious but turned out to be innocent*. The ethics officers make three points:

1. A situation is not always what it seems.
2. The question of how an action will appear is [often] one that an agent should consider.
3. The appropriate thing to do is report the matter to the Ethics Office rather than either to gossip about it or to ignore an apparent misdeed. Then the truth can be discovered and wrongdoing stopped or suspicion dispelled. *Reporting something that turns out to be innocent does not incur a penalty.*

TI gives advice about situations that *looked suspicious but turn out to be innocent.* The ethics officers make three points:

1. A situation is not always what it seems.
2. The question of how an action will appear is [often] one that an agent should consider.
3. The appropriate thing to do is report the matter to the Ethics Office rather than either to gossip about it or to ignore an apparent misdeed. Then the truth can be discovered and wrongdoing stopped or suspicion dispelled. *There is no penalty for reporting something that turns out to be innocent.*

Reporting a matter rather than gossiping about it clearly requires a trustworthy and a risk-free means for handling employee complaints and concerns.

The TI Office offers "ethics quick tests" to assess potential actions: The first four of these are:

1. Is the action legal?
2. Does it comply with our values?
3. If you do it, will you feel bad?
4. How will it look in the newspaper?

These are not strictly speaking tests for whether an action is ethically unacceptable or wrong. In particular, conformity with some companies' values might even lead to unethical behavior if the companies were corrupt. Furthermore, as we saw in the introduction, feelings may be based more on one's personal history than of

*This advice came from Texas Instruments in the 1980s and 1990s during which time TI was a large corporation with two separate divisions, one that made consumer products and one that did work under government contracts.

the ethical acceptability of an act. These four tests are better described as tests of whether an action *needs further ethical scrutiny*. A company's concern with how an action will look (say, in the newspaper) despite its recognition that things are not always as they appear is presumably due to its concern for the company's *reputation*. (We considered advice from Texas Instruments on benchmarking and reverse engineering in Chapter 6. An index of TI advice on other particular topics may be found at http://www.onlineethics.org/CMS/workplace/workcases/ti-ethics.aspx.)

> *How ought one deal with ethical problems in large organizations (given that they may not be what they seem)? How, if at all, does your answer depend on your estimate of the ethical integrity of the organization?*

## The Work Environment and Ethical and Legal Considerations

Why do organizations seek to prevent harassment and expressions of prejudice among their members?

Engineering codes of ethics frequently include prohibitions against prejudicial treatment of others. For example, the IEEE code includes a pledge to "treat fairly all persons regardless of such factors as race, religion, gender, disability, age, or national origin." The ACM code of conduct says "1.4. Be fair and take action not to discriminate." In addition to such ethical requirements, legal constraints and company culture govern the work environment.

Consideration of the law enters a company's assessment of behavior both because of the prudential considerations about avoiding legal penalties and because of the moral authority of law that we discussed in the section on Ethics, Conscience, and the Law in the introduction. Norms concerning the work environment are found both in legislation and in case law. Many legal norms are framed in terms of a person's right not to be subjected to an abusive work environment. Legislation itself is often prompted by a change in public consciousness that follows some extreme events. One such event that was particularly significant for engineering schools in Canada and the United States was the "Montreal Massacre." It reawakened awareness to the problem of the high level of violence directed against women and strengthened public support of gun control in Canada.

---

### The Montreal Massacre

On December 6, 1989, Marc Lepine shot and killed fourteen women at the Ecole Polytechnique in Montreal using a semiautomatic Sturm Ruger Mini-14 rifle. He also wounded thirteen others, mostly women, before committing suicide with the same gun. All but one of the slain women were students in the engineering school. Lepine blamed his own failures on feminists.

Groups at engineering schools in Canada and at some U.S. engineering schools hold candlelight vigils on December 6, between 5 and 6 p.m. – the time the killings took place – to remember those killed and help ensure that such events will not be repeated.

The standards of behavior in ethically concerned organizations, such as universities, corporations, and government agencies, go beyond efforts to comply with the law. The responsible organizations work to promote a positive and mutually respectful work environment. This positive strategy not only helps prevent the development of notorious violations of legal rights but also fosters high morale and helps groups function well.

*Why do organizations seek to prevent harassment and expressions of prejudice among their members?*

### Title VII of the U.S. Civil Rights Act of 1964

What was significant about the Supreme Court's interpretation of Title VII of the U.S. Civil Rights Act of 1964 in its decision on *Meritor Savings Bank v. Vinson*?

The U.S. Civil Rights Act of 1964 grew out of public outrage following the abuse and murders of African Americans involved in the civil rights movement. Title VII of the Civil Rights Act makes it unlawful for an employer to discriminate against any individual with respect to compensation, terms, conditions, or privileges of employment because of that person's race, color, religion, sex, or national origin. (Title IX makes similar provisions for educational institutions.)

### Supreme Court Concurring Opinions

When some justices file *concurring opinions* in a case, they are agreeing with the decision of the majority of the court but give different reasons for coming to that conclusion. This practice reflects the importance of the reasoning *behind* a decision.

Supreme Court decisions have a major effect on the legal system in the United States. Their immediate effect is only part of the story. The reasoning behind each decision often has a significant role in establishing expectations about future legal decisions (including those about where the burden of proof will lie), decisions on matters that were unimagined at the time of the decision.

Therefore, the ability of Supreme Court justices to carefully consider, formulate, interpret, and articulate such reasoning (as well as their record of upholding legal ethics) is very important in their selection. This point is obscured when the media focus only on the question of whether some candidate will be politically conservative or liberal.

In a 1986 decision (*Meritor Savings Bank v. Vinson*) the Supreme Court interpreted this language to prohibit discrimination that caused other sorts of injury as well as economic loss. It held that the phrase "terms, conditions, or privileges of employment" shows that Congress intended "to strike at the entire spectrum of disparate treatment" in employment, and that included subjecting people to a discriminatorily hostile or abusive environment. In that opinion the Court had said that mere utterance of an offensive epithet does not so significantly affect the offended employee's working environment as to violate Title VII. In this case, however, the offensive behavior was "so heavily polluted with discrimination as to destroy completely the emotional and psychological stability" of the workers in question that it was in clear violation of Title VII.

Title VII of the Civil Rights Act and its interpretation in *Meritor* set the stage for the 1993 Supreme Court decision in *Harris v. Forklift*. The decision in *Harris* further clarified what constitutes a work environment so discriminatorily hostile or abusive as to be grounds for legal action.

*What was significant about the Supreme Court's interpretation of Title VII of the U.S. Civil Rights Act of 1964 in its decision on* Meritor Savings Bank v. Vinson?

### U.S. Supreme Court Decision on Harris v. Forklift

How did the Supreme Court's decision in *Harris v. Forklift* extend the interpretation of Title VII of the U.S. Civil Rights Act of 1964 beyond the court's interpretation in *Meritor*?

Teresa Harris had worked as a manager at Forklift Systems, Inc., an equipment rental company, from April 1985 until October 1987. During the time of Harris's employment at Forklift, Charles Hardy, Forklift's president, often insulted her because of her gender and often made her the target of unwanted sexual innuendoes. For example, Hardy told Harris on several occasions, in front of other employees, "You're a woman, what do you know?" "We need a man as the rental manager." At least once, he told her she was "a dumb ass woman." In addition, in front of others, he suggested that he and Harris go to a motel to negotiate Harris's raise. Hardy occasionally asked Harris and other female employees to get coins from his front pants pocket or threw objects on the ground in front of them and asked them to pick up the objects.

In *Harris*, the Supreme Court held that discriminatory behavior that creates a work environment abusive to employees because of their race, gender, religion, or national origin violates the norm of workplace equality set out in Title VII. It does so even *without* evidence that the discriminatory behavior disabled them. It further held that whether an environment is hostile or abusive could be determined *only* by looking at all the circumstances and *not solely* at the degree of disability that results from it. The factors it mentioned are: frequency of the discriminatory conduct; its severity; whether it is physically threatening or humiliating, or merely offensive; and whether it unreasonably interferes with an employee's work performance.

The Supreme Court's decision in case number 92-1168, *Teresa Harris, Petitioner v. Forklift Systems, Inc.* (November 9, 1993)* further refined the criteria for what offensive behavior is grounds for legal action under Title VII. As it had established in the *Meritor* decision, offensive behavior that causes significant psychological injury is "actionable" (i.e., it warrants legal redress). In the *Harris* case, the psychological injury was *not* shown to have *disabled* Teresa Harris as it had disabled the complainants in *Meritor*. On this ground, the District Court held that Harris's injury was not sufficiently serious to be actionable.

The Supreme Court in a 9 to 0 decision overturned the lower court decision, stating that it is sufficient that a reasonable person find the work environment hostile or abusive and the victim perceives the environment to be abusive. The Supreme Court held that a discriminatorily abusive work environment can, and often will, undermine an employee's job performance or keep employees from advancing in their careers, even if it does not cause a nervous breakdown or other severe psychological disability. Discriminatory behavior that creates a work environment abusive to employees because of their race, gender, religion, or national

---

*This decision is available at several places on the Web, including: http://www.oyez.org/cases/ 1990-1999/1993/1993_92_1168/ and http://www.law.cornell.edu/supct/html/92-1168.ZO.html.

origin violates the norm of workplace equality set out in Title VII even without evidence that the discriminatory behavior disabled them. The Court held that whether an environment is hostile or abusive could be determined *only* by looking at all the circumstances and not solely at the degree of disability that results from it. The factors it mentioned are frequency of the discriminatory conduct; its severity; whether it is physically threatening or humiliating, or merely offensive; and whether it unreasonably interferes with an employee's work performance.

In mid-August 1987, after Harris complained to Hardy about his conduct he claimed he was only joking, apologized, and promised to stop. In early September, however, Hardy began anew. On October 1, Harris collected her paycheck and quit. She then sued Forklift, claiming that Hardy's conduct had created an abusive work environment for her because of her gender.

The U.S. District Court for the Middle District of Tennessee had judged that some of Hardy's comments offended Harris, and would offend the reasonable woman, but held Hardy's conduct would not have risen to the level of interfering with Harris's work performance. In a decision delivered by Justice Sandra Day O'Connor, a unanimous court overturned the lower court's decisions. Justices Scalia and Ginsberg filed concurring opinions.

Extreme cases of abuse draw attention precisely because they are extreme. Such cases can leave the mistaken impression that only such extreme behavior is objectionable or illegal, however. *Harris v. Forklift* established that evident harassment violates the law even if it does not psychologically cripple the harassed person.

> *How did the Supreme Court's decision in* Harris v. Forklift *extend the interpretation of Title VII of the U.S. Civil Rights Act of 1964 beyond the Court's interpretation in* Meritor?

## From Overcoming Prejudice to Valuing Diversity

> Why should companies or anyone else be concerned about subtle discrimination or "micro-inequities"?

MIT Ombudsperson Mary Rowe has argued that persistent acts of subtle discrimination, most of which are not amenable to legal control, do greater damage than the clearly offensive but rarer behavior of Hardy in the Forklift case.[27] "Micro-inequities," as Rowe calls them, function "like the dripping of water, random drops themselves do little damage; endless drops in one place can have profound effects." These inequities may take the form of persistent application of negative stereotypes to individuals despite their actual attributes. Rowe suggests using measures such as employee attitude surveys to bring attention to such problems, providing means for individuals to obtain confidential advice and support, and offering management training to overcome subtle discrimination. Her suggestions accord with those of Westin, listed at the beginning of this chapter.

[27] Rowe, Mary P. 1990. "Barriers to Equality: The Power of Subtle Discrimination to Maintain Unequal Opportunity." *Employee Responsibilities and Rights Journal* 3(2): 153–163.

Organizations that plan to thrive in a time of an increasingly diverse workforce often run diversity training workshops for their employees as well as managers. Some go beyond combating prejudice and discrimination to valuing diversity. Valuing diversity is the goal of recognizing and valuing differences rather than just treating everyone the same. Issues of prejudice are most often framed as issues of "discrimination." Discrimination, recognizing the differences between things, is often good. It is a compliment to say a person has "discriminating tastes," for example. *Unjustified* differences in treatment on the basis of race, gender, ethnicity, religion, nationality, sexual orientation, and disability are called "**discrimination**" for short. Discrimination leads us to focus on differences in treatment and ask whether those are justified. An equally important question is whether everyone is treated and expected to act the same way, but in a way that only those in the dominant group finds natural or easy.

A scenario by Joel Palacios MIT '96 aptly illustrates the difficulty for various minorities when only the majority way of doing things is acknowledged.

## Diversity and Barriers to Advancement[a]

Jane and Maria started working for the same company, at the same time and under the same supervisor, Ms. Manager. Jane and Ms. Manager are both European Americans while Maria is a Mexican American.

Soon after they started working, Ms. Manager invited both of them to her traditional Sunday afternoon barbecue, an event that is held biweekly and attended by many professionals in the company. Both Jane and Maria attended the event. Although Jane seemed to have a great time, Maria felt uncomfortable because she was the only minority member present out of about six employees and their families. Her cultural expectations of the event had differed from those of the others. For example, she prepared a dish to share with everyone. Other families had each brought their own food and drinks. She also felt that it was difficult to find common ground with her coworkers outside the world of their profession. Maria decided not to attend any future barbecues because she felt uncomfortable

The supervisor continued to invite both subordinates. Jane attended every time; however, Maria never did, and struggled to come up with reasons why she could not do so. She did not want the supervisor and other employees to take her rejections personally. As time went on, Maria sensed the personal relationship between Jane and Ms. Manager developing into a strong one. Eventually, a year after they had both joined the company, Jane received a good promotion, due in part to a fine recommendation from Ms. Manager.

Maria had the impression that Ms. Manager favored Jane, and that her favoritism had been reflected in her recommendation of Jane. Maria felt that she had been doing superior work and that her contributions to the company were at least as significant as Jane's. Maria became even more concerned about the situation when the new subordinate, hired to replace Jane, turned out to be another European American. A month after this, the new subordinate seemed to be following the footsteps of Jane, developing a strong personal relationship with Ms. Manager.

What should Maria do?

### Getting Started

Subtle discrimination such as that described in this case is very difficult to address. Maria may be able to advance only in a company that is more alert to these issues.

---

[a] This scenario forms part of a project that also contains interviews about how best to deal with the issues. It is available in the Problems section of the Online Ethics Center, http://onlineethics.org.

*Why should companies or anyone else be concerned about subtle discrimination or "micro-inequities"?*

## Organizational Responses to Offensive Behavior and Harassment

What values and priorities underlie organizational responses to offensive behavior and harassment?

Ethically active organizations concern themselves with subtle issues of harassment and work environments as well as with legal issues. These subtler norms are evident in the scoring of two other mini-cases from Lockheed Martin.

Although most complaints of sexual harassment are brought by women, some are brought by men. (**Sexual harassment** is harassment based on biological sex and, as we saw, is prohibited by the Civil Rights Act of 1964. Although harassment of women often contains sexual content, the definition of sexual harassment does *not require* the presence of sexual content in the offensive behavior.) Mini-case 57 describes a situation that would more typically be experienced by males.

> You are a quality inspector. After making your own calculations, you disagree with your supervisor about whether the quality of the item is at an acceptable level. With a rolled up newspaper in his hand, your supervisor swings it in your direction, hitting the back of the chair you are sitting in. What do you do?

The candidate answers supplied are:

A.  Swing back at him
B.  State unequivocally that such behavior is unacceptable in business and advise him you intend to take this matter up with the manager, to whom you both report.
C.  Get up and go straight to the EEO office.
D.  Since the boss says, "I was only joking," you ignore the act.

Not surprisingly, response A receives −10. D receives −5 points, with the emphatic comment, "Intimidation unchecked is intimidation encouraged. Lack of response will encourage this sort of behavior to expand."

B is given the highest score, +10, with the comment, "Not only will this response get the item a third-party inspection, but it will also put your supervisor on notice that you do not accept his action."

The reference to the equal employment office (EEO) in answer C suggests that discrimination may be at work. (In addition to the categories of discrimination mentioned in Title VII, many high-tech companies are also alert to discrimination based on sexual orientation.) The EEO option is scored +5 with the comment, "This is your privilege but it doesn't solve the problem." However, the EEO may be most appropriately consulted if the person suspects discrimination and does not feel comfortable raising that issue with the manager mentioned in B. Companies must be sure their managers can handle issues of discrimination sensitively, if these recommendations are to work.

Mini-case 72 raises some even subtler issues for managers.

When [one] male supervisor talks to any female employee, he always addresses her as "Sweetie." You have overheard him use this term several times. As the supervisor's manager, should you do anything?

The candidate answers are:

A. No, since no one has complained.
B. Yes, talk to the supervisor and explain that, while he may have no sexual intention, his use of "Sweetie" may cause resentment among some of the employees.
C. Yes. Order the supervisor to call an all-hands meeting and apologize for the unintended slights.
D. No, because there is nothing wrong with calling a female employee "Sweetie" or other endearments.

Answer A receives –10 points with the comment, "To some such informalities are, at best, unwelcome, and, at worst, a form of sexual harassment. Action should be taken to correct the situation even without prodding from an employee." Answer D receives a resounding – 20 points and the rebuke, "A manager's role is to assure a productive, professional working environment. Option D means you have abdicated."

Answer C receives 0 points and the comment that there is no evidence of harm done without a complaint and so the response is "premature." (Indeed, making such an example of the perpetrator is an overreaction and for that reason is a bad, rather than a neutral, move. That Answer C is not graded negatively may be overcompensation for the recent past in which managers tended to be too tolerant of such behavior. Some tendency to overcompensate often follows a period of neglect of a problem.)

Answer B receives 10 points with the approving comment, "Acting in a firm, nonjudgmental fashion, you are now doing your job as a manager – proactively, not reactively."

> If some seemingly innocuous behavior is offensive to the person to whom the behavior is directed, the offended party should tell the offending party that the seemingly innocuous behavior is unwelcome. If, *after being told,* the offending party repeats the behavior, then and only then is the behavior harassment.

Notice that the person who is responsible for preserving the working relationship in the *Gray Matters* case we have been discussing is the manager. The behavior in the "Sweetie" case, unlike the behavior in *Harris v. Forklift*, is something a person might do without intending any offense, and the law's ideal "reasonable person" might or might not find it offensive. Some of my male students, after working through the "Sweetie" case, have been anxious that some seemingly innocuous behavior of theirs might be offensive. The rough rule is: if some seemingly innocuous behavior is offensive to the person to whom the behavior is directed, *the offended party should tell the offending party that the seemingly innocuous behavior is unwelcome. If, after being told,* the offending party repeats the behavior, then and only then is the behavior harassment.

The guidance given in the answers to the *Gray Matters* mini-cases shows a company's concern to go beyond respecting employees' legal rights and to promote trust, consideration, and good working relationships within the company.

*What values and priorities underlie organizational responses to offensive behavior and harassment?*

## Ethics in a Global Context

How is one to distinguish mere cultural differences (such as when another culture has *different limits* on some behavior from those customary in the United States) from circumstances in which cultures or societies differ in their toleration of corruption?

During the development of the *Gray Matters* game, the cutoff for the value of gifts that can be accepted from business contacts changed. The cutoff figure used in commentary on earlier mini-cases was $10. The higher figure of $20 appears in some later ones, presumably reflecting inflation. Particular cutoff figures do not carry any moral imperatives behind them, and the lavishness of gifts that may be innocently offered varies significantly from one culture to the next and even from one profession to the next.

U.S. engineers doing business with Japanese firms have often received gifts from those firms that are lavish by U.S. standards. These gifts are normal hospitality by Japanese standards and are given without any expectation that the recipient will do anything improper in return, so they are not bribes. Because they are out of line with what members of U.S. corporations ordinarily give and receive, many U.S. companies doing extensive business with Japanese companies have established practices for dealing with these gifts, such as pooling them and holding an employee drawing for them, or giving them to a charity.

Different cultures have different expectations about how close people ordinarily stand to one another, when, if ever, it is appropriate to look straight into someone's eyes, or whether it is respectful to show the soles of one's shoes to others. Differences in cultural expectations complicate the application of ethical standards across national boundaries, but new associations such as the European Economic Community and the North American Free Trade Association and the growing importance of multinational trade are leading companies to think more cross-culturally about ethics as well as etiquette.

At a minimum, U.S. companies must comply with the so-called Foreign Corrupt Practices Act (FCPA) of 1977 that we briefly discussed in Chapter 2. This act makes it a crime for U.S. corporations to accept or offer payments to foreign governments and political parties in order to obtain or retain business. It does not forbid making minor payments to low-level officials to "grease the skids," although the latter might, depending on the situation, also count as a bribe or as extortion[28] and so be unethical even though legal. Although the example given in

---

[28] For an NSPE BER case that presents a problem of the sort that prompted this legislation see Case 76-6, Gifts to Foreign Officials. The NSPE Board of Ethical Review (BER) cases for 1976–2007 with judgments offered by the BER based on application of the then current NSPE code of Ethics are available in hard copy in the volumes 5–9 of *Opinions of the Board of Ethical Review*, from the National Society of Professional Engineers. See the reference guide with an index

the introduction of paying extortion for the return of one's improperly confiscated belongings would, other things equal, be morally justified, paying extortion to a government to prevent it from terminating one's business relationship in which one has made a heavy initial investment may be *illegal* under the FCPA even if morally justified. Here again we see how the function of the law differs from the moral evaluation of some act.

The FCPA was enacted after it became known that U.S. corporations had paid millions of dollars to foreign governments to obtain or retain business. The intent of the law is not only to prevent U.S. corporations from acting reprehensibly but also to ensure that extortion attempts by foreign officials *do not succeed*. To comply with this law and at the same time avoid insulting potential business partners, some companies require that all payments *be a part of any contract.*[29] Gifts that are expected as a part of normal courtesy in some cultures may also be made a part of the contract.

> The Foreign Corrupt Practices Act was passed after it became known that U.S. corporations had paid millions of dollars to foreign governments to obtain or retain business. The intent of the law is not only to prevent U.S. corporations from acting reprehensibly but also to discourage such attempts at extortion. For this reason, it prohibits certain extortion payments to officials of foreign governments, even if such payments *could* be morally justified.

It may seem that FCPA puts U.S. corporations at a disadvantage in competing with corporations from countries that have no legal strictures against paying bribes, but the FCPA has had some long-term effects that benefit U.S. corporations. For example, the World Bank has prohibitions against making loans that will be used to pay bribes. Therefore, proposals to the World Bank that employ U.S. corporations have a prima facie advantage over plans to use corporations from countries that countenance the payment of bribes.

The NSPE in its 1998 case and judgment, 98-2: Gifts to Foreign Officials – Application of Code of Ethics to Non-U.S. Engineers, judges as *unethical* the actions of a hypothetical international member of the NSPE who makes payments to foreign officials that would be illegal for U.S. firms but that are legal in the engineer's *own* country.

> How is one to distinguish innocent cultural differences (such as when another culture has different limits *on some behavior from those customary in the United States*) from circumstances in which cultures or societies differ in their toleration of corruption?

## Conclusion

In this chapter, we have seen that company values influence both the relationship between engineers and managers and engineers' opportunities to fulfill their responsibilities. High-quality organizational complaint procedures may provide

of cases through 2009 at http://www.nspe.org/resources/pdfs/Ethics/EthicsReferenceGuide.pdf. Cases and opinions from 1976 through 2001 (only) are available at http://www.niee.org/pdd. cfm?pt=NIEE&doc=EthicsCases.

[29] Lytton, William B. 1996. *Combating Corruption in Foreign Markets. The Evolving Role of Ethics in Business: Conference Report.* New York: The Conference Board, Inc.

means for finding good resolutions of disagreements, but are not easy to establish. Establishing the rights, obligations, and responsibilities of employers and professional engineering employees contributes to mutual understanding and enhances trust between those employers and their engineering employees. Professional societies, in cooperation with employers, have established such norms in employment guidelines.

Legal limits on treatment of employees are in force, but these only curtail relatively extreme behaviors. To provide a good working environment that fosters trust and cooperation requires addressing many subtler issues. Farsighted companies and research institutions are concerned with subtler aspects of workplace climate both out of concern for employees and because a good workplace climate fosters high productivity.

# PART 3

# RESPONSIBLE RESEARCH CONDUCT

# 8 Ethics in the Changing Domain of Research

Which, if any, values at stake in research conduct are values that further the pursuit of truth and knowledge?

Why is trust necessary to the flourishing of engineering research?

---

## Doubts about Published Results[a]

You are a computer science graduate student and for two years have been working on an operating system design in Professor Carr's group. Professor Carr has designed a set of novel heuristics for file-system cache maintenance. Carr published performance graphs describing a simulation of a prototype file input/output subsystem in a journal article and included the graphs in the proposal for the group's current grant. The graphs indicate that Carr's heuristic methods will significantly improve file-system cache performance.

You devise a modification to the file-system cache heuristics and ask Carr how to run the simulation code to test the modification. Carr replies that the simulation code had not been used in a long time and had been archived to tape. Carr says it is not worth the trouble trying to remember the archived filenames, because the simulation code was very poor and written in a language that does not run in the group's current computing environment. He tells you to write a new, up-to-date simulator.

As you worked on the new simulator, you ask Carr how to simulate several classes of events, but Carr claims to not remember these details of the old simulator. When you finish building a new simulator, your results are considerably worse than those reported in the performance graphs that Carr published.

You now suspect that Carr did not do a previous simulation and made up the numbers in the performance. Some of your own presentations and papers have been based on Carr's performance data.

What can/should you do?
What, if any, ambiguities do you face?
What are the risks in this situation to yourself or others?

### Getting Started
Graduate programs differ dramatically in the support they provide students in difficult situations like this one. The presence of good support is one thing to look for in a graduate program. How would you go about finding out what support a program offers?

---

[a]Based on a scenario contributed by an MIT graduate student who preferred not to be identified, with new technical particulars added by Prof. Albert R. Meyer, EECS, MIT.

273

Engineering research and research collaborations involving engineers, like other cooperative endeavors, require trust to flourish. As we saw in Chapter 4, **trust** is confident reliance; that is, it has two elements: confidence and reliance.[1] If we have no alternatives, we may continue to rely on things or people, even though we have lost confidence in them, but reliance without confidence leads to a downward spiral. That is unfortunately the situation in some areas of research. For example, I know of instances in which investigators have lost confidence in the fairness and honesty of peer reviewers. Those investigators must continue to rely on proposal submission and journal publication to fund their research and make known their accomplishments. Because of their fear that reviewers might try to steal their work, some have intentionally withheld information or even made misstatements in the manuscripts that they submitted for publication. In the case of misstatements, their intention has been to correct the intentional misstatements only in the final proofs of their articles. Such behavior only hampers the work of honorable reviewers and editors. The investigator who makes misstatements cannot guarantee that she will not become incapacitated before sending back the final proofs. Thus, deceptions of this kind endanger the research record and so make it less trust*worthy*.

As sociologist Niklas Luhmann observed,[2] trust simplifies life. It would be too burdensome to consider all the possible disappointments, defections, and betrayals by those on whom we rely; all possible consequences of those disappointments; and all actions we might take to prevent those disappointments or change their effect. Without trust, research will become ridden with defensive ploys. Blind or naïve trust will not suffice. Trust that is naïve or blind commonly leads to disappointment with an extra sting of shame at having been duped. It is *warranted* trust and trust*worthy* behavior that support enduring trust and cooperation.

> The need for trustworthy professionals in modern society is not captured in the frequent suggestion that trust is necessary because the trusting party cannot control or monitor the trusted party's performance. It would do someone little good to witness everything an engineer did in designing and overseeing construction of a bridge, or even to be able to guide the engineer's actions, unless that person also happened to be an engineer.

The need for trustworthy professionals in modern society is not captured in the frequent suggestion that trust is necessary because the trusting party cannot control or monitor the trusted party's performance. It would do someone little good to witness everything an engineer did in designing and overseeing construction of a bridge, or even to be able to guide the engineer's actions, unless that person also happened to be an engineer. Although most people might be able to recognize some acts of gross negligence, they would not understand the implications of most of what they saw an engineer do and would have no idea of how to improve the engineer's performance.

There are two elements to responsible or trustworthy behavior in professionals: competence and concern. Being incompetent is not by itself a moral failing, although taking work beyond one's competence when doing so puts others at

[1] Baier, Annette. 1986. "Trust and Antitrust," *Ethics* 96: 232–260; reprinted in *Moral Prejudices*. Cambridge, MA: Harvard University Press, 95–129.

[2] Luhmann, Niklas. 1988. "Familiarity, Confidence, Trust: Problems and Alternatives." In *Trust: Making and Breaking Cooperative Relations*, edited by Diego Gambetta. Oxford: Basil Blackwell, 94–108.

risk is one. The trustworthiness of a professional depends upon the professional's integration of competence and concern, however. A modern society devotes resources to the education of its citizens. Those resources enable members of various professions to master bodies of professional knowledge and use their educated discretion to make good decisions in the area of their expertise. The moral concern required of professionals is not merely that they be careful and mean well. They must marshal their expertise to achieve good outcomes *in the special domain of their expertise*, because society entrusts this domain of decisions to the members of their profession. It may be a moral failing for research investigators to litter by negligently disposing of their lunch, but it is a failure in *professional responsibility* if they are negligent about attributing research credit or about the accuracy of the reports they author. There are no good alternatives to having trustworthy professionals, because both individuals and society must rely on the judgment and the discretion of the professional.

> There are two elements to responsible or trustworthy behavior in professionals: competence and concern.

## Values in Science and Engineering

Work in the history and philosophy of science, such as Helen Longino's *Science as Social Knowledge*, has discredited the idea that the methods and judgments of science are value-free, although the place of values in science is not the same as the place of values in, say, politics.[a] Although quite different in most respects, both the first and second editions (1989; 1995) of *On Being a Scientist* recognize the place of values in science (and engineering). (The third edition [2009] is very similar to the second.) All editions discuss two types of values. First are those values involved in differentiating a good explanation or theory from a poor one. These values, often called values "**internal to science**," are values such as simplicity, consistency, and the ability to yield accurate predictions that earlier in this book were characterized as epistemic or knowledge values. The second type is the values that an investigator carries over from other aspects of life into scientific and engineering work – the values "**external to science**."

Another reason that trustworthy investigators are essential for research to thrive is highlighted in Baier's discussion of the moral basis for a decent trust. To be morally decent, she has argued, trust should withstand the disclosure of the basis for that trust. For example, if a research supervisor's trust in her supervisee's honesty is based on the belief that the supervisee is too timid or unimaginative to fabricate data or experiments, disclosure of that belief will give the supervisee an incentive to commit misconduct. Therefore, although closer oversight of research may well reduce dishonesty, the moral climate of scientific research will suffer further if oversight reduces dishonesty *only* by instilling fear of detection. Oversight by supervisors and collaborators should serve two important ends: It should lessen self-deception among investigators, self-deception that may lead them into desperate situations in which they will be tempted to cheat. Moreover, it should foster a full understanding and appreciation of the values that contribute to good science and how those are best implemented in specific research contexts.

[a] Longino, Helen. 1990. *Science as Social Knowledge*. Princeton, NJ: Princeton University Press.

*Which, if any, values at stake in research conduct are values that further the pursuit of truth and knowledge (sometimes described as values "internal to science and engineering")?*

*Why is trust necessary to the flourishing of engineering research?*

## The U.S. Government-Wide Definition of Research Misconduct

What is the U.S. government's definition of research misconduct? Why does it cover just certain types of serious wrongdoing in research and not every act of serious wrongdoing in research?

### A Single U.S. Government-Wide Definition

Previously, two agencies that are the source of most of the funding for science and engineering research, viz., the National Science Foundation (**NSF**) and the Department of Health and Human Services (**HHS**) (which includes the National Institutes of Health [**NIH**]), each had its own (slightly different) definitions of research misconduct. Some other government funding agencies had no definition at all. All, including the Department of Energy (**DoE**), the U.S. Department of Agriculture (**USDA**), the National Oceanic and Atmospheric Administration (**NOAA**), and the National Endowment for the Humanities (NEH), now operate with the same definition. When it was proposed, the authors of the definition stated that they had stopped short of seeking consensus among agencies on a *policy for dealing with allegations of misconduct*, and proposed only a definition of *research misconduct*. The definition is widely referred to as a "policy," however.[a]

Finally, in December 2000, after decades of wrangling with some universities and investigator groups and several abandoned attempts to agree on a definition of research misconduct, the U.S. government officially adopted a definition. (That definition is available on various government agency Web sites including those of the National Endowment for the Humanities [NEH], http://www.neh.gov/grants/guidelines/researchmisconduct.html; the Office of Research Integrity [ORI] of the HHS, http://ori.dhhs.gov/policies/fed_research_misconduct.shtml; and the U.S. Department of Agriculture, http://www.ocio.usda.gov/directives/doc/DR2401-001.htm.) Any U.S. agency that funds scientific or engineering research now uses that definition. The definition itself was thoroughly vetted in the research community, including at a town meeting on November 17, 1999, held at the National Academies building in Washington, DC.

---

[a] See http://ori.hhs.gov/policies/fed_research_misconduct.shtml.

The definition itself (in Section I) reads:

**Research misconduct** is defined as fabrication, falsification, or plagiarism in proposing, performing, or reviewing research, or in reporting research results.
**Fabrication** is making up data or results and recording or reporting them.
**Falsification** is manipulating research materials, equipment, or processes, or changing or omitting data or results such that the research is not accurately represented in the research record.*

---

*The NEH adds a clarification as to what counts as "the research record" to the definition itself, saying it is "the record of data or results that embody the facts emerging from the research, and includes, but is not limited to, research proposals, progress reports, abstracts, theses, oral presentations, internal reports, journal articles, and books."

**Plagiarism** is the appropriation of another person's ideas, processes, results, or words without giving appropriate credit.*

That definition is significantly augmented in Section II, Findings of Research Misconduct, which stipulates:

A finding of research misconduct *requires*[†] that:

There be a *significant departure from accepted practices of the relevant research community*; and the misconduct be committed intentionally, or knowingly, or *recklessly*; and the allegation be proven by a preponderance of evidence.

---

> **Research misconduct** is defined as fabrication, falsification, or plagiarism in proposing, performing, or reviewing research, or in reporting research results.
> **Fabrication** is making up data or results and recording or reporting them.
> **Falsification** is manipulating research materials, equipment, or processes, or changing or omitting data or results such that the research is not accurately represented in the research record.
> **Plagiarism** is the appropriation of another person's ideas, processes, results, or words without giving appropriate credit.
> A **finding of research misconduct** *requires* that:
> - there be a significant departure from accepted practices of the relevant research community
> - the misconduct be committed intentionally, or knowingly, or recklessly
> - the allegation be proven by a preponderance of evidence

As we shall see, the addition of the term "recklessly" is significant.

The responsibility for research integrity has two major components: ensuring the integrity of research results and dealing fairly with others, especially by appropriately acknowledging their work done and protecting the welfare of research subjects. Other professional responsibilities of research investigators include laboratory safety and protection of public health and safety in the conduct of research. Violations of the standards for the treatment of human subjects in experiments (and of animal subjects) are not included in this definition of research misconduct, because standards for the treatment of research subjects *predated* the government-wide definition of misconduct. Such violations carry equally severe penalties, however. This fact illustrates that the failure to count some wrongdoing as research misconduct does *not* mean that the wrongdoing is seen as a less serious offense than research misconduct.

Beginning in the 1980s when research conduct first drew wide attention in the United States, discussions of it often began with consideration of serious wrongdoing in research.[3] This negative approach to the subject of responsible research conduct is in contrast to discussion of engineers' and scientists' other key responsibilities, such as their responsibility for safety. Although some discussions of engineers' responsibility for safety

---

*Bolding added.
[†]Italics added.
[3]Professional responsibility is generally discussed affirmatively, rather than with the principal ways one might be derelict in fulfilling that responsibility. That society has always been concerned about catastrophic accidents and has encouraged engineers to consider their responsibility for safety may partly explain the difference.

## Standards of Proof and Research Misconduct

Different standards of proof are used for different offenses (and the punishments for them). For capital cases in the United States, (cases in which the penalty is execution) the standard is *proof beyond a reasonable doubt.* In other types of cases the standard is a *clear and convincing proof.* For research misconduct the still weaker standard of "preponderance of evidence" is used. The **"preponderance of evidence"** standard requires that it be shown to be *more likely than not* that the accused committed research misconduct.

## Charles Babbage (1871–1971)

Charles Babbage was the inventor of both the "difference engine" and the "analytic engine," the precursor of the modern electronic computer. He founded several professional societies.

begin with the story of a mistake in the form of a notorious accident, suggestion of *deliberate* or *intentional* wrongdoing is rare. Even negligence and recklessness may be absent from the accident story, because accidents may result from highly unusual circumstances or lack of knowledge available at the time.[4] Thus, the mistakes may be innocent mistakes. As Henry Petroski observed, "to engineer is human."[5] Learning from mistakes does not require laying blame for those mistakes, although it is *sometimes* appropriate to do so.

The subject of research misconduct is not entirely new among researchers. As far back as 1830, the English mathematician Charles Babbage wrote an influential book on dishonesty in research.[6] In his book, Babbage defined several terms to describe research misconduct, including one that is still very much in use: "cooking the data." To **cook the data** is to select only those data that fit one's hypothesis and to discard those that do not. Using data solely because they support one's hypothesis would now count as falsification of data, which is a type of research misconduct. "Cooking," however, is a term that investigators also use today, sometimes in jest, when describing methods for data selection that are ad hoc or not fully understood, but that make "messy" or "noisy" data look more conclusive. Such slippery terms signal that the speaker sees something amiss but is unwilling or unable to describe it precisely.

In research, mistakes are controlled through such mechanisms as peer review of research reports prior to publication and the requirement for replication of results. **Replication**, which is the repetition of an experiment to ensure that the experimental procedure yields a consistent result, is not even possible for all types of experiments – such as large-scale clinical studies. Even where possible, they are imperfect.[7]

---

[4] Charles Perrow in *Normal Accidents* (New York: Basic Books, 1984) argues that in highly complex systems some catastrophic accidents are "normal."

[5] Petroski, Henry. 1985. *To Engineer Is Human.* New York: St. Martin's Press.

[6] Babbage, 1830. See discussion of Babbage in C. Ian Jackson's 1992 *Honor in Science* (Sigma Xi, the Scientific Research Society, Research Triangle Park, NC).

[7] See the papers and discussion in the "Peer Review" section of *Ethics and Policy in Scientific Publication*, 1990, by the Committee on Editorial Policy of the Council of Biology Editors (CBE) (John C. Bailar, Marcia Angell, Sharon Boots, Karl Heumann, Melanie Miller, Evelyn Myers, Nancy Palmer, Sidney Weinhouse, and Patricia Woolf). Bethesda, MD: Council of Biology Editors, Inc., pp. 257–284.

Some laboratories replicate their own results, especially major ones, before publishing them in order to catch any mistakes and thus further their reputation as scrupulous investigators and avoid later embarrassment. Even where replication is possible in principle, the experimental procedure may be expensive or dangerous to carry out. Control of both honest errors and deliberate departures from standards of responsible research practice is demanding.

## Is "Scientific Misconduct" "Research Misconduct"?

The term "research conduct" itself needs clarification. "Scientific conduct" has sometimes been used as a synonym for "research conduct," and "scientific misconduct" has been used as a synonym for "research misconduct." As examples, the third edition (2009) of *On Being a Scientist* does so, and a convocation on research ethics that the National Academy of Sciences (NAS) held in June 1994 was called a "Convocation on *Scientific Conduct*" much to the confusion of some attendees who expected an agenda that would include the societal implications of science and engineering.

The misleading breadth of the term "scientific conduct" may go unnoticed, because many scientists of previous generations were slow to recognize any socially and ethically significant results of scientific work other than the growth of knowledge through research. It also shows the lack of an established vocabulary with which to discuss researchers' responsibilities.

Research ethics came to broad public attention only after some flagrant cases of research misconduct (and the institutional mishandling of those cases) came to light. Some scientists – for example, Daniel Koshland, when he was the editor of the influential publication *Science* – responded to the concern about research misconduct by saying that too much was being made of a very few notorious offenses.[8] Some bizarre or flagrant cases *have* been overemphasized. The research community contributes to this distorted emphasis by attending only to such cases and ignoring more common violations of ethical norms in research.

The focus on the major wrongdoing known as "research misconduct" runs the danger of making it seem that research ethics is merely an attempt to hold the line against deliberate deception, rather than a concern to develop, maintain, and transmit standards of research integrity in a context of increasing complexity in research practice. The literature on research ethics is so heavily focused on research misconduct, however, that misconduct and the terms used to describe it provide an obvious starting point. The alternative approach is to focus on the responsibility for *research integrity*. Research integrity has several major components: ensuring the integrity of research results, and dealing fairly with others, especially by appropriately acknowledging their research contributions.

Research disciplines are now reinventing the language to discuss responsible research behavior and departures from it. The need for terms is partly due to a period of silence on the subject following Babbage's early work, but also due to the rapid growth of research and the emergence of new research conditions. Adequate

---

[8] For brief summaries of misconduct cases that occurred at universities and that included both bizarre and typical cases see Allan Mazur, "The Experience of Universities in Handling Allegations of Fraud or Misconduct in Research," *Project on Scientific Fraud and Misconduct: Report on Workshop Number Two*, edited by Rosemary Chalk. Washington, DC: AAAS, 1989, 67–94.

terms are needed to clearly communicate ethical values and to deliberate about ethical standards that are appropriate for novel circumstances and unprecedented situations.

*What is the U.S. government's definition of research misconduct? Why does it cover just certain types of serious wrongdoing in research and not every act of serious wrongdoing in research?*

## Research Misconduct Distinguished from Mistakes and Errors

Why are the incompetent errors that corrupt research results not counted as research misconduct by the U.S. government definition, when engineering societies do hold that practicing engineers have a responsibility to take work only within their competence?

### Plagiarism Is Not Unique to Research

One common attempt to explain this exclusion is to say that "research misconduct" covers only wrongdoing unique to the conduct of scientific research. However, as the late Donald Buzzelli argued, if "research misconduct" covered only wrongdoing unique to research, it would exclude plagiarism. Plagiarism may be of poems, photographs, and other artistic works as well as of research.

As we have seen, "research misconduct" is not applied to all types of wrongdoing done in a research setting, but only certain actions that seriously threaten research integrity. For example, if an investigator takes home pieces of lab equipment for personal use, that act would count as stealing, or at least as misappropriation of property, rather than as research misconduct. Not all failure to follow standards of good research practice counts as research misconduct, even if some failure jeopardizes the research results. For example, failure to run experimental controls (when it would be possible to do so) is certainly a departure from good research practice, but is regarded as a sign of incompetence, rather than misconduct. This point is emphasized in the U.S. government definition of research misconduct where it says, "Research misconduct does not include honest error or differences of opinion." The ethical responsibilities of investigators do *not* include a responsibility to undertake work only within the investigators' competence, although that requirement *does* receive strong emphasis in codes of engineering practice. The difference in responsibilities is explained by the practice of reviewing research reports before publication by reviewers who are selected to be able to judge the worth of the research reported. In contrast, the practice of engineers, like the practice of physicians, is not reviewed with the same regularity, so engineers in practice, like physicians, are charged with protecting others against their own incompetence.

### The Struggle over Misconduct Definitions

Prior to the adoption of the 2000 government-wide definition of misconduct, the NSF[a] and the HHS (NIH) clashed with organizations like the national academies and some professional organizations and universities over the definition of misconduct. This struggle played out in disputes that *appeared* to be about definitions, but had major implications for the amount of discretion the government was allowed in deciding which cases to investigate as possible instances of research misconduct.

In the 1980s, the NSF definition of research misconduct began by specifying fabrication, falsification, and plagiarism as acts that exemplify[b] or constitute research misconduct. However, it went on to say "*or other serious deviation from accepted practices in proposing*, carrying out, or reporting results from activities funded by NSF; or retaliation of any kind against a person who reported or provided information about suspected or alleged misconduct and who has not acted in bad faith."

Research organizations objected to the phrase "or other serious deviation from accepted practices in proposing, carrying out, or reporting results" as vague and liable to include "honest error."[c]

---

[a] National Science Foundation, 2002. "45 CFR Part 689" Federal Register, 67(17) (January 25).

[b] Donald Buzzelli of the Office of the Inspector General at NSF argues that "falsification, fabrication and plagiarism" were intended as *examples* of "serious deviations from accepted practice" rather than defining instances of such deviation, a point that the "other serious deviation" clause simply spells out. See comments by Donald Buzzelli on the definitions of research misconduct in the Research Ethics section of the Online Ethics Center.

[c] I am indebted to the late Donald E. Buzzelli for bringing this point to my attention. The struggle is described in greater detail in the first edition of *Ethics in Engineering Practice and Research*.

As we saw in Chapter 1, professions generally seek autonomy so it is not surprising that the research community is wary of government regulation. To be relatively free of such regulation the research community needs to further develop its own ability to regulate itself. It is in everyone's interest that it does so, but that self-regulation is not well developed. Normal mechanisms of research practice had been used to accomplish the task, but studies have shown that such mechanisms are not sufficient even to purge corrupted results from the literature in a reasonable time.[9]

---

*Why are the incompetent errors that corrupt research results not counted as research misconduct by the U.S. government definition, when engineering societies do hold that practicing engineers have a responsibility to take work only within their competence?*

## Recent History of Attention to Research Misconduct

How does the development of attention to research conduct by the research community compare with the development of attention to standards of professional responsibility in engineering practice by the engineering community?

The research community's understanding of responsible research conduct and of the importance of professional responsibility in research has developed rapidly since the mid-1980s. The ethical standards for research conduct received

---

[9] Kiang, Nelson. 1995. "How Are Scientific Corrections Made?" *Science and Engineering Ethics* 1(4) (October): 347. Guertin, Robert. 1995. "[Commentary on] How Are Scientific Corrections Made? (by N. Kiang)," *Science and Engineering Ethics* 1(4) (October): 357. Pfeifer, Mark P. and Snodgrass, Gwendolyn L. 1990. "The Continued Use of Retracted, Invalid Scientific Literature," *Journal of the American Medical Association* 263(10): 1420–1423.

## Is Research Misconduct "Fraud"?

In the 1980s the term "fraud" was widely used to describe research misconduct. For example, in 1985 the U.S. Congress passed a section of the Public Health Service (PHS) Act titled "Protection Against Scientific Fraud."[a] "Fraud" or "scientific fraud" is still sometimes used, especially in popular writing. However, as the National Academy of Sciences Panel on Scientific Responsibility and the Conduct of Research pointed out, "fraud" is a poor term for research misconduct.[b] It is misleading for two reasons: First, the legal definition of "fraud" requires that some party be injured by the fraudulent action. In addition, the legal notion of fraud has three basic elements:

1. The perpetrator makes a false representation;
2. The perpetrator *knows the representation is false or recklessly disregards whether it is true or false*; and,
3. The perpetrator *intends to deceive* others into believing the representation.

Instances of research misconduct commonly stem from an attempt to "cut corners" to confirm a result that the perpetrators *deeply believe to be true* (and for which they make a *deceptive* argument).

The first edition (1989) of *On Being a Scientist* claimed that "[t]he acid test of fraud is the intention to deceive." The second and third editions do not use the term "fraud."

---

[a] For a discussion of section 493 of the Public Health Service Act and subsequent response to it, see Semiannual Report to the Congress, Number 9 (April 1, 1993–September 30, 1993) by the Office of Inspector General, National Science Foundation.

[b] National Academy of Sciences Panel on Scientific Responsibility and the Conduct of Research. 1992. *Responsible Science: Ensuring the Integrity of the Research Process*, Vol. I, Washington, DC: National Academies Press, 25.

little *public* discussion in the United States or in the twentieth century until the 1980s. The discussion began by focusing not on professional responsibility and trustworthiness, but on controlling research misconduct (then commonly and mistakenly called "scientific fraud"). At that time, extreme, flagrant, and sometimes bizarre cases of "research misconduct" (in the technical sense of falsification, fabrication, or plagiarism) and the mishandling of those cases came to light. The quarter-century discussion of ethics in the conduct of research contrasts with the much longer history of discussion of other professional responsibilities of engineers.

After the 1985 publication of Sigma Xi's *Honor in Science*, which sharply criticized Robert Millikan for a lie about his data selection in his 1913 paper on electron charge, others became embroiled in the controversy about how to interpret Millikan's action. This discussion of ethics in research, like many others in the 1980s, was very polarized. Discussants tended either to take the position that Millikan's action was in no way objectionable or to claim that it constituted falsification of his results (a species of research misconduct). (I argue for a different interpretation of the Millikan case in another section of this chapter.)

In the 1980s, a surprising number of scientists responded defensively to evidence of cases of flagrant misconduct and were reluctant to acknowledge the need for greater attention to research integrity. One can get a flavor of those times by reading the first edition (1989) of the National Academy of Sciences' (NAS) *On Being a Scientist*.[10] It was a first attempt by the NAS to contribute something to the education of young investigators about responsible research conduct. However, it neglected to address issues of how to interpret the actions of any *successful* scientist whose research conduct had been questioned. For example, although that first edition contains a picture of a crucial page in Millikan's laboratory notebook that *shows* data points he dropped, nowhere does it discuss Millikan's actions.

---

[10] Committee on the Conduct of Science. 1989. *On Being a Scientist*, first edition. Washington, DC: National Academies Press.

The tone of 1988 congressional oversight hearings chaired by John Dingell that investigated how research institutions were responding to misconduct allegations reinforced the belief of many scientists that both they and research itself were under siege.

Despite the embattled stance of many investigators, the research community had no choice but to acknowledge the clear mishandling of allegations of research misconduct by many universities, detailed in Allan Mazur's 1989 report.[11] For example, John Darsee, a cardiologist and clinical investigator, fabricated data in more than a dozen research papers, and at least forty-five abstracts listed faculty members as coauthors on articles and abstracts without their knowledge or consent, but no effective action was taken against him. Darsee moved from Emory University to Harvard, where he continued the same practices until he was finally caught fabricating data in 1981. Robert Slutsky was an extremely prolific investigator at the University of California, San Diego, writing 160 papers in seven years. He too added coauthors to his papers without justification. After a reviewer questioned the duplication of data in two of his papers, he abruptly resigned. Only then was an investigation launched. Twelve of his published papers contained fabricated results and another forty-eight of his papers were questionable. In the 1980s, a common explanation in scientific circles for these acknowledged instances of research misconduct was that they were due to a very few rogue investigators, most of whom were mentally ill.

Whatever else happened, research institutions did need to develop better misconduct procedures. In order to continue to receive government research funding, especially NIH funding, most research universities at least did begin to establish or improve their procedures for handling allegations of misconduct.

In the early 1980s, Walter Stewart and Ned Feder documented lax behavior by many *coauthors* of John Darsee, behavior that allowed Darsee to deceive the scientific community in a long list of publications. Because some of those coauthors threatened to sue *Nature* if it published the exposé, publication of the Stewart and Feder article was delayed until 1987.[12] Even afterward, the scientific community has been slow to absorb the lesson that *broader lapses of professional responsibility* by those who would themselves never commit research misconduct nonetheless may set the stage for misconduct by others. In 1996, Francis Collins, the head of the National Institutes of Health's Human Genome Project, reported that a junior researcher in his lab (his graduate student) had fabricated data in five papers coauthored with Collins. Many in the scientific community accepted Collins' judgment that he could not have prevented the fabrication or detected it earlier, except via the unacceptable alternative of double-checking everyone's work.[13] Such a quick dismissal of the responsibilities of coauthors, especially senior coauthors, contrasts with the much more nuanced judgment by the committee at Bell Labs that investigated research misconduct

---

[11] Mazur, Allan, *op. cit.*

[12] Stewart, Walter and Feder, Ned. 1987. "The Integrity of the Scientific Literature." *Nature* 325 (January 15): 207–214.

[13] See for example, Eliot Marshall, "Fraud Strikes Top Genome Lab," *Science* 274(5289): 908–910, November 8, 1996.

there in 2002.[14] After absolving the perpetrator's coauthors of any complicity in research misconduct, they went on to raise the difficult and much subtler issue of the professional responsibility of coauthors for work that bears their names (and in this case contained fabricated or falsified data).

> In 2002, a research misconduct investigation committee at Lucent Technologies raised the difficult and much subtler issue of the professional responsibility of coauthors for work that bears their names.

That case at Bell Labs and another case in the same year at Lawrence Berkeley National Laboratory made investigators abruptly aware that research fields in the physical sciences were also vulnerable to research misconduct and that very promising and talented investigators could commit research misconduct. (We briefly consider those cases in the next section.) Furthermore, these cases showed that research misconduct was not confined to trainees and mentally ill underlings. The two investigators who were found guilty of research misconduct in those two cases were "rising stars" in highly visible areas of research.

*How does the development of attention to research conduct by the research community compare with the development of attention to standards of professional responsibility in engineering practice by the engineering community?*

## Distinguishing Falsification from Legitimate "Data Selection"

What are the criteria by which falsification of data is distinguished from legitimate data selection (i.e., legitimately treating some data points differently from others, perhaps even to the extent of disregarding them)?

**Data selection** is the differential treatment of data. When *made according to legitimate criteria*, data selection is an indispensable part of science. It is legitimate to discard some data if that "run" or sample is contaminated – for example, because you dropped the sample on the floor – or if statistical methods that are applicable to the sort of data you have collected warrant discarding some "outliers." To ensure the legitimacy of the criteria used for selection, those criteria should be *explicitly stated*.

> When *made according to legitimate criteria*, data selection is an indispensable part of research. To ensure the acceptability of the criteria used for selection, those criteria should be explicitly stated.

Changing one's data merely to fit one's expectations or preferences is falsification. Changing the value of data is absolutely prohibited, so no question arises of when it is justified to do it. However, excluding some data points or smoothing the curve plotted from the data may be either *justified* data selection or unjustified "cooking." The crucial justification will depend on the characteristics of the data, such as how noisy the data are. Some data selection is carried out by software statistical packages used to "crunch" the data. It is

[14]This 2002 report is available at http://publish.aps.org/reports/.

important to understand the *criteria* that one's data must meet to make the use of a given software package appropriate.

> *What are the criteria by which falsification of data is distinguished from legitimate data selection?*

## Robert Millikan's Treatment of the Data for Determination of Electron Charge

Suppose that in conducting an experiment you sometimes see something strange occurring in your experiment. You have some ideas about the factors that may be confounding your observations. Those ideas have enabled you to improve your experimental setup and reduce the frequency of strange behavior, but you have not yet eliminated all the episodes of strange behavior. For those experimental runs in which you recognize the strange behavior, the data that you obtain look quite different from the data you obtain when nothing looks strange to you. Your reasoning about the phenomena based on the data that you obtained from the experiments that did *not* look aberrant to you has been crucial to developing your understanding of the underlying phenomena, an understanding that has had some independent confirmation. Therefore, you have confidence in your intuition that the data obtained from the aberrant-looking experimental runs are not indicative of the natural phenomena that you are seeking to investigate. How do you describe your method?

The selection and presentation of data are a professional responsibility and require the exercise of judgment. Discretion/judgment is required to recognize sources of "noise" (i.e., extraneous influences on observations of the phenomena under investigation) and to apply statistical methods to deal with noisy data, even where the source of the noise is unknown. Making the required judgments is therefore more complex than simply reliably recording one's observations. Self-deception is more of a risk when one must exercise discretion, however.

The complexities involved in data selection are well illustrated in the story of Robert Millikan, who in 1923 won the Nobel Prize for his work establishing that the electron carries a characteristic amount of charge rather than carrying a varying amount of charge. The story of his data selection to which we shall presently turn is interesting for its bearing on both scientific method and research ethics. It provides both a historically interesting example of the exercise of intuition in science and an example of the evolution of standards of scientific practice in early twentieth-century physics.

As we saw in the introduction, **intuition** is the ability to immediately recognize what is going on in a situation. In contrast, the ability to infer what is going on from other, independently identified evidence or premises is called "*reasoning*." The ability to recognize something without being able to articulate the basis for one's recognition is familiar in everyday life. One example we considered earlier was that of recognizing an acquaintance at a great distance just from the person's walk, without being able to say what it is that is distinctive about that walk. Describing some of Millikan's data selection as the operation of "intuition" rather than as "reasoning" suggests that Millikan was not able to articulate all the features of his observations that made him think something was amiss with some of his experimental observations. He was often able to think of reasonable explanations of why "things went wrong" with the experimental situation, however. Among the hypotheses that he offered for what "went wrong" in some experimental runs were that "two drops stuck together" or that "dust" interfered. Some of these hypotheses helped him improve conditions in subsequent experimental preparations.[15]

In practical work, such as engineering practice or clinical medicine, one's professional standing is based on the successful outcome of one's practice, rather than solely on the quality of the reasons one can give for professional decisions. For example, engineers are credited for the quality, safety, and reliability of the products they have designed. Physicians are credited for the health outcomes of their patients. In most areas of professional practice, reliance on intuition and professional experience is well accepted. It is because of the importance of developing the practitioner's "eye," intuition, or ability to give accurate or insightful judgments that education in practical scientific areas includes internship or apprenticeship experience that allows the student to "get a feel for" what is important and how experienced professionals go about their work. If a practitioner turns out to be wrong very often, others stop seeking that person's professional opinion.

Research has a much more difficult time dealing with judgments based on intuition and experience, as contrasted with those based on reasons that can be fully articulated. This is largely due to the role of peer evaluation in deciding which research results are worthy of publication. In evaluating a research report, the investigators' reasoning, and not just their conclusions, is important. Saying "I discarded all the data taken when there was something funny going on in the experiment" may describe the operation of true insight, but it is not likely to be very convincing to reviewers. Reliance on intuition leaves one especially vulnerable to self-deception, but research cannot entirely dispense with intuition either, as the case of Robert Millikan illustrates.

> In evaluating a research report, the authors' reasoning, and not just their conclusions, is important.

The full story of Millikan's research makes fascinating reading. It has been thoroughly researched and engagingly presented by science historian Gerald Holton in two articles.[16] Holton convincingly argues that Millikan's intuition was part of what made him a better researcher than his rival, Felix Ehrenhaft. In particular, Millikan's ability to recognize and select the data that most accurately reflected the underlying phenomenon was what put him ahead of Ehrenhaft, who indiscriminately used all of his data and therefore

[15] Holton, Gerald. 1978. "Subelectrons, Presuppositions, and the Millikan-Ehrenhaft Dispute," first published in *Historical Studies in the Physical Sciences*, 9, pp. 161–224 and reprinted in his 1978 book *The Scientific Imagination*, pp. 60–61. My page references to Holton's paper are to the reprint in *The Scientific Imagination*.

[16] See Holton, 1978, and "On Doing One's Damnedest: The Evolution of Trust in Scientific Findings," Chapter 7 in Holton's 1995 book, *Einstein, History, and Other Passions*, New York: American Institute of Physics. Holton himself did not comment extensively about Millikan's moral lapse. Instead he was content to display the discrepancy in detail, not only in quotations from Millikan's work but also in a reproduction of two pages from Millikan's laboratory notebook dated March 15, 1912. These pages are shown in Figure 8.1. Among the clearly written comments are "Beauty" on the left-hand page by the data taken from one drop and, on the right-hand page, "Error high will not use" by another drop. (These pages are also reproduced in the first edition [1989] of *On Being a Scientist*.)

came to the conclusion that the electron could hold varying amounts of charge. Figure 8.1 shows pages from Millikan's notebook with his annotations indicating his estimates of the worth of specific data points.

As those who have repeated Millikan's oil drop experiment know, it is often evident that "something funny" is going on – for example, two oil drops may stick together and behave in ways that are different from the behavior of a single oil drop. If an investigator has an *independent* basis for believing that some data are flawed – a basis other than that the values obtained are not the ones expected – then the investigator has some justification for excluding those data points. Of course, it is important to be even-handed in excluding data that are suspect, discarding both those that do and do not support one's hypotheses. For example, if a piece of experimental equipment was discovered to be malfunctioning, an investigator might discard all data back to the last time the instrument was tested and found to be in good working order. Of course, that might also mean discarding a great deal of data.

A present-day evaluation of Millikan's work is complicated by the fact that today's standards for data selection were just developing in Millikan's time. Indeed, Millikan helped develop them. The evidence that he was operating with different methodological criteria is part of what makes the story of his research significant for the history of science and the history of research methodology in particular. For example, Holton quotes passages from Millikan's 1910 paper in which Millikan frankly reports attitudes toward data handling that sound outlandish by contemporary standards:

[I]n the section entitled "Results," Millikan frankly begins by confessing to having eliminated all observations on seven of the water drops. . . . A typical comment of his, on three of the drops, was: "Although all of these observations gave values of e within 2 percent of the final mean, the uncertainties of the observations were such that I would have discarded them had they not agreed with the results of the other observations, and consequently I felt obliged to discard them as it was." Today one would not treat data thus, and one would surely not speak about such a curious procedure so openly.[17]

That Millikan was so open about the methods he used demonstrates how far he was, in 1910, from attempting to deceive anyone. Further, that his paper containing these comments was published in a very prestigious journal, the *Physical Review*, shows that his description of his data handling did not strike his contemporaries as deviant. Standards were developing, however. Three years later, when Millikan published a major paper on the character of electron charge based on his oil drop experiment, he seems to have become self-conscious about the issue of selecting data. In this 1913 paper he writes, in italics,

*It is to be remarked, too, that this is not a selected group of drops but represents all of the drops experimented on during 60 consecutive days.*[18]

[17] Holton, Gerald. 1994. "On Doing One's Damnedest."
[18] Quoted by Holton, 1978, p. 63, from Robert A. Millikan, "On The Elementary Electrical Charge and the Avogadro Constant," *Physical Review*, 2, 1913, pp. 109–143.

Two Pages from Millikan's Notebook Showing His Evaluation of His Data Points (Courtesy of the Archives, California Institute of Technology)

Regrettably, Millikan's statement is false, which he must have known when he made it. It is instructive to examine both this moral lapse on Millikan's part and the heated debate and curious silences about this lapse that have occurred within the scientific community since 1981.

Granted that Millikan was wrong to lie, why did he do it? There is no evidence that Millikan lied about other matters. Indeed, the passage quoted from Holton shows Millikan to have exhibited a praiseworthy openness about his methods

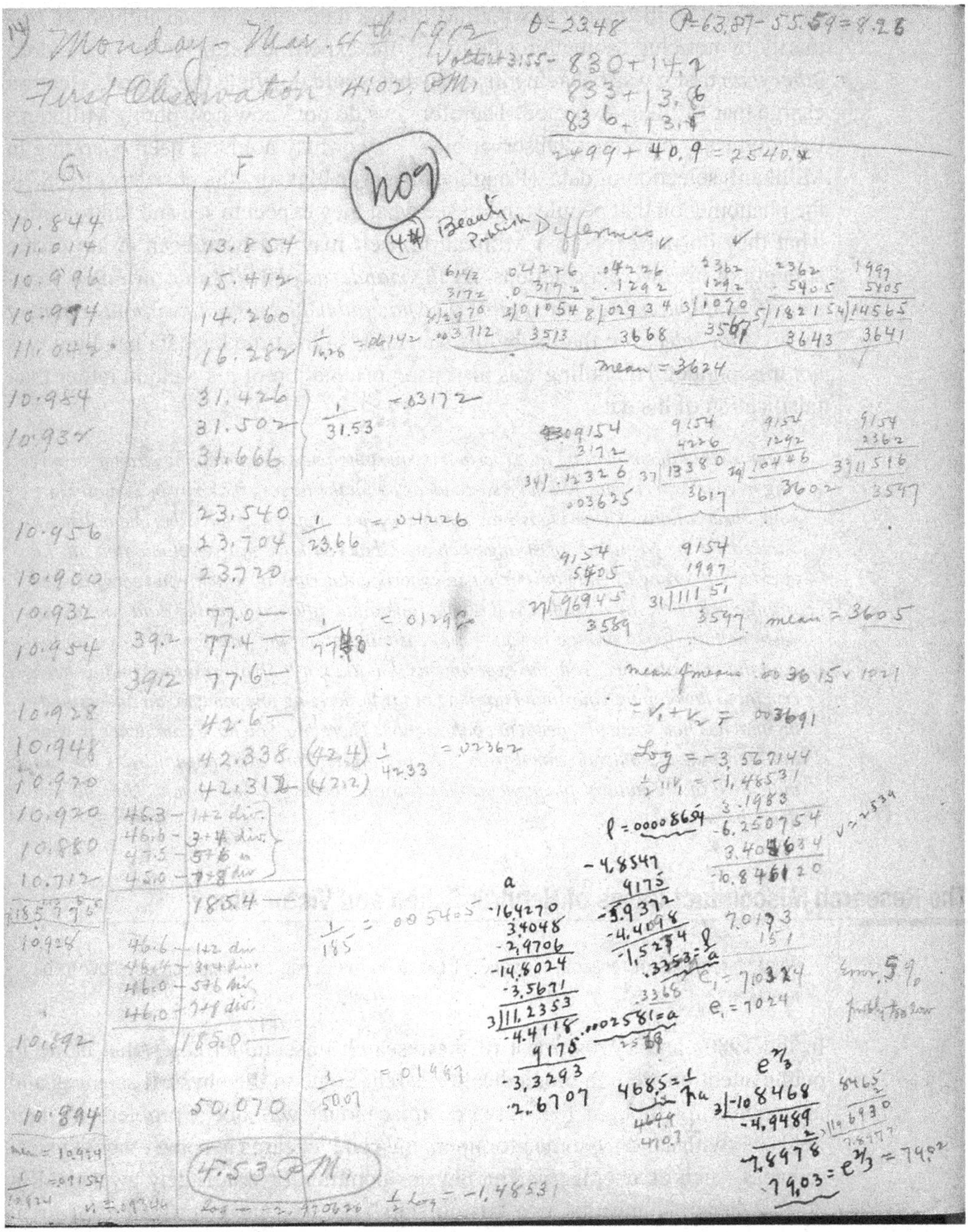

**Figure 8.1** (continued)

(even if the methods themselves look peculiar by present-day standards). Perhaps in the years between 1910 and 1913 Millikan had become aware of the emerging opinion that researchers ought to give reasons for discarding data. Perhaps someone challenged him on the passage from his 1910 paper confessing his method. In the research for his 1913 paper Millikan had used methods for data selection that he could not fully explain. By today's standards, Millikan would have been expected to give reasons for discarding some readings. What we do not and cannot know is the extent to which Millikan's data selection was influenced primarily by noticing "something funny" in the experimental behavior of the drop *other than* that it was behaving in a way that would not yield the value of electron charge that Millikan expected. Therefore, we do not know how purely Millikan's intuition was operating. "Observer bias" may or may not have been operating in Millikan's selection of data. (Roughly, **observer bias**, or "the observer effect," is the phenomenon that people tend to see what they expect to see and fail to notice what they do not expect.[19]) Millikan himself may not have been aware of the strength of his own expectations. *By the standards of Millikan's own time, however, it would have been acceptable if he had published his data without comment on his data selection*; that is, by the standards of his time his data handling was *not* misconduct. His failing was his misrepresentation of his method rather than falsification of his data.

> *Suppose that in conducting an experiment you sometimes see something strange occurring in your experiment. You have some ideas about the factors that may be confounding your observations. Those ideas have enabled you to improve your experimental setup and reduce the frequency of strange behavior, but you have not yet eliminated all the episodes of strange behavior. For those experimental runs in which you recognize the strange behavior, the data that you obtain look quite different from the data you obtain when nothing looks strange to you. Your reasoning about the phenomena based on the data that you obtained from the experiments that did not look aberrant to you has been crucial to developing your understanding of the underlying phenomena, an understanding that has had some independent confirmation. Therefore, you have confidence in your intuition that the data obtained from the aberrant-looking experimental runs are not indicative of the natural phenomena that you are seeking to investigate. How do you describe your method?*

## The Research Misconduct Cases of Hendrik Schön and Victor Ninov

What stereotypes of research misconduct cases do the Schön and Ninov cases show to be mistaken?

In the 1980s and 1990s, most of the research misconduct cases that came to public attention were in biomedical research. Some in the physical sciences and in engineering thought that research misconduct was not a problem in their fields, notwithstanding some prominent misconduct cases in some other physical sciences, such as chemistry. The physics community was rudely awakened in 2002, when it found that two "rising stars" (i.e., young investigators of great promise) had each committed research misconduct and done so in cutting-edge areas of research conducted at two of the most prestigious research facilities in the United States. Although mental illness was used to explain the guilty party's behavior in some early misconduct cases,[20] there was no evidence of mental illness here.

---

[19] Social psychology also has a narrower meaning for this term. See http://psychology. about.com/od/aindex/g/actor-observer.htm.

[20] Racker, Efraim. 1989. "A View of Misconduct in Science," *Nature* 339(May): 91–93.

## The Misconduct of Hendrik Schön at Bell Labs

An investigation committee was formed in May 2002 by Lucent Technologies/Bell Labs to investigate "the possibility of scientific misconduct, the validity of the data and whether or not proper scientific methodology was used in papers by Hendrik Schön, et al., that are being challenged in the scientific community."[a]

By June 20, 2002, allegations of misconduct had been made about twenty-five of Schön's papers (all published with coauthors). The committee selected twenty-four of these papers for detailed examination and grouped them into three classes:

+ *"Substitution of data* (substitution of whole figures, single curves and partial curves in different or the same paper to represent different materials, devices or conditions)

+ *Unrealistic precision of data* (precision beyond that expected in a real experiment or requiring unreasonable statistical probability)

+ *Results that contradict known physics* (behavior inconsistent with stated device parameters and prevailing physical understanding, so as to suggest possible misrepresentation of data)"

... The Committee requested primary (raw) data files for some of the papers but was unable to examine them because they no longer exist[ed]."

Although the committee found Hendrik Schön to be "a hard working and productive scientist," it found there was much to criticize in Schön's behavior and classified some of his actions as research misconduct.

The committee found that Hendrik Schön alone had performed all device fabrication (i.e., device construction), physical measurement, and data processing in the work in question (with minor exceptions) and that no coauthor or other colleague had participated beyond providing starting materials, nor had any coauthor or other colleague witnessed the most significant physical results. Schön had not systematically maintained proper laboratory records for the work in question. Furthermore, he had deleted virtually all primary (raw) electronic data files, "reportedly because the old computer available to him lacked sufficient memory." The devices with which one might have confirmed the claimed results were not available to the committee, having been damaged or discarded. "Finally, key processing equipment no longer produces the unparalleled results that enabled many of the key experiments. Hence, it is not possible to confirm or refute directly the validity of the claims in the work in question." However, the committee found compelling evidence "that manipulation and misrepresentation of data occurred. In its mildest form, whole data sets were substituted to represent different materials or devices. Hendrik Schön acknowledge[d] that the data are incorrect in many of these instances. He state[d] that these substitutions could have occurred by honest mistake. The recurrent nature of such mistakes suggests a deeper problem. At a minimum, Hendrik Schön showed reckless disregard for the sanctity of data in the value system of science. His failure to retain primary data files compound[ed] the problem. More troublesome [to the committee were] the substitutions of single curves or even parts of single curves, in multiple figures representing different materials or devices, and the use of mathematical functions to represent real data. Hendrik Schön acknowledge[d] these practices in many instances, but state[d] that they were done to achieve a more convincing representation of behavior that was nonetheless observed." The committee found "these practices ... completely unacceptable" and concluded that they represented research misconduct. Indeed, the committee judged that "of the twenty-four Final Allegations examined, Hendrik Schön committed scientific misconduct in sixteen.... Of the remaining eight, two were judged to have no clear relationship to publications, while six were troubling but did not provide compelling evidence of scientific misconduct."[b]

---

[a] The Lucent Technologies 2002 *Report of the Investigation Committee on the Possibility of Scientific Misconduct in the Work of Hendrik Schön and Coauthors* is available on the American Physical Society's Web site, http://publish.aps.org/reports/lucentrep.pdf.

[b] *Ibid.*, 12. Note that despite the 2002 date, the committee uses the term, "scientific misconduct" rather than "research misconduct."

---

## The Misconduct of Victor Ninov at Lawrence Berkeley National Laboratory

In July 2002, officials at Lawrence Berkeley National Laboratory concluded from their investigation that research leading to the 1998 claim to have discovered two new elements was riddled with research misconduct by Victor Ninov, an investigator at the lab who had been regarded as an expert on the physics of heavy elements.

Although the discovery of two new elements (a supposed "super-heavy" element, called "118," for the atomic number of this supposed particle, and its decay product called "116") had been greeted with much acclaim,[a] other investigators were not able to replicate the experimental results on which the claim was based. The failure to replicate Ninov's results led officials at Lawrence Berkeley National Lab to undertake their own investigation of the research. Prior to that investigation Ninov's colleagues and coauthors had left it to Victor Ninov alone to deal with the raw data, because only Ninov knew how to run the computer programs that analyzed the data.

The investigation found evidence in a computer log file that data had been cut and pasted and numeric values had been changed. Ninov claimed he was innocent and pointed out that others had access to the computer that contained the data.[b]

Lawrence Berkeley National Laboratory fired Ninov, reprimanded others for not being sufficiently vigilant, and issued a news release withdrawing the discovery.

---

[a] A 1999 news story about the article in *Physical Review Letters* announcing the discovery of 118 (an article that has since been retracted) may be found at http://physicsworld.com/cws/article/news/3027. The news story also reports the pride that Bill Richardson, then head of the Department of Energy (which runs the Lawrence Berkeley labs), took in the supposed discovery.

[b] Johnson, George. 2002, October 15. "At Lawrence Berkeley, Physicists Say a Colleague Took Them for a Ride," *The New York Times*, F1. This article is available at http://www.nytimes.com/2002/10/15/science/at-lawrence-berkeley-physicists-say-a-colleague-took-them-for-a-ride.html.

---

That two such promising investigators would commit research misconduct seems inexplicable to many. Some people have hypothesized that each felt work pressure, but pressure does not explain why someone would commit misconduct in research that was sure to receive scrutiny and therefore virtually certain to be detected eventually. It is foolhardy as well as unethical to attempt to convince other investigators of a nonexistent relationship in a "hot" area of research, because others would attempt replication or other experiments that would reveal the truth. Therefore, it is likely that Schön and Ninov began their slide down the slippery slope of research misconduct by attempting to make the data look more convincing as evidence for a phenomenon that *each sincerely believed existed in nature*. This hypothesis is supported by the fact that Schön explained his having substituted single curves or even parts of single curves in multiple figures representing different materials or devices and using mathematical functions to represent real data by saying "that these practices were used for the purpose of achieving a *better and/or more convincing representation of the observed phenomena.*"[21] The investigation committee concluded that such practices constituted *research misconduct*, however.

> *What stereotypes of research misconduct cases do the Schön and Ninov cases show to be mistaken?*

## Fabrication: From Hoaxes to "Cutting Corners"

> What does James Urban's fabrication of data illustrate about the role of recklessness (as contrasted with an intent to deceive) in research misconduct?

In the 1980s famous hoaxes, such as the Piltdown man hoax, were included in such discussions of research misconduct as Broad and Wade's book, *Betrayers of the Truth* and in the PBS NOVA video, *Do Scientists Cheat?* The term "**hoax**" has the connotation of fooling others for the sake of doing so rather than for some other end, such as appearing to be a productive researcher. Perpetrators of hoaxes are often anonymous. "Piltdown man" was a hoax, created by passing off a combination of human and ape bones as the remains of a single humanoid "missing link." Such famous hoaxes in science are *intentional* deceptions, but are rather different from most cases of research misconduct. Hoaxes are rare in engineering, medicine, or the natural sciences. The few hoaxes that I have encountered in these fields have been hastily concocted ruses, such as rumors about computer viruses, "Trojan horses" and "email bombs."

Attempts to mislead others about the character of underlying phenomena are rare, even in cases where misconduct is found. In a more common scenario, the perpetrator believes in the truth of a certain conclusion about natural phenomena and has at least some grounds for that belief, but acts to deceive others about the nature and strength of the evidence for that conclusion. An especially clear illustration is the fabrication of data in a research article that James Urban submitted to a journal.

### The Fabrication of Data by James Urban

In the early 1990s, James Urban, a post-doctoral fellow at Caltech, was found to have fabricated data in a manuscript he *originally submitted* to the journal *Cell*. He claimed that the data reported in the *published version* of the paper were genuine, however. The data in the published version were certainly different from those in the manuscript originally submitted to *Cell*. Some of Urban's lab books

---

[21] Lucent Technologies, 2002, 12. Italics added.

were missing and so could not be examined. (He said that they were lost in a subsequent move across the country.)

Urban *did not deny the charge of fabrication*, but he did deny any *intent to deceive*. Clearly, he did intend to deceive the editor and reviewers for *Cell* into thinking that he had obtained experimental results that he had not in fact obtained; that much intent to deceive is implied by the term "fabrication." One official close to the case said that Urban believed he knew how the experiment would turn out and, because of the pressure to publish, tried to "speed" the review process by fabricating the data in the original manuscript. The official was convinced that Urban would not have published without having first inserted data he had actually obtained experimentally.[a] Therefore, the point of Urban's denying an *intent* to deceive was that he did not intend to deceive others *about the natural phenomena*. Apparently, Caltech interpreted Urban's intentions in fabricating data as Urban described, because they found him guilty of "serious misconduct" but not of "fraud," which Caltech distinguished from "serious misconduct" and regarded as a graver charge. The *Cell* article was retracted.

[a] Roberts, Leslie. 1991. "Misconduct: Caltech's Trial by Fire," *Science* 253: 1344–1347 (September 20); p. 1346.

This example further illustrates that fabrication of research results, although a serious misrepresentation, often does *not* represent as *true* a conclusion about some phenomena that the perpetrator *knows to be false or has no sound basis for thinking true*. An investigator committing misconduct may have some scientific evidence for a conclusion but seeks to *deceive others about the strength of the evidence* for that conclusion.

Fabrication of *plausible* findings in which one firmly believes, no less than fabrication of results to convince others of what one knows to be false, thwarts the progress of scientific research and confuses the application of research results for human benefit. It is a reckless act, rather than an intentionally deceptive one, however. If the accusation of intent to deceive is raised, it is important to ask whom the perpetrator is intending to deceive *about what*, because although all deception may need moral justification, deceptions differ in their gravity.

Recklessness in research is similar to recklessness in other contexts, such as driving too fast to keep control of one's vehicle. **Recklessness** is manifest in taking serious risks that, ethically speaking, *one ought not to take*.* It shows a disregard of major values and standards and is, therefore, irresponsible. Not only is recklessness, rather than actual fraud, at the heart of much research misconduct[22] but "reckless research" also underlies many of the lesser departures from responsible research conduct that the NRC panel that authored *Responsible Research* (among others) calls "**questionable research practices**," but which might better be called "ethically objectionable research practices."

*The corresponding legal notion of recklessness is the equivalent of the legal notion of gross negligence. It is thus an extreme form of negligence, a concept that we considered in Chapter 1.

[22] In addition to the sources already cited, the reader will find descriptions of misconduct cases frequently written up in *Science* and in journals of the disciplines in which they occurred. The Office of the Inspector General of NSF issues a semi-annual report to the Congress, which includes a section on oversight dealing with misconduct cases and other audits and inspections conducted by that office.

> **Recklessness** is manifest in taking serious risks that, ethically speaking, *one ought not to take*. It shows a disregard of major values and standards.

Prior to the 2000 adoption of the government-wide definition of research misconduct, such misconduct was taken to require the *intent* to deceive others into believing true something the perpetrator knew to be false, or at least had *no* good reason for thinking true. (As we have seen now reckless as well as intentionally deceptive action is sufficient for an act to be research misconduct.) Because "intent" is difficult to prove, some previous findings of research misconduct, such as that against Thereza Imanishi-Kari,[23] were overturned on appeal.

In nonresearch contexts describing an act as "**cutting corners**" may mean just taking some shortcut, but in connection with research practice the phrase is regularly used to describe practices that, unlike honest mistakes or even careless mistakes, are *knowingly* undertaken *violations of the standards of good research practice*. Even though it is not the researchers' intention to put bogus results into the literature or to misappropriate credit, cutting corners often risks doing so. Eleanor Shore, who served as the dean charged with oversight of research ethics at Harvard Medical School, observed that personal expediency is often the motive for such acts.[24] In the name of expediency, perpetrators may recklessly disregard accepted standards of research practice and so put at risk the integrity of research results or the proper assignment of credit for research.

> In research contexts, the term "cutting corners" is regularly used to describe practices that, unlike honest mistakes or even careless mistakes, are *knowingly* undertaken *violations of the standards of good research practice*.

Notice that some violations of accepted standards of research practice do *not* put at risk the integrity of research results or fair crediting of research contributions. Such acts are often seen as unprofessional behavior because they are seen as lessening the dignity of research as a profession or in some way undermining the professional autonomy of investigators, but they are *not* seen as violations of research ethics. An example is the announcement of research findings in a press conference rather than by publishing in a peer-reviewed publication.[25] The argument for doubting that research integrity will be compromised thereby is that although a publication-by-press-conference may confuse the public, other researchers will not regard such findings in the same light as results published in peer-reviewed journals.

Some practices that imperil research integrity do not violate standards of research practice within a given field of research because they have yet to be generally recognized as inferior. The earlier discussion of Robert Millikan's data selection methods illustrates how standards of research practice change.

Consider how the definition of recklessness fits the James Urban case discussed earlier in this section. Let us accept Urban's account of what he did: He did do the experiments in question; they are accurately reflected in the published (and subsequently withdrawn) version of his paper; he did not intend to put fraudulent

---

[23] See the first edition of this book for a detailed discussion of that case.
[24] Shore, Eleanor. 1995. "Effectiveness of Research Guidelines in Prevention of Scientific Misconduct," *Science and Engineering Ethics* 1, 4(October): 383.
[25] This is a practice that the NAS panel classifies as a "questionable research practice" *op. cit.* p. 28.

data into the literature but only to shorten the delay between obtaining research results and publishing them.

Suppose, however, that the experiments that he conducted had given him a significantly different result than what he had projected. Would he have had the moral courage to withdraw the paper once accepted, at the risk of offending the editors of a prestigious journal like *Cell?* Suppose that he had died or become incapacitated and so was not able to complete the experiments, or an earthquake or other accident had disrupted his laboratory. In addition to avoiding certain "delays" built into the standard publication procedures (and so giving him an unfair advantage over any competing investigators), his action also endangered the integrity of his research results, because in many circumstances, exceptional efforts would have been required to publish only genuine results. In some circumstances, he would not have been able to act at all. His case illustrates the wisdom of the inclusion of "recklessly" in the 2000 U.S. government-wide definition of research misconduct.

Recklessness in misconduct cases is further illustrated by a university's finding of misconduct against a researcher accused of plagiarism. (Plagiarism, the appropriation of another's ideas or writings and representation of them as one's own, will be discussed in detail in Chapter 9.) The accusation was made against an investigator in chemistry for using text from published articles, without attribution, in his grant proposal to NSF. The investigator had copied the work of others verbatim into his notes without quotation marks or attribution. As a result he could not distinguish his own work from that of others when he came to use his notes in writing a grant proposal. Although this was *not* the perpetrator's *deliberate misrepresentation* of another's work, it was not simply a careless mistake either. Dropping some quotation marks in transcribing some notes would have been a careless, perhaps even a negligent mistake. In this case, however, there were no quotation marks to lose. The failing was one of recklessness even if it was not the intentional or calculated theft of ideas and words. His university found him guilty of research misconduct (but not plagiarism) and said that the subject had displayed "a reckless disregard for appropriate procedures of scholarship" and had "knowingly and repeatedly [engaged] in a pattern of research note taking that given enough time, was inevitably going to produce precisely the situation that arose with his NSF grant proposals."[26]

> Reckless research takes risks that are irresponsible and that *one ought not to take.*

*What does James Urban's fabrication of data illustrate about the role of recklessness (as contrasted with an intent to deceive) in research misconduct?*

## Self-Deception in Research Misconduct

*What is self-deception and how does it compare with observer bias?*

The introduction discussed self-deception as the failure to spell out, even to oneself, what one is doing, in circumstances under which it would be normal to

[26]Office of the Inspector General, National Science Foundation. 1992. Semiannual Report to the Congress, No. 7, April 1, 1993–September 30, 1993, 37.

do so. This characterization[27] makes it clear both why self-deception is regarded as a moral failing and why self-deception, like ordinary deception, is less than honest.

The moral (and cognitive) failing of self-deception is different from the psychological fact of **observer bias**. Psychology has studied the normal human tendency to see what one expects to see. Research methods have been refined to control for that bias. For example, in research in which the effect of some variable is being tested in a clinical setting, a double-blind method is used. In a **double-blind** clinical study neither the patients who are study subjects, nor the observer who records the data, are aware of which patients are part of the experimental group and which are members of the control group. Today, when we recognize at least some sources of observer bias, to fail to control for such bias *would* be negligent and perhaps self-deceptive, *not* because the bias itself is self-deception, but because believing that one is immune from such bias would be.

The 1989 edition of *On Being a Scientist* confused observer bias with self-deception and so makes it appear that those previous generations of scientists whom we would now say were influenced by observer bias were all self-deceived. However, where *no one* recognizes the phenomenon of observer bias, an individual investigator is not self-deceived for failing to control for it.

A clear example of self-deception in the recent history of research is the case in which someone on the Pons and Fleischmann team changed the observed value of a signal line on the γ-ray spectrum to a value appropriate to the neutron production he interpreted it to indicate.

---

### Self-Deception in a Report of Cold Fusion

On March 23, 1989, Stanley Pons and Martin Fleischmann announced a remarkable breakthrough – the accomplishment of nuclear fusion (of deuterium) at *room temperature* in their University of Utah laboratory. Their evidence for this "cold nuclear fusion" was a graph that recorded a supposed gamma-ray emission when a cube of palladium had "melted and partly vaporized."[a] An electric current had passed through a palladium electrode and a platinum electrode in a beaker of heavy water and lithium. The gamma-ray energy showed a peak at 2.22 MeV, the unique amount of energy released when a neutron is captured by a proton. (Neutron capture is possible only if the neutron has been generated.) Some scientists were skeptical.

In fact, the research and results that Pons and Fleischmann and their supporters found proved to have many discrepancies. One critic was Richard Petrasso. Petrasso's group requested that Fleischmann et al. show the full gamma-ray spectra that he observed. It revealed that the signal line had an energy of 2.496 MeV, not 2.22 MeV. (In addition, the line lacked a "Compton edge," a specific pattern that should have been evident at 1.99 MeV. The Compton edge results from the detector's reaction to gamma rays.) Petrasso also pointed out that the gamma-ray line reported by Fleischmann et al. was two times smaller than the resolution the

---

[27] For example, see the first edition (1989) of *On Being a Scientist*, which treated self-deception as a manifestation of the fallibility of human perception, reasoning, and foresight.

measuring instrument would permit. Other scientists showed that the pressure claimed by Fleischmann et al. was a miscalculation.[b]

---

[a]Taubes, Gary. 1993. *Bad Science: The Short Life and Weird Times of Cold Fusion.* New York: Random House, 4.

[b]*Ibid.,* 3–4, 44–45, 142–3, 167–174, 253–4. See also the exchange between Fleischmann and Pons and Richard Petrasso published as "Measurement of Gamma-Rays from Cold Fusion," *Nature* 339 (June 29, 1989): 667–668.

---

Had cold fusion as originally envisioned proven possible,[28] it would have shown a relatively inexpensive route to releasing energy from fusion to be possible. This would have had enormous economic (and military) implications. In the case of research with implications as momentous as cold fusion, no one would expect to engage in falsification and have the act go undetected, because the results would be scrutinized in excruciating detail. Changing that value makes sense only if one assumes that the investigator (or the team) was fully confident that he had observed cold fusion, that he believed he knew what the data should be and deceived himself into believing that he was correcting, rather than falsifying his data in changing the value of the signal line. No charges of research misconduct were brought, but the research community came to view the Pons and Fleishmann report of cold fusion as an odd aberration from good scientific practice.

*What is self-deception and how does it compare with observer bias?*

## Honesty about Method and Results Central to Research Integrity

Why is ensuring the integrity of data/results more than a matter of following rules about how to keep one's lab notebook or data record?

---

### Group Misconduct Regarding Laboratory Procedure[a]

You are a graduate student working as part of a group on a large project. The results from your group experiments are used for other experimental work. Your faculty supervisor, the principal investigator (PI) for the project, wants you to use a new procedure in your experimental work. She expects the new procedure to yield results that are better suited to the conditions of the other experimental work. The other members of your group do not want to change the procedure they have been using; the new one requires significantly more work. They believe the PI will not notice if the old procedure is used.

You rely on the group for assistance in your own thesis work, but if you go along with the decision to use the old procedure, the quality of the data will most likely be inferior, you will mislead the PI, and perhaps the whole scientific community.

You argue for using the new procedure and informing the PI that the work will just have to take longer – information that she is not likely to receive well. The rest of the group is not persuaded. What should you do and how can you go about it?

---

[a]Based on a scenario by Arun Patel and Ravi Patil (MIT '92).

---

[28]See David L. Goodstein, 1994, "Whatever Happened to Cold Fusion?" *Engineering & Science,* Fall, 15–25; reprinted from *The American Scholar,* 63:4 Autumn 1994.

Data now come in many forms – no longer just observations to be recorded in laboratory notebooks, but, for example, photographs and micrographs as well.* Using a bound notebook with numbered pages for laboratory observations may have been a good standard practice when all data were in the form of written observations, but that norm is not applicable to data in the form of photographs or computer printouts. Furthermore, safeguarding the integrity of results is as much a matter of being truthful about one's methods as accurately reporting one's data.

> Safeguarding the integrity of results is as much a matter of being truthful about one's methods as accurately reporting one's data.

As science and engineering have become more specialized, the need for collaboration among investigators with different expertise has increased. Researchers often have only a very general idea of the standards of research practice applicable in other disciplines. The need for large-scale studies has produced new collaborative arrangements among many individuals, which may involve many institutions. Such large collaborations can create new occasion for error, confusion, and misrepresentation.[29] In some cases, this has led to questionable findings and eroded public confidence in the value of research.

*Why is ensuring the integrity of data/results more than a matter of following rules about how to keep one's lab notebook?*

## Factors That Undermine Research Integrity

What factors might contribute to an investigator's willingness to start down the slippery slope of falsification or fabrication of data or methods?

Various explanations have been offered for research misconduct. Some early writers like Broad and Wade in their 1982 book, *Betrayers of the Truth*,[30] presented falsification and fabrication of results as a problem with a long history in science. Broad and Wade were rightly criticized, however, for failing in some cases to distinguish between dishonesty and the use of methods (such as the data selection by Robert Millikan) that would fail to be acceptable by today's standards, but were acceptable in earlier periods.

A 1994 National Research Council study argued that cutbacks in research funding for the biological and biomedical sciences had disproportionately deprived

---

*Falsification of photographs is a particular problem for research. Currently, editors at some journals, such as *The Journal of Cell Biology*, estimate that 20 percent of the articles their journal (unknowingly) accepts have at least one image that has been inappropriately manipulated. Hany Farid, "Seeing is not Believing, Doctoring digital photos is easy. Detecting it can be hard." *IEEE Spectrum*, August 2009, 44.

[29] Norman, Colin. 1988. "Stanford Inquiry Casts Doubt on 11 Papers," *Science* 242 (November 4): 659–661.

[30] Broad, William J. and Nicholas Wade. 1982. *Betrayers of the Truth*. New York: Simon and Schuster.

young investigators of research funds.[31] The threat of the loss of their careers placed able young investigators under exceptional pressure. The shortage of jobs for PhDs in physics in the last decade may have created similar pressures to "cut corners."

Some graduate students have reported feeling driven to falsification or fabrication by pressure from their research supervisors to find experimental confirmation of the supervisor's own theory.[32]

The number of graduate students per faculty research supervisor has grown dramatically in some fields, which raises serious questions about the quality of research supervision and mentoring for those students. The lack of faculty supervision is further complicated by the presence of post-docs in some fields. The presence of post-docs has sometimes meant that they are the primary recipients of faculty supervision and graduate students are more dependent on supervision by post-docs. The supervision of graduate students by post-docs may or may not receive faculty oversight.

Some features of undergraduate education in science and engineering may be inadvertently fostering bad research conduct. Academic integrity surveys at research universities show that an alarming number of science and engineering students – the majority at some universities – admit to falsifying their lab reports.[33] The explosive growth in the number of scientific investigators after WWII has made it difficult for new social controls to emerge rapidly enough to replace those that functioned when the community of investigators was smaller so "everyone knew each other." Other changes, especially

- The prospect of major financial gain for investigators in life science fields that had not previously known it
- The dramatic increase in
  - Articles with multiple authors and
  - Interdisciplinary collaborations,

have required investigators to address new types of problems of fairness and research oversight despite an absence of consensus about norms for addressing those problems.

*None* of these explanations ethically *justifies* research misconduct or the more general irresponsible research conduct, but identification of factors that foster it

---

[31] National Research Council Committee on the Funding of Young Investigators in the Biological and Biomedical Sciences. 1994. *The Funding of Young Investigators in the Biological and Biomedical Sciences.* Washington, DC: National Academies Press.

[32] For example, see the case of the student who faked data for a version of the volley theory of audition in Nelson Kiang, 1995, "How are Scientific Corrections Made?" *Science and Engineering Ethics* 1, 4, (October): 347–356.

[33] See for example, Elizabeth W. Davidson, Heather E. Cate, Cecil M. Lewis, Jr., and Melanie Hunter, 2000, "Data Manipulation in the Undergraduate Laboratory: What Are We Teaching?" *Investigating Research Integrity Proceedings of the First ORI Research Conference on Research Integrity* available at http://ori.hhs.gov/documents/proceedings_rri.pdf (pp. 27–34 in the hard copy). I have had access to confidential documents from several universities that show the same pattern.

may help the research community and individual investigators recognize temptations to cheat and take action to remove or resist those temptations.

*What factors might contribute to an investigator's willingness to start down the slippery slope of falsification or fabrication of data or methods?*

## The Emerging Emphasis on Understanding and Fostering *Responsible* Conduct

What advantages are there to fostering responsible research conduct rather than simply identifying and punishing research misconduct?

A broader concern with fostering responsible conduct, and not merely detecting and punishing research misconduct, is now emerging. The concern with the broader aspects of responsible conduct has several sources. As the 1992 report of the MIT Committee on Academic Responsibility found,[34] charges of misconduct are prone to arise in settings where instances of other wrongdoing, abuse, and conflict have been left unresolved. The correlation between misconduct charges and poor research environments suggests that better responses to subtler problems of research conduct can reduce the incidence of misconduct charges. As those who have been through misconduct investigations can testify, misconduct investigations are time-consuming and impede research because they frequently impound data, and are emotionally harrowing, whatever the outcome. More fundamentally, trust is essential to the maintenance of the research enterprise. This latter point was articulated in the mid-1990s in several influential publications, including an opinion piece in *Science* by Bruce Alberts and Kenneth Shine,[35] then presidents of the National Academy of Sciences and the Institute of Medicine, and in the third edition (2009) of *On Being a Scientist*.[36] As the preface to the second edition of *On Being a Scientist* put it:

> The scientific enterprise is built on a foundation of trust. Society trusts that scientific research results are an honest and accurate reflection of a researcher's work. Researchers equally trust that their colleagues have gathered data carefully, have used appropriate analytic and statistical techniques, have reported their results accurately and have treated the work of other researchers with respect.[37]

---

[34] Committee on Academic Responsibility Appointed by the President and Provost of MIT. 1992. *Fostering Academic Integrity*. Boston: Massachusetts Institute of Technology.

[35] Alberts, Bruce and Kenneth Shine. 1994. "Scientists and the Integrity of Research," *Science* 266 (December 9): 1660.

[36] Committee on Science, Engineering, and Public Policy of the National Academy of Sciences, National Academy of Engineering, and Institute of Medicine. 1995. *On Being a Scientist*, second edition. Washington, DC: National Academies Press.

[37] *Ibid.*, ix. Available online at http://www.nap.edu/catalog.php?record_id=12192. The previous edition (second, 1995) is also available at http://books.nap.edu/books/0309051967/html/index.html. The second edition, like the third, emphasizes fostering trust.

## When Is Trust a Good Thing?

As Annette Baier has argued, not all trust relationships are worth creating or maintaining. She argues that to be "morally decent" trust should *withstand the disclosure of the basis for that trust*. Therefore, if a research supervisor's trust in her supervisee's honesty is based on the belief that the supervisee is too timid or unimaginative to fabricate data or experiments, disclosure of that belief will give the supervisee an incentive to cheat. Therefore, although closer oversight of research may well reduce dishonesty, the moral climate of scientific research will suffer further, if oversight reduces dishonesty *only* by instilling fear of detection.

A corollary is that a culture of suspicion and disappointment undermines confidence in the results on which one builds, clouds the joys of discovery, spoils the pleasures of teamwork, and destroys many daily satisfactions of research investigation as well as complicating many research activities. The destruction of the existential pleasure of investigation receives surprisingly little attention, except tangentially in discussions of the high attrition or even suicide rates among trainees in some laboratories and departments.[38]

There is no intrinsic scarcity of enjoyment or satisfaction that people may take in mastery of the skills and acquisition of the virtues needed to conduct research. However, such pleasure is eroded by evidence that others are exploiting one's trust to get a competitive advantage in seeking external rewards, such as status and money. If the pleasure in doing research is eroded, only such intrinsically scarce external rewards will remain as incentives. In that case, competition for those external rewards will become more cutthroat as the fear of detection becomes the only check on "cutting corners" in pursuit of those external rewards.[39]

*What advantages are there to fostering responsible research conduct rather than simply identifying and punishing research misconduct?*

## Responsible Authorship, Reviewing and Editing

What responsibilities does an author of a research article have?

In the fifth section of the introduction we reviewed five of the fourteen guidelines for authors that the ACS gives to those seeking to publish in any of the ACS journals (including *Chemical and Engineering News*). Those guidelines specified the obligations of authors to

- Refrain from plagiarizing*
- Present an accurate account of their research

---

[38] See for example, Alison Schneider, "Harvard Faces the Aftermath of a Graduate Student's Suicide," *The Chronicle of Higher Education*, October 23, 1998, A12.

[39] Roberts, Kavussanu, and Sprague reported on the effect of two research climates: One emphasized the acquisition and exercise of mastery (mastery of a field, becoming a proficient investigator); the other emphasized getting research results. They found that the "mastery" environments were more supportive of the intellectual and professional development of trainees than were the "results" environments. Roberts, Glyn C., Maria Kavussanu, and Robert L. Sprague. 2001. "Mentoring and the Impact of the Research Climate," *Science and Engineering Ethics* 7: 525–537.

*The ACS added an explicit prohibition of plagiarism only in its 2010 revision, which accounts for that prohibition appearing as item thirteen in its list of obligations of authors. The ACS models its guidance on refraining from plagiarism on guidance contained in *Authorial Integrity in Scientific Publication* from the Society for Industrial and Applied Mathematics (SIAM). (The ACS credits SIAM.) *Authorial Integrity in Scientific Publication* is available at http://www.siam.org/books/plagiarism.php. (See the later discussion of the responsibility for animals used in experiments.)

- Use journal space wisely
- Reveal any hazards in the conduct of their experiments (and so protect the safety of other investigators who might seek to replicate their experiments)
- Identify all sources of information contained in their research report that are not common knowledge [for the readers of the ACS publication to which a manuscript is being submitted] and refrain from revealing any information obtained from confidential sources
- Reveal any financial or other conflicts of interest (i.e., competing financial or other interests that might be affected by publication of the article)

Other obligations of authors set out in these ACS guidelines address the subjects of:

- Providing sufficient detail about research to allow others to replicate the work
- Citing previous publications that were influential in determining the nature of the research described in the submitted manuscript and the sources of any research materials supplied by a nonauthor.
- Refraining from personal attacks in criticizing others' work.
- The distinction between those who qualify for coauthorship and those whose contributions to the reported research should be included in the acknowledgements section.
- The responsibility and accountability of all authors for the research results reported in the submitted research report.
- The special obligation of the author who submits an article [to an ACS publication] to ensure that all coauthors have received a draft of the manuscript and agreed to be authors, and that only those qualified to be authors are listed.
- An author's obligation to refrain from submitting the same research for publication more than once (which is often called "**duplicate publication**" or, less appropriately, "self-plagiarism") and inform the editor [of the ACS publication] and supply that editor with copies of any related manuscripts that the author has under editorial consideration or in press.
- The ACS strictures against duplicate publication are less severe than those of some biomedical journals, however. Thus, the ACS guidelines say: "It is generally permissible to submit a manuscript for a full paper expanding on a previously published brief preliminary account (a "communication" or "letter") of the same work. However, at the time of submission, the editor should be made aware of the earlier communication, and the preliminary communication should be cited in the manuscript." The ACS guidelines on prior publication are like those found in engineering fields where electronic preprints and the like are common.
- Conforming to institutional requirements as defined by one's institutional animal care and use committee (IACUC) in conducting research with animals (see the later discussion of the responsibility for animals used in experiments).

The Schön and Ninov cases discussed earlier do reveal the research community's emerging awareness of how the responsibility of coauthors for an accurate account of research does involve some oversight of the work of their coauthors.

---

### Professional Responsibilities of Hendrik Schön's Coauthors

In its report examining misconduct in Hendrik Schön's research the investigation committee found "all coauthors of Hendrik Schön in the work in question completely cleared of scientific misconduct.... In addition to addressing the question of scientific misconduct, the Committee also addressed the question whether the coauthors of Hendrik Schön exercised appropriate professional responsibility in ensuring the validity of data and physical claims in the papers in question. By virtue of their coauthorship, they implicitly endorse the validity of the work. There is no implication here of scientific misconduct; the issue is one of professional responsibility."

The committee found this issue to be extremely difficult, in part because the research community has not carefully considered the issue nor developed clear, widely accepted standards of behavior. "In order to proceed, the Committee adopted, for working purposes, a minimal set of principles that it feels should be honored in collaborative research. At its core, the question of professional responsibility involves the balance between the trust necessary in any collaborative research and the responsibility all researchers bear for the veracity of the results with which they are associated." It did not adopt the commonly held view that (unless each coauthor specifies the nature of her contribution in a footnote) each coauthor is responsible for the *entirety* of the collaborative research, but rather thought that there were differences in extent of responsibility of coauthors that varied with those investigators' "expertise, seniority and levels of participation" in the research.

For each coauthor, the committee considered the nature of their participation in the research and what the committee saw as their differing degrees of responsibility. "The Committee concluded that the coauthors of Hendrik Schön in the work in question have, in the main, met their responsibilities, but that in one case (that of the most senior and powerful coauthor) questions remain that the Committee felt unqualified to resolve, given the absence of a broader consensus on the nature of the responsibilities of participants in collaborative research endeavors."[a]

Subsequently in 2002, Schön's coauthors retracted eight articles in *Science* that Schön had coauthored. This was the largest block retraction of articles by the journal *Science*.[b]

The following year, also at the request of some of Schön's coauthors, *Nature* retracted seven articles that it had published by Schön et al.

*Science* and *Nature* are two of the most prestigious research publications.

---

[a] The Lucent Technologies 2002 *Report of the Investigation Committee on the Possibility of Scientific Misconduct in the Work of Hendrik Schön and Coauthors* is available on the American Physical Society's Web site, http://publish.aps.org/reports/lucentrep.pdf.

[b] Associated Press. 2002. "Science Star's Papers Retracted," *Wired*, October 31, at http://www.wired.com/science/discoveries/news/2002/10/56125.

---

### Professional Responsibilities of Ninov's Coauthors

The report of the committee that investigated research misconduct in Ninov's research did not differentiate among Ninov's coauthors and implicitly criticized all of them. The report said that the committee found it "incredible that not a single collaborator checked the validity of Ninov's conclusions of having found three element 118 decay chains by tracing these events back to the raw data tapes."[a]

*Physical Review Letters* at first resisted Lawrence Berkeley Lab's request for a retraction, heeding Ninov's argument that retraction ought to await the results of further experiments, but when investigations at Lawrence Berkeley were complete, the journal printed a retraction.

---

[a] Johnson, George. 2002, October 15. "At Lawrence Berkeley, Physicists Say a Colleague Took Them for a Ride," *New York Times*. This article is available at http://www.nytimes.com/2002/10/15/science/at-lawrence-berkeley-physicists-say-a-colleague-took-them-for-a-ride.html.

---

*What are the responsibilities of an author of a research article? If you would argue that the responsibilities of coauthors are not all the same, state the criteria that you would use in differentiating their professional responsibility for the integrity of work they coauthor.*

## Conflicts of Interest in Authoring, Editing, or Reviewing Research

What conflicts of interest, other than conflicts that may arise from the *financial* interests of an author, editor, or reviewer, may arise in the process of publishing research?

In the introduction, we saw that a party has a **conflict of interest** or is in **a conflict of interest position** when that party

- Is in a position of *trust* that requires the *exercise of judgment on behalf of others* (people, institutions, etc.)
- Has interests, obligations, or responsibilities of the sort that *might* interfere with the exercise of such judgment, and
- Having those interests is neither *obvious* nor *usual* for others in the same position of trust.

Financial conflicts of interest – conflicts of interest that occur when the agent's interests are financial – are common in academic life. Many universities ask members of their faculties to fill out a financial conflict of interest form every year disclosing any financial interests that the faculty members or their immediate families have that might be affected by decisions that the faculty member is entrusted to make. Financial conflicts of interest also arise in authoring research when an investigator has financial interests (such as stock or company ownership, consulting relationships) that may be affected by publishing the results of some research. You may have heard of some recent notorious examples of concealed and corrupting conflicts of interest in publications of studies of drug efficiency or other therapies, but only since World War II has much biological and biomedical research had major financial implications for the faculty members conducting the research. Engineering has seen such research for a longer period, because even before World War II members of engineering faculties often started their own companies or engaged in extensive consulting (up to "one day per week" at many private engineering schools and less at public universities). Engineering schools, therefore, have had more experience and a longer history of dealing with such financial conflicts of interest, and some instituted measures for dealing with it long before recent government regulations on financial conflict of interest.

Beginning in the 1980s, financial conflict of interest began to receive public attention and was the subject of government attention in congressional investigations and legislation.[40] The Food and Drug Administration Amendments Act of 2007* restricts investigators' participation on advisory committees for the U.S. Food and Drug Administration if they, or members of their immediate family, have a financial interest that could be affected by the advice they may give as members of such committees. (The minimum financial interest that must be reported gradually decreases from 2008 to 2012.)

The ACM/IEEE–CS Joint Task Force on Software Engineering Ethics and Professional Practices[41] gives the following guidance about conflicts of interest:

> Software engineers shall maintain integrity and independence in their professional judgment. In particular, software engineers shall, as appropriate:
>
> . . . .
>
> 4.05. Disclose to all concerned parties those conflicts of interest that cannot reasonably be avoided or escaped.
>
> 4.06. Refuse to participate, as members or advisors, in a private, governmental or professional body concerned with software related issues, in which they, their employers or their clients have undisclosed potential conflicts of interest.

Many journals require that authors who have interests that might compete with their obligation to present a clear objective account disclose those interests when they publish their research. We saw earlier that the American Chemical Society (ACS) advises authors in its journals that they have an ethical obligation to:

> reveal to the editor and to the readers of the journal any potential and/or relevant competing financial or other interest that might be affected by publication of the results contained in the authors' manuscript. Sources of funding of the research reported should be clearly stated. In addition, all authors should declare

---

[40] *Conflicts of Interest in Engineering Research* by Mark Frankel available at https://www. citiprogram.org/members/learnersII/moduletext.asp?strKeyID=4E5DDF40-2BE0-4C32-ACA9-2B2D4F3D96E4-3551988&module=12875#9 (requires login).

*This is Public Law 110-85. It is available at http://frwebgate.access.gpo.gov/cgi-bin/getdoc.cgi?dbname=110_cong_public_laws&docid=f:publ085.110.pdf.

[41] ACM/IEEE Software Engineering Code of Ethics and Professional Practice, Version 5.2, is available at http://www.acm.org/about/se-code#full.

1. The existence of any significant financial interest (>\$10,000 or >5% equity interest) in corporate or commercial entities dealing with the subject of the manuscript;
2. Any employment or other relationship (within the past three years) with entities that have a financial or other interest in the results of the manuscript (to include paid consulting, expert testimony, honoraria, and membership of advisory boards or committees of the entity).

The authors should advise the editor in writing either that there is no conflict of interest to declare, or should disclose potential conflict of interests that will be acknowledged in the published article, whether by insertion of a footnote, or incorporation of a sentence or paragraph in the "acknowledgments" section, or by other format of disclosure to the reader as specified by the journal.[42]

Such a financial conflict of interest might also exist for an editor or reviewer who is editing or reviewing a manuscript authored by other investigators.

As important as financial conflicts of interest are, another sort of conflict of interest exists in publication, which is subtler but which can have a major impact on research conduct. This sort of conflict of interest exists for journal editors and reviewers whose own research or career interests may be affected by publication of research by other parties with whom they are in competition for research funds, talented graduate students, etc. The ACS *Ethical Guidelines to Publication of Chemical Research* list as the fourth obligation of reviewers:

A reviewer should be sensitive to the appearance of a conflict of interest when the manuscript under review is closely related to the reviewer's work in progress or published. If in doubt, the reviewer should return the manuscript promptly without review, advising the editor of the conflict of interest or bias. Alternatively, the reviewer may wish to furnish a signed review stating the reviewer's interest in the work, with the understanding that it may, at the editor's discretion, be transmitted to the author.

Conflicts of interest in reviewing are more likely to be experienced or observed by engineering trainees than are financial conflicts of interest, because trainees (and their immediate families) are less likely than senior researchers to have significant financial interests that might conflict with giving a complete and accurate report of their research. (I *have* heard concerns expressed by engineering trainees about their research supervisors' mishandling of their own [the supervisors] conflicts of interest in reviewing manuscripts, however.) Editors need to select reviewers who know something about the research area being reported, so appropriate reviewers may have some conflict of interest. Reviewers who receive a manuscript authored by

> A conflict of interest exists for journal editors and reviewers whose own research or career interests may be affected by publication of research by other parties with whom they are in competition for research funds, talented graduate students, etc.

---

[42] This is item 12 in the list of obligations of authors in the latest version (2010) of the ACS's *Ethical Guidelines to Publication of Chemical Research*. Those guidelines may be accessed at http://pubs.acs.org/page/policy/index.html.

an investigator who is their competitor should do their best to fairly review the competitor's work. Curtailing corruption on the part of reviewers largely falls to the editors of research journals. Journal editors are becoming much more aware of their responsibilities for the integrity of published research, and organizations such as the Council of Science Editors (CSE) – see http://www.councilscienceeditors.org/ – are helping editors to recognize and fulfill these responsibilities.

---

**Evaluating a Reviewer's Claim of Prior Discovery**

As the editor of JAAA you ask Prof. Sharp to review a manuscript submitted to your journal by Prof. Wright, because you know that Sharp works in the same subject area.

Sharp takes the full time allowed for review of Wright's article and, at the end of the review period, sends a letter recommending that JAAA reject the work by Wright, on the grounds that it is not novel. Sharp claims that the submitted manuscript simply repeats the work that Sharp has published in another journal, but Sharp does not give a full citation of that work.

What do you do?

---

*What conflicts of interest, other than conflicts that may arise from the financial interests of an author, editor, or reviewer, may arise in the process of publishing research?*

## Responsibilities in the Supervisor–Trainee ("Mentor–Mentee") Relationship

What makes the relationship between a research supervisor and that supervisor's trainee (sometimes called the "mentor–mentee" relationship) especially deserving of attention in research ethics (i.e., more deserving of attention than other relationships among research investigators)?

The relationship between a trainee (graduate student or post-doc) and her research supervisor is one in which the trainee acquires most of her practical knowledge about the conduct of research, including the ethical aspects of research conduct. For that reason alone, it would be a crucial concern for research ethics.

The supervisor–trainee relationship is actually very complex, and it changes over time because at least the trainee (and sometimes the supervisor) develops considerably over the course of the relationship as the trainee matures as a research investigator, and so becomes more of a colleague and less like a student. (In fields in which a period of post-doc training is typical, the graduate student becomes more like a post-doc – and indeed may receive much of her supervision from a post-doc, rather than from her faculty advisor – rather than becoming a colleague of her faculty advisor.) The changing nature of the relationship is just one of the complicating factors. Another is that the research supervisor is typically the trainee's principal source of information and evaluation of

### When Is a Research Supervisor a Mentor?

The term "mentor" is taken from *The Odyssey*. Mentor was the name of Odysseus' trusted guide and counselor. In the context of professional ethics the term carries the connotation of someone who helps a younger person to understand and negotiate ethical and practical challenges in developing as a professional. Some research supervisors act as mentors to their supervisees. Others confine themselves to teaching research methods only. Still others by their behavior teach negative lessons and are sometimes called "**toxic mentors.**"

the quality and significance of the trainee's own research. If the trainee has concerns about the fairness of the credit that the supervisor assigns that trainee for the trainee's contributions to their joint research, those concerns will not be easily resolved by hearing only the supervisor's assessment of the importance of the trainee's contributions. (Some engineering graduate programs seek to ensure that their graduate students have ready access to a second opinion by assigning to each student a departmental advisor who is different from the student's research supervisor.) Gender and ethnic differences can further complicate the relationship.

> The supervisor–trainee relationship is a crucial subject in research ethics because the research supervisor is typically the trainee's principal source of information and evaluation of the quality and significance of the trainee's own research. If the trainee has concerns about the fairness of the credit that the supervisor assigns that trainee for the trainee's contributions to their joint research, those concerns will not be easily resolved by hearing only the supervisor's assessment of the importance of the trainee's contributions.

Carl Djerassi's 1991 novel, *Cantor's Dilemma*, addresses many issues of trust in the supervisor–trainee relationship (and illustrates some common gender differences in supervisors). Because Djerassi had a distinguished career in chemical research, the details of the story reflect many aspects of actual research practice as they exist in the physical sciences.

The 2008 film *Dark Matter*, www.darkmatterthefilm.com, deals with the relationship between a theoretical physicist supervisor, Jacob Reiser (played by Aidan Quinn) and a brilliant graduate student from China, Liu Xing (played by Liu Ye). The cultural differences between the supervisor and trainee receive extensive attention in this film. The film should be useful in stimulating discussion of these differences. Because the filmmaker, Chen Shi-Zheng, does not have extensive research experience, his exercise of "poetic license" in crafting the drama may mislead some viewers about the relationship between research supervisors and their trainees, however.

Partly because the story is told from Liu's perspective, some subtleties of the norms that govern the supervisor–trainee relationship are passed over. The impression the film gave me of Jacob Reiser was of a supervisor who was motivated by the desire to have his students elaborate his theory, rather than prepare them to make their own contributions. Although that motivation is common among supervisors, the actual situation is also usually strongly influenced by the nature of the funding that a research supervisor can obtain for that supervisor's graduate students. Commonly the supervisor will have grants (or perhaps contracts) that pay the student's stipends. In return, the graduate students carry out research related to the purposes of the grant (or contract). Therefore, the situation of a student in engineering or the natural sciences is quite unlike that of a graduate student in the humanities. In the humanities, a student's thesis research is typically not supported by the supervisor's grants, and the thesis topic is more a matter of the student's own choosing. In the situation of the typical graduate student in engineering or science, the supervisor's responsibilities and obligations vis-à-vis the funding source limit the student's choice of research topics. Therefore, a supervisor's expectation to have a hand in choosing the graduate student's thesis topic derives in part from the supervisor's responsibilities vis-à-vis the funder of the research.

Sound reasons for the supervisor's shock and anger at the student publishing on his own are not shown in the film, so the impression is left that the only explanation is injury to the supervisor's vanity. Later in this chapter when we consider a supervisor's responsibility for the quality of her trainee's research we will revisit this issue. In the film *Dark Matter*, graduate student Liu did try unsuccessfully to show his paper to Reiser, although Liu did not tell Reiser of his plan to publish.

A faculty member in a science department gave me the following scenario based on his own belief that he and other faculty members were often remiss in responding to their trainees' need to publish. Therefore, he thought they should discuss this situation in which a trainee fails to receive feedback on a manuscript to be submitted for publication.

---

### Timely Publication[a]

Cory has been a graduate student in Prof. Harried's lab for the past 6 years and expects to graduate soon. Cory is waiting for Harried to review and critique a manuscript before Cory submits it for publication. Cory submitted the manuscript to Harried more than a year ago, and repeatedly has asked Harried about it. The manuscript is still sitting on Harried's desk. Other labs around the country have been working on this sort of problem. If the manuscript isn't sent out soon, someone else is likely to publish the findings first.

---

[a] I wrote this scenario based on the suggestions of a faculty member who preferred not to be named. It is available at http://www.onlineethics.org/cms/16230.aspx.

---

Subsequently a post-doc read the Timely Publication scenario on the Web and asked the OEC Ethics Helpline what she could do in a very similar situation with her previous supervisor (at a university she had now left). Several of us who were staffing the OEC Ethics Helpline suggested that she write a very polite letter to the professor in question. We suggested that she emphasize that although she would prefer to publish with the former supervisor, because the standards for authorship at the target journal required involvement of all authors in writing or reviewing the manuscript submitted for publication, then if the former supervisor did not have time to give to the article, she would have to publish alone. (A variant scenario for post-docs is also posted at http://www.onlineethics.org/cms/16230.aspx.)

*What makes the relationship between a research supervisor and that supervisor's trainee especially deserving of attention in research?*

## Human Research Subjects/Participants

### Historical Background
What events in the twentieth century led to the adoption of new standards for the treatment of human research subjects/participants in the United States and elsewhere?

Research with human subjects/participants is less common in some engineering disciplines than in psychology and biomedicine, but it occurs often in some newer engineering fields, such as biomedical engineering and in studies of human factors.

In 1946, after the discovery of brutal medical experiments carried out by the Nazis[43] the Nuremberg code required informed consent for human experimentation.[44] The requirement was refined in the Helsinki declarations issued by the World Medical Association in 1962 with subsequent revisions in 1964 and 1975 and 1989. To meet the "**informed consent standard**" for experimenting on human beings requires that a person who is to become a subject/participant in an experiment must first be given full information about the experiment, and freely consent to participate (i.e., agree to participate without being in any way coerced).

## Human Subjects or Human Participants

Until recently when research experiments were conducted on people, they were referred to as "subjects" or "research subjects." In response to the objection that this made the people sound like mere material for the investigators to use and failed to recognize that in the typical case, the people had given their informed consent and *decided* to participate in the research, some propose calling these "participants."

To differentiate them from the investigators who collaborated in the research (and whose obligations and responsibilities are discussed here) I call them "subjects/participants."

The informed consent standard was only gradually adopted in the United States, however, and it took some shocking cases in this country to demonstrate the need for reform. In 1966, a well-respected clinician and investigator, Henry Beecher, published an article in the prestigious *New England Journal of Medicine*, in which he reported the unethical treatment of human research subjects in many premier institutions in the United States.[45] Subsequently, the NIH developed the first Public Health Service Policy on the Protection of Human Subjects. At first this policy had only limited application. Later the policy was expanded to apply to all human subjects/participants in research conducted or supported by what was then the Department of Health, Education and Welfare.[46]

One of the most infamous experiments in the United States was the Tuskegee Syphilis Study, a study that the Public Health Service had conducted from 1932–1972.[47] This study had started innocently enough as a treatment program for syphilis, although in the 1930s, before the discovery of penicillin, treatments were largely ineffective. When it became clear that resources were not available to treat all those who had syphilis, the project became a study of *untreated* syphilis. Later, *even after the discovery of an effective treatment*, namely penicillin, the shocking decision was

---

[43] Caplan, Arthur L. 1992. *When Medicine Went Mad: Bioethics and the Holocaust.* Totowa, NJ: Humana Press.

[44] The requirement of informed consent now applies to behavioral research as well as medical research and research in engineering and the natural sciences. Many of the classic experiments in psychology involved deception and even clear harm to subjects and would not be allowed today.

[45] Beecher, Henry K. 1966. "Ethics and Clinical Research," *New England Journal of Medicine* 274(24) (June 16): 1354–1360.

[46] Levine, Robert J. 1986. *Ethics and the Regulation of Clinical Research*, second edition. Baltimore, MD: Urban & Schwarzenberg.

[47] Brandt, Allan M. 1985. *No Magic Bullet: A Social History of Venereal Disease in the United States since 1880.* New York: Oxford University Press.

Although the informed consent standard had been adopted in some countries after World War II, the informed consent standard was only gradually adopted in the United States, and it took some shocking cases in this country to demonstrate the need for reform in U.S. research practice.

made to continue the study and even to prevent subjects from getting treatment when they gained access to it by other means. Those from whom effective treatment was knowingly withheld were male African American sharecroppers. The medical community went for decades without raising questions about articles with titles like "The course of untreated syphilis in Negro men." It is hard to imagine that journal readers would have withheld comment if the articles had been titled: "The course of untreated syphilis in college students." The tolerance of the medical community for the continuance of this research evidences racism and a willingness to exploit the powerless. Because Tuskegee, the institution that conducted the study, is a historically black institution, not only whites were at fault, however.

Another set of experiments dating from World War II that have more recently come to light are a series of radiation experiments funded by the Department of Energy (DoE). The motivation for some of these studies was to learn more about radiation injuries for the sake of workers who had been exposed to radiation in weapons work during World War II. Other DoE-funded radiation studies simply used radioactive tracers, for example, to conduct nutrition studies. Some experiments did meet today's standards for the treatment of human subjects. In others, the patients were unharmed but their informed consent was never obtained. Still others were extremely damaging to subjects, including tests that irradiated the testicles of prisoners or subjected patients supposedly dying of cancer to massive doses of radiation. (Some of the irradiated patients turned out not to have fatal cancer.[48])

The radiation experiments that seriously harmed patients violated not only the informed consent standard but the standard that predated it as well. As we saw in the introduction, before the adoption of the informed consent standard, the informal rule for ethical experimentation was for investigators to first do to themselves anything to which they proposed to subject others – an inverted golden rule. The radiation experiments were typical of experiments in earlier decades, in that patients and prisoners were usually used in experiments that were unethical even by that earlier standard. Today, regulations on human experimentation are intended both to implement the informed consent standard and to provide special protections for vulnerable populations who have been the most subject to abuses in the past. The DoE radiation tests have also been the grounds for lawsuits against hospitals and universities.[49]

*What events in the United States in the twentieth century led to the adoption of new standards for the treatment of human research subjects/participants?*

### Current Requirements Governing Human Subjects/Participants

What is an "IRB" and what is the rationale for requiring the IRB's approval of any research with human subjects/participants?

[48] Mann, Charles C. 1994. "Radiation: Balancing the Record," *Science* 263(5146): 470–474.

[49] Wheeler, David L. 1995, October 13. "Making Amends to Radiation Victims," *The Chronicle of Higher Education.*

Today in the United States, the ethical and legal right to refuse to participate in experiments is coming to be regarded as an absolute right, like the right to refuse medical treatment. These rights are reflected in law and regulation.

Any institution, such as a university, government laboratory, or hospital, that receives U.S. government funding for *any* of the research it conducts must follow government regulations on the use of human subjects/participants in *all* of its research with human subjects/participants.

## Law and Regulation

Regulation has the force of law. The principal difference between law and regulation is that updating regulation does not require new legislative action. It works this way: Legislation gives a certain governmental entity the power to regulate some matter. That entity can update the regulations governing the matter in question through a process that, although complex, does not require new legislation.

**Institutional review boards (IRBs)** are a key support for the exercise of that right. In the United States, any institution receiving government funds must have an IRB that reviews all research protocols for experiments involving human subjects. (If investigators are applying for government funding for a particular study, that study usually must be approved by the IRB *before* the research can be funded.) The membership of such boards shows little turnover, and its accumulated experience prepares it to recognize dangers that an individual experimenter might overlook. It is also in a position to judge when some requirement for human subject protection should be changed or waived. (We will explore this point further in the next section.)

In experimentation on children or others not fully competent to consent for themselves, IRBs are alert to the danger that those who give consent as their "proxy" may have motives that conflict with protecting the interests of their wards. For guardians to be paid for the research participation of the children in their care would create such a conflict of interest, for example. In many cases the assent of the child subject as well as the subject's guardian is required to continue the experiment. The guardian is able to give fully informed legal consent, but experimenters may also be required to stop the experiment if children or adults with mental retardation do not assent – see the subsection on the common rule, later in this chapter for discussion of a child's assent.

Because experience has shown that institutionalized subjects, such as patients or prisoners, have reduced options and are liable to be influenced by what they think authorities want, special restrictions are placed on the recruitment of such people to avoid subtle coercion.

Other protections and assurances are specific to the situation and vulnerabilities of the subjects. Students in psychology courses used to be regularly required to be subjects in the psychology experiments of their professors. This situation continued into the 1980s. That practice is now recognized to be coercive, so that students must be offered alternatives to test participation.

The requirement of something like informed consent for *treatment* is much older than for experiment; treatment without consent has long been understood in legal terms as a "battery." In both cases, the mechanism of informed consent involves the patient or subject/participant receiving an explanation and signing a written form with the required information.

## Two Types of Informed Consent

Informed consent for medical treatment is rather different from informed consent to participate in research, although in each case the procedure involves signing a consent form. In the case of treatment, the informed consent procedure is meant to be part of a larger enterprise of shared decision making between patient and provider. Therefore, a patient may change a consent form before signing it, for example, to specify that only the named surgeon, and not the surgeon's students or "associates" will perform the surgery. In contrast, research subjects cannot modify a research protocol. They can only decide to participate or decline to participate.

Biomedical research frequently involves human subjects/participants. In the United States, the approval process for biomedical devices requires a lengthy (and costly)[50] review process by the Food and Drug Administration (FDA). Such a process is required for drugs and devices, although not for all experimental treatments. For example, the artificial heart required FDA approval, although transplantation of a baboon heart did not. In 1994, the FDA halted development of a new device for emergency resuscitation of heart attack patients because testing of the device requires informed consent.[51] This consent could not be obtained from patients actually *having* heart attacks.

*What is an "IRB" and what is the rationale for requiring the IRB's approval of any research with human subjects/participants?*

### Human Subjects/Participants in Product Testing

Why is the recruitment of human subjects for product testing not covered by government regulations regarding informed consent?

Testing of most products is not subject to the regulations regarding the use of human subjects in research, because manufacturers do not receive government funds for any of their testing. An extreme example of injury in product testing occurred in October 1991 when an aircraft manufacturer, McDonnell Douglas, was reported to have caused serious injury to senior citizens recruited to test an evacuation procedure. According to the *Wall Street Journal*, McDonnell Douglas was seeking to demonstrate that it could safely increase the seating of an all-economy-class aircraft from 293 seats to 400. In response to criticisms that the use of employees and family members of employees as test subjects biased the test results, the company recruited senior citizens. Nothing resembling informed consent was obtained from these subjects/participants. Failure to obtain informed consent was *not* itself prohibited by law, although the action certainly appears reckless to anyone familiar with ethical standards for human experimentation.

Even absent any familiarity with those standards, the company violated ordinary standards of prudence and responsibility when, after eleven people had been injured in the morning evacuation drill, it reportedly proceeded to conduct an afternoon evacuation drill without warning the participants of possible injury. In all, forty-four people were reportedly injured, including eleven who were taken

---

[50] For example, Trimedyne, after spending $2 million to develop a device to use a laser to vaporize fatty deposits in coronary arteries, waited from 1983–1993 for FDA permission to use in diseased *leg* arteries. Finally it received permission only for "no risk, no benefit" use in leg arteries (in a patient who was to have artery grafts for other reasons).

[51] Winslow, Ron. 1994. "FDA Halts Tests on Device That Shows Promise for Victims of Cardiac Arrest," *Wall Street Journal*, Wednesday, May 11, B8.

to the hospital, six of whom had broken bones. One sixty-year-old woman was paralyzed from the neck down due to a fracture of her spine.[52] The injured parties were able to sue for damages, of course, but the point of informed consent requirements is to *prevent* such occurrences.

*Why is the recruitment of human subjects for product testing not covered by government regulations regarding informed consent?*

### The Common Rule for the Protection of Human Subjects/Participants in Research

What is meant by "minimal risk"? What reasons can you think of for giving this notion such a prominent place among the criteria for judging when the design of a proposed research study adequately protects any human subjects/participants in the study?

The U.S. government has adopted a single policy governing the use of human subjects/participants in research, which "applies to all research involving human subjects conducted, supported or otherwise subject to regulation by any federal department or agency" that conducts or funds research. This policy is widely referred to as "the common rule." You may read the entire common rule at http://www.nsf.gov/bfa/dias/policy/docs/45cfr690.pdf. Here I briefly summarize that rule.

The common rule defines **minimal risk** (to a subject/participant in research) as meaning that "the probability and magnitude of harm or discomfort [to the research subject/participant] anticipated in the research are not greater in and of themselves than those ordinarily encountered in daily life or during the performance of routine physical or psychological examinations or tests."[53] The rule then specifies certain sorts of information as necessary for someone to give "informed consent" for participating in any research study. The sorts of information it specifies are these:

(1) A statement that the study involves research, an explanation of the purposes of the research and the expected duration of the subject's participation, a description of the procedures to be followed, and identification of any procedures that are experimental;

(2) A description of any reasonably foreseeable risks or discomforts to the subject;

(3) A description of any benefits to the subject or to others that may reasonably be expected from the research;

(4) [If the research is about some treatment, a] disclosure of appropriate alternative procedures or courses of treatment, if any, that might be advantageous to the subject;

(5) A statement describing the extent, if any, to which confidentiality of records identifying the subject will be maintained;

(6) For research involving more than minimal risk, an explanation as to whether any compensation and any medical treatments are available if injury occurs

---

[52] Nazario, Sonia L. 1991. "McDonnell Douglas Jet Evacuation Drills Leave 44 Injured." *Wall Street Journal*, Wednesday, October 30, A3–A4.

[53] The Common Rule for the Protection of Human Subjects, p. 4, accessed at http://www.nsf.gov/bfa/dias/policy/docs/45cfr690.pdf.

and, if so, what they consist of, or where further information may be obtained;

(7) An explanation of whom to contact for answers to pertinent questions about the research and research subjects' rights, and whom to contact in the event of a research-related injury to the subject; and

(8) A statement that participation is voluntary, refusal to participate will involve no penalty or loss of benefits to which the subject is otherwise entitled, and the subject may discontinue participation at any time without penalty or loss of benefits to which the subject is otherwise entitled.[54]

The common rule then lists additional elements of information that are to be provided to each subject/participant about any research to which such information applies. For example, a subject who is pregnant or may become pregnant may need to be told that the experimental treatment may carry unknown risks to the embryo or fetus.

The rule also allows that an IRB *may* approve a consent procedure that excludes or alters some of the requirements that it has set for informed consent or even waive the requirement to obtain informed consent provided the IRB finds and documents that

- The research involves no more than minimal risk to the subjects.
- The research is on public benefit programs and it could not be conducted without the alteration or waiver.
- The waiver or alteration will not violate the subject/participant's rights or harm that person, and the research could not be conducted without the waiver or alteration.

The usual means of documenting the informed consent procedure is by a written consent form (approved by the IRB) that is signed by the subject or the subject's legally authorized representative. An IRB may waive the requirement to obtain a signed consent form for some or all subjects, however, if it finds either

- That the research presents no more than minimal risk of harm to subjects and involves no procedures for which written consent is normally required outside of the research context.
- That the only record linking the subject and the research would be the consent document and the principal risk would be potential harm resulting from a breach of confidentiality. In which case each subject will be asked whether the subject wants documentation linking the subject with the research, and the subjects can choose whether to sign before participating.

The informed consent procedure is straightforward when the subject/participant is competent to give consent. However, there are special categories of subjects/participants for whom obtaining informed consent is problematic:

- Children
- Prisoners

[54]*Ibid.*, 10.

- People with reduced mental capacity such as those with brain injury or mental retardation
- People with mental illnesses that interfere with their judgment or even their interest in securing their own well-being
- People with less education or who are unaccustomed to having their rights recognized and so are unaccustomed to asserting those rights, or are otherwise less likely to asserting them, such as the frail, elderly, poor, or members of certain minority groups

Research that involves the use of children or involves prisoners as research subjects/participants is thoroughly regulated, perhaps because both children and prisoners were groups that had been exploited in earlier testing. See Title 45 of the Code of Federal Regulations, Part 46, Protection of Human Subjects, Subpart C, Additional Protections Pertaining to Biomedical and Behavioral Research Involving Prisoners as Subjects, and Subpart D, Additional Protections for Children Involved as Subjects in Research.[55]

Experimentation on prisoners is effectively limited to research that answers questions about incarceration – addressing prisoners as institutionalized persons (rather than research that would inform treatment of nonprisoner populations). Restrictions on research with prisoners were formulated with recognition of their diminished freedom in general and their fewer alternatives to participation in particular.

In addition to the informed consent of the child's parents or guardian, the child's assent is required for research, if the child is capable of giving it. Although the child may be too young to appreciate all the information that the parents or guardian may have weighed in giving "informed consent," the child may be able to assent. **Assent** roughly means that the child *expresses the willingness* to participate. The absence of the child's objection to participating is not enough to constitute assent.

Regulations regarding research with human subjects/participants and special regulations regarding children and prisoners have been so thoroughly thought out as to disallow many research studies that might exploit them. However, research with people who have mental illness or reduced mental capacity is less regulated and calls for more judgment in deciding how best to protect their welfare and rights.[56] The situation in "The Patient Who Wants to Withdraw," although a problem for a medical resident rather than an engineer, illustrates some of the difficulties in following the procedures set out for the adult patients who are competent to give (or withdraw) their consent at any time.

---

[55] Code of Federal Regulations, Title 45, PUBLIC WELFARE, PART 46, Protection of Human Subjects, Revised January 15, 2009. Subpart D, Additional Protections for Children Involved as Subject in Research, available at http://www.hhs.gov/ohrp/humansubjects/guidance/45cfr46.htm.

[56] See "The Ethics of Research with Human Subjects Who Are Mentally Ill" at http://www.onlineethics.org/CMS/2963/modindex/mentres.aspx and "The Ethics of Research with Subjects Who Have Dementia" at http://www.onlineethics.org/CMS/2963/modindex/ADreseth.aspx for more on these complexities.

## The Patient Who Wants to Withdraw

You are the psychiatric resident on-call tonight and your responsibilities include all admissions to the adult locked units as well as covering the current inpatient population. The head nurse pages you for help in resolving a very complicated issue involving an agitated patient.

Upon arriving to the ward, you are informed by the nurse that the patient, Cory Lee, had been admitted to the ward that afternoon after experiencing an exacerbation of her psychotic symptoms that included paranoia and auditory hallucinations. She was currently enrolled in a research study involving a new oral antipsychotic drug that her personal psychiatrist is conducting. Lee was now in week two of the study following a washout period of her previous psychotropic (i.e., a period in which she received no antipsychotic medication). She had agreed to participate in the study because, despite experiencing reasonable control of her psychotic symptoms, the side effects of the medication that she was taking were intolerable. She developed severe psychiatric symptoms during this second week and her psychiatrist decided to admit her and continue the study.

As the day progressed, Lee became increasingly agitated, and was unable to respond to staff's request to return to her room. Lee refused to follow staff's instructions unless the staff could convince her doctor to stop giving her that "pill from Hell." Her psychiatrist was informed of the patient's request to stop taking the experimental drug, but the psychiatrist insisted that the patient's current condition precluded her from making a rational decision. No other neuroleptic medications were to be given to this patient because doing so would nullify the study. The staff was unable to pacify the patient and the situation escalated further. The nurse requests that you "Do something!" You determine that Lee, though agitated, is willing to discuss her concerns.

During her conversation with you, her volume of speech is loud and, at times, frightening, but she does not threaten violence. Her thoughts are organized and, despite her admitting to hearing voices during the conversation, the content is clear. She says she would prefer the side effects of her "old faithful drug" rather than endure these present symptoms. You are aware that competent research subjects are free to opt out of research studies at any time.

What is your responsibility in this matter?

What factors help determine issues of competency when there are active psychotic symptoms?

What, if anything, can the resident do if s/he disagrees with the recommendation of the patient's psychiatrist or that of the attending physician in the hospital?[a]

---

[a]Stuart Youngner, M.D., collaborated in the construction of this discussion case.

*What is meant by "minimal risk"? What reasons can you think of for giving this notion such a prominent place among the criteria for judging when the design of a proposed research study adequately protects any human subjects/participants in the study?*

## Responsibility for Experimental Animals

What justification is there for setting standards for the treatment of nonhuman vertebrates used in experiments? Evaluate the adequacy of that justification (i.e., say why you think that justification is or is not convincing and why).

What is the justification for treating the mice in the walls of one's building very differently from those used in experiments?

Nonhuman animals can replace human subjects/participants in certain experiments that would cause more than minimal harm to research subjects, but doing so subjects those experimental animals to the harm in question. One alternative to the use of living experimental subjects in experiments that cause those subjects more than minimal harm is to forgo any research that causes serious harm to subjects.

## Physiology Becomes an Experimental Science

Before Bernard's work, the view called "vitalism" was widely popular. According to it, an *élan vital, vis vitali*, or vital force was present in living things. This force produced unpredictable responses to stimuli, so that even if one prepared two experiments in exactly the same way, the life force could produce different outcomes. As Lord Kelvin had put it, "The influence of animal or vegetable life on matter is infinitely beyond the range of any scientific inquiry hitherto entered on."

Research using animals as subjects became common after Claude Bernard (1813–1878), the French physiologist, founded physiology as an experimental science. Bernard conducted his experiments at home, and his wife, whose ample dowry funded his work, became active in the cause of "anti-vivisection," as objection to such experiments was then called. So, ethical controversy surrounded animal experimentation from the first. The membership of People for the Ethical Treatment of Animals (PETA), one of the groups most critical of the use of animals in experiments, has grown rapidly since the 1970s, however.

### Using Animals in Medical Experiments

Suppose that certain experiments are proposed to aid the development of better means of coping with extreme pain in humans. These experiments would involve performing a variety of procedures that would be very painful to the experimental subjects. The subjects would have to be vertebrates to obtain information that could apply to humans. Species closer to humans in evolutionary terms might give more meaningful results, but it is not known exactly what traits of laboratory animals are most relevant. Because the information about neural response would be crucial, subjects would receive no pain medication. The subjects would be temporarily paralyzed to keep them from flailing about.

What, if any, species would it be ethically acceptable to use in such experiments? Is it morally relevant whether the subjects are mammals? If so, why?

Would the intelligence of members of that species be morally relevant to the decision, and if so, how?

Would the presence or absence of a complex social system in which members care for other members of the species be a morally relevant factor to consider?

Would it be morally relevant that one candidate species resembled humans more closely?

Would it be relevant that some particular individuals had once been human pets? If so, how would the ethics of using such animals compare with those of using animals bred for experimental use?

A comparable experiment to the previous one was a study of head trauma to restrained baboons conducted at the Experimental Head Injury Laboratory at the University of Pennsylvania. The Society for Neuroscience estimates that there are about 2 million cases of head trauma each year causing 50,000 deaths. Experiments aimed at improving treatment for people with head injury frequently involve subjecting experimental animals to head injury. The Pennsylvania study became well known when members of the Animal Liberation Front, a clandestine group, broke into the Experimental Head Injury Laboratory and stole a videotape record of the experiments. They gave the videotape to PETA which then circulated a twenty-five-minute selection of excerpts.[57]

The question of whether or under what circumstances it is permissible to experiment on animals for human benefit involves *the moral standing of the experimental subjects*, the harm or discomfort that the experiments will cause them, and the benefit to be derived for humans. (The treatment of farm animals raises some similar issues, but usually does not arise directly in the work of engineering and science, as does the use of animals in experiments.) As we saw in the introduction, for a being to have moral standing means that such an individual's well-being must be considered for its own sake. Not all beings that have moral standing need have the *same* moral standing. The existence of laws against cruelty to animals evidences the widespread conviction that animals (especially vertebrates) have moral standing, but if they were thought to have the same moral standing as humans, then having work animals would be regarded as slavery and it is not. That certain beings have moral standing does not require that they have the same moral standing as people, nor that they be said to have rights, but only that their good must be considered for its own sake.

The NIH requirement for Institutional Animal Care and Use Committees (IACUCs) to review any experiment that uses nonhuman animals as subjects shows that government regulation and popular sentiment agree that at least *certain* sorts of animals, especially mammals, birds, and other vertebrates, do have moral standing. An **IACUC** is "a self-regulating entity that, according to U.S. federal law, must be established by institutions that use laboratory animals for research or instructional purposes to oversee and evaluate all aspects of the institution's animal care and use program."[58] IACUCs function much as IRBs do in that they are guided by federal regulations,[59] but each IACUC and IRB is empowered to make case-by-case decisions as to whether to reject, accept, or require modifications in the research protocols it reviews.

The ethical question that is still disputed is *how high* is the moral standing of nonhuman animals (or of specific subtypes of animals, say vertebrates, because only certain types of animals are covered by U.S. government regulations on the use of animals in research). The answer to this question will be very important in deciding what, if any, negative consequences to animals (or at least vertebrates) are morally acceptable in order to achieve benefits to others, especially humans. If certain animals have the *same* moral standing as people, then because animals are *incapable* of giving informed consent, experimentation looks like unjustified risk shifting. Animal rights theorists who hold that some or all nonhuman animals

---

[57] Tannenbaum, Jerrold and Rowan, Andrew. 1985. "Rethinking the Morality of Animal Research," *Hastings Center Report* (October), 32–36.

[58] See definition at the IACUC Web site, http://www.iacuc.org/aboutus.htm.

[59] IACUC.org supplies links at http://www.iacuc.org/guidelines.htm. The Canadian regulations may be directly accessed at http://www.ccac.ca/en/CCAC_Programs/CCAC_Programs_ETC.htm and those for the United States at http://www.nal.usda.gov/awic/pubs/noawicpubs/educ.htm. Regrettably, the U.S. document has few internal links and requires scrolling through the document.

do have the same standing as people, therefore, hold that the use of animals to benefit people is unjustifiable.[60]

If some being is itself a *moral agent*, that fact is generally taken to show that the being has very high moral standing, because being a moral agent is one criterion of being a person. Some other criteria commonly given as criteria for having high moral standing are rationality or at least intelligence, sentience (the ability to experience pain and pleasure), and being alive.

As we saw earlier, there has been a tendency for people to accord higher moral standing to those who are *like themselves*. Unless the similarities are *morally relevant ones* this tendency is merely a self-serving bias.

Many of the specific responsibilities for the welfare of experimental animals depend upon more than the moral standing of the species or individual animal. Notice that what is permissible treatment of wild mice is very different from the permissible treatment of experimental animals. Why are the mice in the walls (whether they are wild strains or feral laboratory strains) treated so differently from the mice in the cages?

We saw earlier that responsibility stems not only from knowledge but also from a position of relationship and control. Therefore, parents have responsibilities for their children regardless of whether they know much about rearing children. The difference in the treatment of the mice in the walls and the mice in the cages is another instance of responsibility arising from control. Experimental animals are under the experimenters' control. This position of control carries with it special responsibilities that do not apply in the case of others of the same species. In a similar way, a pet owner might be liable under legislation against cruelty to animals for doing to a pet something that the same person would be allowed to do to an animal pest.

> The requirements on the treatment of human and animal subjects reflect both the moral standing of people and animals, and the special responsibilities one takes on by establishing a relationship of some control with respect to these individuals.

The extermination of pests finds justification in the assumption that they have lower moral standing than humans and in the threat they pose for spreading disease and destroying property. The moral standing of some pest species, especially vertebrates, is often argued to require using methods that force wild pests to relocate elsewhere, or at least kill them in ways that minimize their suffering.

*What (ethical) justification is there for standards for the treatment of nonhuman animals used in experiments? Evaluate the adequacy of that justification.*

*What is the justification for treating the mice in the walls of one's building very differently from those used in experiments? Evaluate the adequacy of that justification.*

---

[60] In contrast to their view, Peter Singer, one of the most famous theorists arguing that animals deserve better treatment, holds that *certain* use of animals in research *is justified when the benefits to humans are sufficiently great and not obtainable by other means*. Singer, Peter. 1985. *In Defense of Animals*. New York: Blackwell; Singer, Peter. 1977. *Animal Liberation*. New York: Avon Books; Singer, Peter. 1990. "The Significance of Animal Suffering," *Behavioral and Brain Sciences* 13(1): 9–12.

# Raising Ethical Concerns in Research

What can you do if you have questions about whether some act or course of action is responsible research conduct?

Much attention has been given to creating safe and reliable ways of raising concerns about possible research misconduct (i.e., about acts of falsification, fabrication, or plagiarism in research) and about the abuse of human research subjects/participants. The best ways of raising concerns about other departures from responsible research conduct will vary more with the specific research context, especially for students and other trainees, who need a knowledgeable and trustworthy advisor to help them think through all the ramifications of those issues. It would help trainees if all research supervisors were ethically aware and knowledgeable about ethical standards and modes of practice in their own organization, but that is not always the case.

Policies to protect good-faith "whistleblowers" (i.e., complainants) have been instituted to ensure that research misconduct can be reported. "**Good faith**" in this context means that the complainant had sound reasons for thinking that research misconduct had occurred and was not simply trying to create difficulties for the person accused of perpetrating misconduct. Being a good-faith complainant does not mean that the complainant is necessarily *right* in thinking that misconduct occurred. (The faculty of one university further clarified a good-faith complainant as one who had reasons for thinking that misconduct had occurred and had not disregarded evidence that misconduct had not occurred.)

## Inquiries and Investigations of Misconduct

The inquiry stage is somewhat similar to a grand jury that decides whether to issue an indictment in a criminal case, in that the inquiry panel decides whether there is enough evidence that research misconduct did occur to proceed with a full investigation. At some institutions, the inquiry panel will conduct any subsequent investigation. At others, the investigation is handled by an entirely different group, often one that handles all investigations into research misconduct at that institution.

Alerting a person who has committed research misconduct (especially falsification or fabrication) has often led that person to destroy records in an attempt to cover the misconduct. It is often advisable if one believes that wrong has been done to go first to the supposed wrongdoer to see if that person can explain or correct the apparent wrong. Doing so is not the recommended procedure in cases of suspected research misconduct, however. Instead, one is advised to contact a staff member of the research standards office (who may have any of several different titles from "research integrity officer" to "vice president for research"). Each university that receives government funding for *any* of the research it conducts must have such a person.

You may take your concern directly to that person and be protected, as long as you are acting on the basis of sound reasons. If you are still a student, you may wish to have some advice about whether the evidence you have has some more likely explanation than research misconduct. For advice, you might seek out a very senior member of the faculty whom you respect and *trust*. (A senior person would be more likely to be immune from pressure to keep the matter silent or to use the accusation to further some other agenda.) Suppose there is no such person. You or a friend could anonymously present the situation as a *hypothetical* situation to someone who is knowledgeable both about current procedures for handling accusations of research misconduct and research in your discipline or field and ask how that person would interpret the situation.

If the evidence you present speaks for itself – for example if you present evidence in the form of documents showing a publication or grant proposal from the accused and also provide the document from which some or all of the text or ideas were plagiarized – then the inquiry and investigation can proceed without your further involvement. If, however, part of the evidence is what you witnessed, you can expect to meet with at least the group that conducts the inquiry into the research misconduct.

> *What can you do if you have questions about whether some act or course of action is responsible research conduct?*

# 9 Responsible Authorship and Credit in Engineering and Scientific Research

What are the ways in which credit can be given for research contributions when writing a research report?

## Backward-Looking Responsibility

In contrast to forward-looking responsibility, backward-looking or retrospective responsibility is responsibility for some act that has already occurred. As we saw in the introduction, causal responsibility is retrospectively assigned or identified, but causal responsibility need not carry ethical significance. (It certainly carries no ethical significance when the causal agent – such as a storm – is not a moral agent.) Blame or credit is also assigned retrospectively. In the assignment of praise or blame, retrospective responsibility *is* ethically significant.

Backward-looking responsibility assumes accountability. One is blamed for a bad outcome for which one is responsible. For example, the party responsible for a tunnel collapse would be blamed for it.

Conversely, one is praised or credited for positive results. Thus, someone who was responsible for an ingenious experimental design would be credited.

Forward-looking responsibility specifies *the good result that is to be achieved*, such as "the responsibility for *the integrity of the research record*." To fulfill a responsibility one must figure out what to do or avoid doing to achieve the specified ends.

Credit for research contributions is assigned in three principal ways in research publications: by authorship (of the research being published), citation (of previously published or formally presented work), and via a written acknowledgment (of some contribution to the present research).

Good research practice also requires fulfilling responsibilities. As was discussed in the introduction, fulfilling responsibilities typically requires both creativity and more exercise of judgment than do fulfilling obligations, respecting others' rights, or following rules of the form "Do X" or "Do not do Y." Responsible authorship requires the concept of forward-looking responsibility. Forward-looking responsibility specifies the *end (i.e., the good result) that is to be achieved*, such as "the responsibility for *the integrity of the research record*." To fulfill a responsibility one must figure out what to do or avoid doing to achieve the specified ends.[1]

Whereas the statements of ethical *obligation* specify what acts one is obliged to perform or refrain from, when choosing a course of action to fulfill an ethical *responsibility*, one typically must consider multiple criteria. Typically, some courses of action, even if they would achieve a desired end, would be unacceptable because of other ethical considerations.

---

[1] Although "responsibility" can be used as a synonym for "obligation" – "the responsibility *to do* something" just means "the obligation to do that thing" – there is no way of stating the responsibility for some *outcome* in terms of obligations. There is no expression "obligation for." Achievement of an outcome requires judgments about *what actions* will best achieve those ends without causing major negative side effects in the research context under consideration.

*What are the ways in which credit can be given for research contributions when writing a research report?*

## Citation and Acknowledgment

Under what conditions should a person's contribution to some research that is being reported be credited via citation and under what conditions should it be credited via an acknowledgment?

Here are some widely accepted criteria for citation of research publications or presentations.

- If in doing the research on which you are reporting you drew on any results or ideas that appeared in previously published or formally presented work – citations of talks and papers presented at disciplinary meetings are appropriate whether the papers were published in a volume of conference proceedings or not.
- The Bibliography or List of Works Cited for your report should be sufficiently complete to allow readers to understand where the reported research fits in the development of engineering and scientific research.
- The Bibliography or List of Works Cited should include any foundational research contributions that are not common knowledge *for the readership of the publication for which you are writing*.
- Theses in university libraries are appropriate for citation, as are some unpublished reports that are nonetheless generally obtainable (e.g., by writing to the issuing laboratory).
- Unpublished work, such as private correspondence, is cited only when no readily obtainable written sources are available. Appropriate crediting of such sources may be better handled with an acknowledgment. In either case, because an unpublished view is attributed to another person, the author should obtain the approval of the person cited or acknowledged, because it may not be a view that the person wishes to take publicly.
- Some, *but not all*, engineering society publication guidelines explicitly say that authors are obligated to obtain *permission* for such acknowledgments.

> If in doing the research on which you are reporting you drew on any results or ideas that appeared in previously published or formally presented work, be sure to cite those sources.

Citation of a person's publication does not make the person cited *accountable* for the works in which her work is cited, and so citation never requires permission of the part(ies) whose works are cited. Those authors *already* bear responsibility for their published work. Therefore, citation is the one form of crediting that never carries accountability with it.

Professional societies have offered some guidance on citation practices. Beginning in 1985, the American Chemical Society (ACS) issued several versions of its Ethical Guidelines to Publication of Chemical Research to guide editors, authors, and reviewers for any of its numerous publications. (Their publications include some, such as *Chemical and Engineering News*, that are explicitly addressed to

> Citation of a person's publication does not make the person cited *accountable* for the works in which her work is cited, and so citation never requires permission of the part(ies) whose works are cited. Those authors *already* bear responsibility for their published work. Therefore, citation is the one form of crediting that never carries accountability with it.

engineers.) The ACS guidelines have been taken as a model by several other engineering and scientific societies. Thus, publication guidelines from four engineering societies – the American Society of Civil Engineers (ASCE), the American Society of Mechanical Engineers (ASME), the American Institute of Aeronautics and Astronautics (AIAA), and the American Institute of Chemical Engineers (AIChE) – are all modeled on the ACS guidelines and contain passages that are virtually identical.[2] Most of these societies acknowledge the ACS guidelines. In the most recent revision (2010), the ACS guidelines include as the ninth of the fourteen obligations listed for authors:

> An author should identify the source of all information quoted or offered, except what is common knowledge. Information obtained privately, as in conversation, correspondence, or in discussion with third parties, should not be used without explicit permission from the investigator with whom the information originated. Information obtained in the course of confidential services, such as refereeing manuscripts or grant applications, should be treated similarly.[3]

Contributions to the reported research that are neither sufficiently significant to qualify a person to join the authors in writing up the research nor contained in a citable source should be recognized in the *acknowledgments*. (Slide presentations may include acknowledgments on a slide with the names of all contributors and the nature of their contributions.)

> Contributions to research that are neither sufficiently significant to qualify a person to join the authors in writing up the research nor contained in a citable source should be recognized in the acknowledgments.

A person whose research contribution is acknowledged in a report of the research is accountable only for the specific contribution for which the person is acknowledged, *not* the whole report. The ethical guidelines of some engineering and scientific societies do require permission (of the person acknowledged) for an acknowledgment in an article in their journals, however, presumably because the representation of what the acknowledged party is said to have done should be approved by that party. Even when permission is *not* required by the journal in which the publication is to appear, it is prudent and considerate to at least inform and perhaps obtain the permission of anyone acknowledged prior to publication of the manuscript. (It would be awkward if some investigator were acknowledged for making a

---

[2]The ASME credits the ACS. The ASCE and AIAA credit the American Geophysical Union publication guidelines, which in turn credit the ACS publication guidelines. Some of these engineering society guidelines contain only sections on obligations of authors and obligations of reviewers, whereas the ACS also begins with the obligation of (its) editors and after 1989, a section on publishing outside of the technical literature. The ACS guidelines have also been used as a model by scientific societies, for example, by the Optical Society of America (OSA) and the American Geophysical Union (AGU) as those societies fully acknowledge.
[3]ACS, 2006.

contribution that that investigator believed to be a bad idea and a misunderstanding of what she had suggested.)

*Under what conditions should a person's contribution to some research that is being reported be credited via citation and under what conditions should it be credited via an acknowledgment?*

# Authorship

### Qualifications for Authorship

What are the criteria that qualify a person for authorship of a research report/article, as contrasted with acknowledgment for contributing to the research?

A person is eligible for authorship of a research report, when (at least) both of the following conditions are met:

- The person has made a major contribution (in such areas as research design, theoretical development, development of a prototype, analysis and interpretation of data) to the research reported.
- That person reviews and approves the final manuscript.

Usually an author also contributes to the writing or critical revision of the manuscript as well, but as we shall see, the engineering community has not reached agreement on whether that is necessary for authorship. Although some authorities require participation in both writing and critical revision, most engineering societies do not.

Authorship makes one accountable for the work that one has authored or coauthored. Unless an author's contribution is explicitly stated to have been limited to a certain area, such as "X contributed the statistical analysis for this work," *the default expectation* is that author is responsible for the *entire work*.

Because each author must take responsibility for the quality and integrity of at least some aspects of the research report and thus answer for any deficiencies in that aspect of the work, all must approve the final version before it appears to the public. Author accountability is also the reason that the author list should include no fictitious names. For the same reason, "gift" or "**ghost**" or "honorary" **authorship** (the listing of someone as an author who does not contribute to the research and hence does not qualify for authorship, although he or she may have some other connection to the research) is unacceptable. A person who does not know or does not agree with the content of a published paper cannot take public responsibility for it.

> A person is eligible for authorship of a research report, when
>
> - the person has made a major contribution (in such areas as research design, theoretical development, development of a prototype, analysis and interpretation of data) to the research reported
> - that person at the least reviews and approves the final manuscript
>
> Authorship makes a person accountable for the report.

One notorious abuse that the prohibition of "ghost authorship" is intended to prevent is that of a company paying an investigator to publish under her own name a ghost-written article reporting research favorable to the company's interests.

In writing a *research* report, authors may receive extensive editorial help that improves the writing of that report, although such help might not be acceptable for all types of writing. The acceptability of such editorial help for *research* reports is due to the fact that the criteria used to judge a research report are such epistemic values as truth and fruitfulness of the inquiry reported. Good writing always makes it easier to disseminate one's work, but in a research report, the quality of the writing is secondary to the quality of the research. An editor is not a ghostwriter, and editing of research articles does not make authors into ghost authors.

Special situations may arise that require some modification of authorship rules. One that has been thought through is the death of one of the coauthors prior to publication. Obviously, if the person has died before the final version is complete, that person cannot review and approve the article, but does deserve credit for her research contribution. The previously mentioned AIChE guidelines state as the eleventh of the twelve obligations of authors:[4]

> The coauthors of a paper should be all those persons who have made significant scientific contributions to the work reported and who share responsibility and accountability for the results. Other contributions should be indicated in a footnote or an "Acknowledgments" section. An administrative relationship to the investigation does not of itself qualify a person for coauthorship (but occasionally it may be appropriate to acknowledge major administrative assistance). *Deceased persons who meet the criterion for inclusion as coauthors should be so included, with a footnote reporting date of death.* No fictitious name should be listed as an author or coauthor. The author who submits a manuscript for publication accepts the responsibility of having included as coauthors all persons appropriate and none inappropriate. The submitting author should have sent each living coauthor a draft copy of the manuscript and have obtained the coauthor's assent to coauthorship of it. (Italics added.)

The IEEE, in its Publication Guidelines, gives criteria for authorship that are superficially similar to those specified by the International Committee of Medical Journal Editors (ICMJE) in its "Uniform Requirements for Manuscripts Submitted to Biomedical Journals," in that it states what seem to be three criteria as necessary for authorship. Its second requirement is significantly different from that of the ICMJE, however.[5]

The IEEE affirms that authorship credit must be reserved for individuals who have met each of the following conditions:

a. Made a significant intellectual contribution to the theoretical development, system or experimental design, prototype development, and/or the analysis and interpretation of data associated with the work contained in the manuscript

b. Contributed to drafting the article *or reviewing* and/or revising it for intellectual content (Italics added)

c. Approved the final version of the manuscript as accepted for publication, including references[6]

[4] AIChE. *Ethical Guidelines for AIChE Publications*.

[5] Retrieved from http://www.icmje.org/ethical_1author.html, dated October 2009 (last visited November 8, 2010).

[6] Presently, the IEEE criteria for authorship are found only in section 8.2.1 of the IEEE's *PSPB Operations Manual*, Amended November 16, 2007. The entire *PSPB Operations Manual* is available as a 875kb pdf at http://www.ieee.org/portal/cms_docs_iportals/iportals/publications/

In including "reviewing" in the second requirement, the IEEE blurs the distinction between the second and third requirement, because reviewing is obviously necessary for meaningful approval of an article. (The ICMJE had listed as its second requirement, "drafting the article or *revising it critically* for important intellectual content;" "revising it critically" is *not* implicitly included in approving the final version of the manuscript. [Italics added.])

The engineering community generally recognizes the first and third of the requirements listed by the IEEE, but at present, has not reached a consensus that more is required for authorship. I know of several cases in which coauthors did not contribute substantially to either writing the article or critically revising it, because they are not fully fluent in the language in which the research report is written.

*What are the criteria that qualify a person for authorship of a research report/article, as contrasted with acknowledgment for contributing to the research?*

### Responsibilities of Authors
What are the responsibilities of an author of a research article or other research report?

Most engineering societies, including ASME, embrace the notion that:

> An author's central obligation is to present a concise and accurate account of the research, work, or project completed, together with an objective discussion of its significance.[7]

> The default expectation is that each author is accountable for the entire work. Therefore, in the absence of a statement about the scope and limits of what each author contributed, all authors are accountable for both the integrity and the competence of the research reported.

Because the default expectation is that each author is accountable for the entire work, in the absence of a statement about the scope and limits of what each author contributed, questions about either the integrity or the competence of the research reported are ones that *all of* the authors are expected to be able to answer. Because of special difficulties in answering for the work of coinvestigators in other fields or disciplines, specification of the nature of one's contribution is especially appropriate in interdisciplinary research. Journals must be willing to give space for such specifications for it to be a practical possibility, however.

*What are the responsibilities of an author of a research article or other research report?*

### Categories of Authors and Their Special Obligations and Responsibilities
What special categories of authors carry special responsibilities?

All those on the author list of a research report are presumed to have fulfilled the obligations of authorship and to be prepared to take responsibility for the article or other publication, either the portion for which they have explicitly identified as their work, or, in the absence of such a specification, the entire work.

PSPB/opsmanual.pdf (last downloaded November 8, 2010). Because the IEEE has removed some valuable ethical guidelines, placing their expectations for responsible authorship in the *PSPB Operations Manual* may show the same reluctance to openly guide the members of the profession on the ethics of their practice.
[7] ASME. 1999. "Ethical Obligations of Authors."

For coauthors of articles in engineering and science journals, subcategories apply and are ethically significant because of the special obligations and responsibilities that fall to coauthors in those categories.

**Lead author** – The lead author is the author, if any, who is principally responsible for the work, the one who made the greatest intellectual contribution. The lead author bears responsibility for the whole research report, even if some other coauthors explicitly state that they take responsibility for only some aspect of it.

**Submitting author** – The submitting author is the author who submits the manuscript for publication and usually is the author who deals with the journal and its editors after submission. This author has a special responsibility to see that all the authors have fulfilled the criteria for authorship and that all have read and approved the final version of the work. If there are special journal requirements, such as the requirement that all authors sign a form saying that they have read and approve the final version, the submitting author must see to it that those are met. The submitting author is often the lead author or the leader of the research team, but need not be. If a senior investigator were publishing an article with one of her trainees, the senior investigator might submit the article, because she knows more about dealing with journals or the specific journal in question. On the other hand, the senior investigator might ask the trainee to handle the submission (to give the trainee experience) even if the senior investigator were the lead author as well as the team leader.

**Corresponding author** – The corresponding author is the person whom interested individuals may contact about the article after it is published. The corresponding author will typically receive most of the reprints of the authored article and answer requests for reprints. Often an asterisk indicates the corresponding author by her name in the author list.

A coauthor may become the corresponding author simply because that person has the most predictable mailing address in the immediate future (if, for instance, all of the other authors are changing institutions in the near future). In a *few* fields, the corresponding author is assumed the leader of the research team. (See the entry for "last author," in the next subsection for more information about the expectations of a leader of a research team.)

**Senior author** – This term is ambiguous and sometimes indicates the lead author and sometimes the most senior of the authors (i.e., the one with the highest academic rank, most prestigious job title, or greatest reputation within the field). The panel at Lucent Technologies' Bell Labs that investigated the possibility of research misconduct in the work of Jan Hendrik Schön concluded that while only Schön was guilty of research misconduct, perhaps the most senior of his coauthors, precisely because of his seniority, ought to have done more to forestall Schön's research misconduct. Although the power of coauthors does seem relevant to the exercise of responsibility, as of 2010, there is no consensus about how to assess the responsibility of coauthors for misconduct committed by a colleague. The suggestion of the panel at Lucent Technologies' investigation panel would make the category of senior author a category of authorship that carries special responsibilities for the integrity of research.

*What special categories of authors carry special responsibilities?*

### Plagiarism
Why is plagiarism such a major issue for research conduct (and academic integrity)?

As we saw in the last chapter, plagiarism is generally understood to be the appropriation of another person's ideas, processes, results, or words without giving appropriate credit. Plagiarism is a type of "research misconduct." It can be of graphical representations (such as photographs, tables, or charts) as well as text, and of electronic media as well as hard copy (such as books, journals, conference proceedings) and multimedia presentations. (Plagiarism can be of artistic works as well as reports of research, but the distinctive standards for plagiarism of artistic works need not concern us here.) Plagiarism is also what most colleges and universities regard as the most serious misappropriation of credit of which their students or faculty members may be found guilty.

The IEEE gives substantially the same definition.[8] Nonetheless, the IEEE also distinguishes what it calls five "levels" or "degrees" of plagiarism to guide IEEE editors and officers in deciding how to respond to charges of misappropriation of credit in IEEE publications. These levels of the extent of plagiarism are useful for delineating what *counts* as unfair in the assignment of credit in contemporary U.S. research (and academic) ethics. They are:

1. Uncredited Verbatim Copying of a Full Paper.
2. Uncredited Verbatim Copying of a Large Portion (greater than 20% and up to 50%) within a Paper,
3. Uncredited Verbatim Copying of Individual Elements (Paragraph(s), Sentence(s), Illustration(s), etc.)
4. Uncredited Improper Paraphrasing of Pages or Paragraphs. [Instances of improper paraphrasing occur when only a few words and phrases have been changed or when the original sentence order has been rearranged; no credit notice or reference appears with the text.] (Parenthetical explanation quoted from the original.)
5. Credited Verbatim Copying of a Major Portion of a Paper without Clear Delineation [Instances could include sections of an original paper copied from another paper; credit notice is used but absence of quotation marks or offset text does not clearly reference or identify the specific, copied material;[9] or, for an *extended* quotation, as of the IEEE delineations quoted here, failure to set the quoted text apart from the original text, e.g., by indentation and change of font. Extended quotations are commonly indicated by these means rather than by quotation marks.] (The first parenthetical remark in the original. The second is added.)

---

[8] The definition the IEEE gives in its *PSPB Operations Manual*, available at http://www.ieee.org/portal/cms_docs_iportals/iportals/publications/PSPB/opsmanual.pdf, section 8.2.1 B, "Responsibilities of Manuscript Authors," is "the use of someone else's prior ideas, processes, results, or words without explicitly acknowledging the original author and source."

[9] *IEEE PSPB Operations Manual*, section 8.2.4 D, "Guidelines for Adjudicating Different Levels of Plagiarism."

Plagiarism to any extent is a violation of the standards of fair crediting. Those who feel hampered by a lack of facility with English or whatever language in which they are publishing do better by having an editor go over their research reports than by quoting or paraphrasing large portions of text from other reports.

> Plagiarism is generally understood to be the appropriation of another person's ideas, processes, results, or words without giving appropriate credit. Plagiarism that meets this definition counts as a type of "research misconduct." Plagiarism applies to graphic representations (such as photographs, tables, or charts) as well as text, and to electronic media as well as to hard copy (such as books, journals, conference proceedings) and multimedia presentations.

In the previous chapter we considered the role of trust in fostering cooperative endeavors such as engineering research. Trust that one's research contributions will be fairly credited is central (along with trust in the honesty of research reports) to the maintenance of the trust necessary for research to flourish. Plagiarism to any extent violates that trust and so undermines that cooperation. The examples given at the beginning of Chapter 8 of self-protective actions that research investigators may take when their trust in being properly credited breaks down illustrate the corrosive effect that violations of standards of fair crediting have on the research environment.

The ethical consideration about the wrongness of plagiarism (notwithstanding the conventionality of particular standards of what actions count as plagiarism or even the cultural dependence of the expectation of individual credit) cannot be reduced to legal considerations of copyright violations, although some engineering authorities, such as the IEEE,[10] have mistakenly asserted that plagiarism is a form of copyright violation. As discussed in Section 7 of the introduction to this book, copyright protection is *not* the legal equivalent of a prohibition of plagiarism. Property rights are a different matter from credit. Thus, an author may (and often does) transfer the copyright of a journal article to the publisher of that journal, but nonetheless retains the credit for having authored the article.

*Why is plagiarism such a major issue for research conduct (and academic integrity)?*

### Fair Sharing of Credit among Coauthors
How does one go about ensuring the fair sharing of credit among collaborating investigators?

As coauthorship becomes more common, it becomes more important that collaborators have a discussion early in their collaboration as to how they will handle authorship decisions, including what and where to publish and which of them is to be credited in what way on which publications. Although the projected research may hold some surprises that will require some rethinking of the initial agreements, the early discussion of authorship issues will prevent misunderstandings. These discussions may be more difficult or complex when collaboration involves multiple disciplines, but they are all the more important

---

[10]See the fourth item on slide 6 of the *IEEE IPR Office Tutorial on Plagiarism* (undated), available at www.ieee.org/portal/cms_docs_iportals/iportals/publications/rights/plagiarism.pps and last downloaded on November 9, 2010, which begins, "It should also be noted that plagiarism is a type of copyright infringement...." Plagiarism covers misappropriation of ideas and text in the public domain, however, and neither is covered by copyright.

because of those complexities (such as different journal practices in the journals of the involved disciplines).

### Credit among Coauthors[a]

Pat is an engineer and a member of a research and development team composed of engineers from several fields of engineering, including materials science and engineering, chemical engineering, and biomedical engineering, which is working with some molecular biologists and physicians to develop a new implantable drug. The team members have agreed on the default expectation (the expectation to be fulfilled unless justification is given for behaving differently) that the order of the names on the author list of coauthored papers will reflect the practices of the discipline of the publication for which the paper is written. For example, normally any paper for a chemical engineering journal would have the chemical engineer as the first author and she would initiate the writing of that paper with whichever collaborators had relevant knowledge and handle its submission. Any publication for a prestigious publication that crosses disciplines, such as *Science* or *Nature*, will normally list the authors in alphabetical order. In all publications, the work of any team members who are not coauthors will be cited or acknowledged wherever relevant.

How adequate and responsible are these agreements about credit?

What alternative expectations, if any, do such teams commonly work out ahead of time and how do they compare with the ones stated? (Any research team should discuss publication and credit early in the planning and conduct of a project.)

---

[a]This scenario is substantially the same as one by Caroline Whitbeck and has the same title in the Ethics Case Discussions, located at http://www.ethicscasediscussions.org/ (last visited November 9, 2010).

It is responsible and admirable that the research team anticipated and discussed authorship complexities. Their agreement yields general guidelines, however, and should not be expected to dictate a strict specification of authorship and order of authors for each publication resulting from the collaboration.

Do you know of any conventions and requirements that are specific to your field (such as special requirements of the journals in which you seek to publish) or any special considerations arising from the *interdisciplinary* collaborations in which engineers in your field commonly participate?

In some fields the *order* in which the authors are listed is taken to be significant, and may indicate the extent of contribution of each. In other fields, the listing may simply be alphabetical (by family name). If the author list is long, alphabetical listing may be the only practical way of handling the ordering of authors. If the ordering is alphabetical, it signifies nothing about the coauthors' relative contributions.

The publication guidelines for some journals, such as the IEEE, say that once the article has been submitted for publication, the order of authors must not change without permission of all living authors. The IEEE further stipulates that although coauthors may remove their names after that time, no author's name is to be removed *by others*, and *no authors names may be added after submission.*[11]

---

[11]Some engineering societies have commented directly on these issues so it should clearly be understood as at least a condition of publication in the journals of those societies. For some others, it may be the articulation of a commonly accepted norm that other societies have not thought to make explicit. Therefore, when publishing in some journal that does not articulate a rule that is stated by others, it is prudent to consult the editor of the journal in which one is publishing before violating that rule.

Determination of the order of authors or of which coauthors fall into the special categories discussed earlier requires judgment. It is a complex matter and one that is made even more complex if multiple fields and disciplines, with different conventions about authorship, are involved in the research being reported.

**First author** – This term is most frequently used to mean the lead author. It is so used because a common convention is to indicate the lead author by placing that person's name first in the author list. One reason that this practice is common is that an article with three or more authors is often referred to using only the name of the first author followed by "et al." In such a case, the article will come to be known by the name of the first author in the list.

Some journals or fields that use author position to signify differential credit fine tune the signification of contribution even further: if an article has two authors, the authors may list their names on the same line (to indicate equal contribution) or they may list the first's above the second (to indicate that the first author is the lead author).

**Last author** – In some of the fields in which the order of the authors *does* indicate contribution, the last author position simply means the author who made the least important contribution.

In other fields in which the order of the authors indicates importance of contribution, especially fields that have been strongly influenced by medical research traditions, the last author position is reserved for the *leader of the research team* that carried out the research. The team leader need not be the lead author; another member of the team may have made the greatest contribution to this specific *research reported*. The leader of the research team is typically the person who planned the research program of which the research reported is a part and so may have the most comprehensive vision of where this research fits in the advance of engineering knowledge.

The head of the laboratory, if any, in which the research was carried out often, but not always, provides the leadership for the research team. Being the head of the laboratory itself only signifies having an administrative relationship to the research and so does not by itself justify being included on the author list, according to currently accepted criteria for authorship.

*How does one go about ensuring the fair sharing of credit among collaborating investigators?*

### When Supervisors and Their Supervisees Share Authorship

What makes the situation of sharing authorship between a research supervisor and one or more of the supervisor's trainees more complex than other coauthorship collaborations?

Trainees depend on their supervisors to help them learn how to judge the importance of various contributions to research and what constitutes fairness in crediting research contributions, as well as for learning the conventions for assigning credit in their field. Therefore, credit issues between research supervisors and their trainees (graduate students and post-docs) must be handled somewhat differently from credit between two mature investigators. The conversation among peers about authorship at the beginning of a collaboration will be very different from the trainee's inquiry into the criteria that the supervisor uses in deciding authorship, differential credit among authors, or when a trainee should present a conference paper.

Supervisors bear a responsibility for educating the trainee in the principles, rules, and criteria used in judging whether in general research conduct is responsible and, in particular, whether credit allocation is fair. Supervisors need to explain their practices and the values underlying them to their supervisees, especially if their supervisees are to have respect for and confidence in standards of fairness. Telling supervisees only after the fact that they have violated some standard is a poor substitute for preparing them for their responsibilities as an investigator.

In principle, engineering may be in a better position to educate the next round of researchers than many of the sciences. In the sciences, a lab is generally the domain of a single faculty member, but in engineering, a lab is often shared by many faculty members who have pooled their resources to acquire expensive equipment. This creates a larger community to educate trainees in research practice. Some departments have gone further and require that every graduate student have, in addition to her research supervisor, a different member of the faculty as a departmental advisor. This way a student has additional opportunities to receive guidance about research conduct and its underlying values.

The research community has paid more attention to research ethics since the late 1980s, and, as the U.S. definition of research misconduct illustrates, formulated some of the existing rules and criteria for responsible research conduct since that time. As a result, some experienced investigators may find either that they are not abreast of current resources and guidelines, or they are not fully articulate about the (appropriate) standards and criteria they have internalized. Furthermore, they may not be aware of points at which their practices differ from those of their colleagues, and so be unprepared to explain to their students just which differences are within the acceptable range, which are outside of it, and why. If you find yourself with such a supervisor, it may be best to send her your questions in an email, so she will have time to think through her answer.

For example, around 1990, I recall an investigator who said his research group all knew that he hated to write up his research and so his trainees knew they could secure first author position (which in his field is interpreted as lead author) by writing the first draft of the manuscript. (Of course, a trainee would have to be intimately involved in some research project to *be able* to write the first draft.) None of his departmental colleagues had that practice, but they agreed that his practice was *within the acceptable range of practices*, given that everyone in his group knew what it was. That department then went to some effort to see that all the graduate students learned more about the criteria used by their own research supervisors, so that other students working for other supervisors would not expect that they would necessarily secure first author position by writing first drafts and feel unfairly treated when they did not.

Trainees need to understand the expectations *within their own fields*. These are best learned from their own supervisors. One such expectation is whether graduate students are encouraged to publish on their own in their particular field. This expectation can vary even from one subfield or "area" to the next.

Trainees, especially entering graduate students, may have difficulty judging why a faculty member is the lead author on a joint publication for which they may believe that they have put in the largest effort. It is important to ask your supervisor about anything that seems amiss, but for a start, note that time put in is *not* a good measure of contribution to research, especially if the time was spent in doing something that could be effectively accomplished by someone with little engineering background. Intellectual contribution is most important and that contribution may occur *before* the graduate student receives a research problem. Finding an interesting engineering problem and knowing what the research problem a student with a certain background is likely to be able to complete within a reasonable amount of time may require more intellectual labor than actually carrying out the research. Students are well advised to learn as much as possible about the place of the work assigned to them in the larger scheme of the supervisor's research as well as about the supervisor's expectation for the work assigned to them. It is also good to know the supervisor's criteria for the selection of target journals, conferences, or other venues for publication.

Finally, conflicts and misunderstandings about authorship may arise within engineering laboratories. These are reduced if trainees and supervisors have a dialogue about credit and the supervisor's crediting practices *early in their relationship*. Engineering research supervisors typically have many potentially competing responsibilities. These responsibilities include (in no significant order):

- For the advance of knowledge in their field of engineering
- For the education of their trainees
- For the wise and appropriate use of grant funding
- To their institutions and for various work assigned to them
- *Vis-à-vis* their collaborators and to investigators in their field.

Those are weighty responsibilities and help explain why professors are notoriously absent-minded. If you, as trainee, think you have not been credited for a research contribution, it is wise to begin by considering that your professor may have simply forgotten what you did. Raising the subject of the research on which you worked and asking your supervisor about the supervisor's estimate of the significance of your contribution are more prudent than either assuming that your contribution was of little value or that you are being treated unfairly.

## What about My Contribution?

For the first year of your graduate studies, you worked with Professor One on the Hot Research project. By the end of the first year, you not only became proficient at operating the complex experimental apparatus but also made a small but notable refinement in the approach to the segment assigned to you. At the end of the first year, Professor One went on leave for a semester and you started working on a different project with another supervisor in the same lab.

In the term following Professor One's return from sabbatical, another student, who is still working on Hot Research, tells you that he and Professor One are coauthoring a paper that incorporated your refinement.

Are there any ambiguities in the situation?

What, if anything, can and should you do?

Notice both what you do not know and where you have only hearsay evidence. In particular, notice what you do not know about what Professor One has in mind. Is One really planning to publish? If so, does One remember your contribution?

If so, is One planning to include you as an author or thank you in an acknowledgment? How can you best find out? How will you phrase what you say in a way that does not prejudice the matter?

## Crediting Trainees Who Help Write Grant Proposals

You are a Ph.D. student working with Professor Pi. You are near the completion of your project and have prepared a paper for publication. Professor Pi has hired another Ph.D. student, Kino, who will continue on the same project after you graduate. Professor Pi would like to obtain further grant funding for that project and prepares a grant proposal with the help of Kino. Professor Pi has an electronic version of your paper and copies most of the figures and about half the text in the grant proposal from your paper. You have presented some of the work reported in your paper at a conference. That presentation is cited in the grant proposal, but only in the Background and Significance section. You are concerned that whoever will read that proposal may attribute to Kino all the work presented in the Progress Report section because Kino is the person for whom the funding is sought. You also worry that you will be submitting exactly the same figures and text when you publish your paper.

### Getting Started

We saw that the current U.S. government-wide definition of research misconduct applies to research reported in proposals so some ethical responsibilities are similar, but proposals are confidential documents and are not immediately available to the interested public as are published reports.

Would the fairness issues be any different if the first student's work were the student's thesis rather than a (publishable) research paper?

Would the situation be different if the advisor had made a presentation of the lab's work to an industry group, rather than summarizing it in a proposal to a government-funding agency? (One faculty member has told me that she includes a slide with the names of all the students who worked on a project in every project presentation she makes.)

What would be appropriate ways of crediting a trainee contributor to a grant proposal? Would it be different if the trainee were listed as a member of the research team to be supported if the work is funded?

What is the range of acceptable variation among faculty members in the way they credit their trainee's work? (One faculty member, when writing a grant proposal that uses the work of a student who will not be getting any financial support from a grant, lists the name of the student among the collaborators [for whom no funding is sought].)

Grant proposals are something of a hybrid. On the one hand, they commonly contain reports of research (either research that leads up to the research proposed, or pilot testing of the hypothesis proposed). Therefore, many of the same strictures apply to them as to other research reports, including that fabrication, falsification, and plagiarism in them count as a serious breach of standards of responsible research behavior. On the other hand, they are confidential documents, not publicly available research reports, so unlike both publications and theses,* they *cannot* be cited. The information contained in them could not be counted as

---

*A thesis has an intermediate status as a research publication. Theses, like technical reports issued by laboratories, *can* be cited in other research reports. Once a thesis has been placed in the library of the degree-granting institution, interested parties may obtain copies of the thesis. Placing a thesis in a university library does *not* count as a prior publication so the thesis author may later submit all or part of that thesis for publication.

a prior publication of some work. Nonetheless, the fact that in writing a grant proposal using the work of someone external to the research group without attribution *would* count as plagiarism should remind everyone that it is important to properly credit the work of those *within* the research group.

> *What makes the situation of sharing authorship between a research supervisor and one or more supervisees more complex than other coauthorship collaborations?*

# Responsibility for Research Quality

### Authors' Responsibility for the Quality of Their Research/Reports
How is an author's responsibility for the quality of their research and research reports understood?

Although honest mistakes are not counted as research misconduct or as any other type of *ethical* violation, there are definite sanctions for doing poor research and writing poor reports. Put positively, there are definite rewards, usually career rewards, for doing good research and writing it up so that others may recognize and benefit from it. Good research and good research reports have epistemic or knowledge value.

> *How is an author's responsibility for the quality of their research and research reports understood?*

### Supervisors' Oversight of the Research of Their Trainees
How does the quality of the work of graduate students and other trainees in engineering reflect on their research supervisors?

Research supervisors have both a responsibility to uphold research standards and help their trainees to meet them. "When May a Student Publish?" illustrates how lack of appreciation of the supervisor's responsibility for the quality of research put into the literature by her students might lead a student to rash action. A computer science professor, Albert Meyer (at the suggestion of another), constructed this problem situation for discussion with graduate students. The aim was to help students think beyond credit to the *quality control responsibilities* of a research supervisor.

## When May a Student Publish?[a]

On the basis of outstanding undergraduate performance, Terry landed a first-year Research Assistantship in Professor Grimm's group within the Large Laboratory for Better Theory. Terry has not yet felt ready to say much to Professor Grimm or the other group members, but has thoroughly studied Grimm's recent CACM paper on parametric cache management and absorbed all the discussions at group meetings. Near the end of the year, Terry discovers a connection between the Algebraic Geometry he has been taking as a minor subject and parametric cache management. Terry quickly writes up a short paper on "Geometrically Parameterized Cache Management" to make the deadline for the annual ACM/IEEE Cache Management Symposium.

In the paper, Terry cites Professor Grimm's published work and several publicly available technical reports of other group members, and carefully acknowledges the contributions from group presentations by Grimm and two other group members. Because the basic idea of Terry's paper comes from the algebraic geometric connection that she alone recognized, Terry feels it is appropriate to be the sole author.

Rushing to the group printer to get the final draft of her paper, Terry meets Pat, one of Grimm's senior students. Pat tells Terry that she noticed that Terry's paper is to be submitted to the ACM/IEEE Symposium and thinks Terry should clear the submission with Professor Grimm. Unfortunately, Grimm is out of town for the rest of the week, and the symposium deadline is tomorrow.

What should Terry do?

What are the risks to Terry or to others in this situation?

---

[a]This scenario was written by Professor Albert Meyer based on a suggestion by Professor David K. Gifford. It is part of a collection of scenarios in the Online Ethics Center for teaching research ethics through faculty-student discussions of problem situations in research and may be accessed at http://www.onlineethics.org/CMS/2963/modindex/resethpages/publish.aspx.

A third computer science faculty member in the same department said (with grim humor) that he would kill any of his students who published without showing him the manuscript first. The third computer science faculty member reasoned that the student's institution and topic area would quickly identify the student author as one of his, and if the publication were junk, his, the supervisor's, reputation would suffer.

In some disciplines or fields it is not *possible* for a student to publish alone; journals in those fields do not accept submissions from graduate students. In other fields sole publication by students is encouraged. *In any case, work should be reviewed first by one's supervisor.* A supervisor's busy schedule may lead to other sorts of delays, however.

Trainees and supervisors often have different interpretations of the issues in this situation, with trainees considering only the question of credit, but supervisors thinking also of quality control of the research that the research group produces. Research supervisors are supposed to give trainees comments and criticisms about their work. A trainee is making a mistake not to take full advantage of the opportunity to get that feedback. The question here is not simply whether students publish as sole authors in a particular field, but whether students should seek to publish anything without first showing the work to their supervisors. Were a trainee to publish something that turned out to be of very poor quality, it could reflect badly on the supervisor and other supervisees of that supervisor.

*How does the quality of the work of graduate students and other trainees in engineering reflect on their research supervisors?*

### Criteria for Deciding What Credit Trainees Merit

What criteria do research supervisors, especially *your* research supervisor, use in evaluating trainees' research contributions?

The criteria that supervisors legitimately use in deciding whether their trainees should be single authors or coauthors are the same as for peer authors, that is, the criteria for authorship discussed earlier. Some examples of criteria that supervisors legitimately use in deciding the differential credit (which may be indicated by the order of authors or by which one becomes the corresponding author, or other conventions used for differential credit in the supervisor's particular field) when publishing with one or more trainees or when instructing or encouraging several trainees to publish together are:

1. The engineering importance of the contribution of each author (including the author-supervisor, if any),
2. The engineering importance of the contribution of each trainee-author (and placing the supervisor last),
3. Where a team of trainees regularly collaborates on research and lead authorship is made to rotate among the trainees, by whose turn it is to be lead author,
4. By the importance of some contribution to the research productivity of the group. (Recall the example of the faculty member discussed earlier, who gave first author position to whichever graduate student wrote the first draft.)

It can also be legitimate for the supervisor to use alphabetical order, but that ordering gives no information about differential credit.

Criteria used to determine differential credit also commonly play a role in deciding which of several trainees who have worked on a project might present the paper at a conference (where trainees are permitted to give presentations). In the latter case, questions of the trainee's skill at public speaking and maturity to handle possibly hostile questions are also frequently considered.

Some examples of *il*legitimate criteria are the following:

1. Which trainee is a personal friend, lover, or relative of the supervisor
2. Which trainee is connected to another powerful person in the supervisor's field or institution
3. Which trainee will stay with the supervisor's group longer, rather than taking her acquired knowledge elsewhere after graduation

*What criteria do research supervisors, especially your research supervisor, use in evaluating trainees' research contributions?*

## Subsidiary Obligations of Authors

What "goods" or values in addition to research integrity and fairness in apportioning credit are ones that authors of research report should pursue?

In addition to obligations and responsibilities that are essential to further the goals of preserving the integrity of the research record and fairly crediting research investigators for their work, authors are generally recognized to have other obligations to avoid overburdening others or wasting their efforts. These obligations and responsibilities are reflected in the titles of the five subsections that immediately follow.

### Do Not Fragment Your Research Reports

What is the point of writing reports that contain a full report of one's research (for a given audience); that is of refraining from fragmenting one's research into a larger number of articles?

One cluster of an author's subsidiary obligations concerns the proper use of journal resources and the time and efforts of journal editors and the reviewers they ask to review manuscripts. Guidelines for the ethical publication of research may remind would-be authors to use journal space "wisely and economically,"[12] and warn authors not to waste journal space, complicate literature searches, obscure the importance of their research, and inconvenience readers by "fragmenting" their research reports. Fragmenting research means dividing research reports into what has sometimes been called the "least publishable unit."

The temptation to fragment one's research reports may arise from situations in which one's research productivity is judged (e.g., by grant proposal reviewers or promotion and tenure committees) by the sheer number of research publications or the number on which one is the lead author. Where research productivity is so judged and investigators feel themselves to be in fierce competition for funds, jobs, or promotions, they may be tempted to appear more productive by fragmenting their research to increase the number of their (refereed) publications. However, simply counting the number of an investigator's publications, while easy to do, is widely recognized to be a poor way of assessing that investigator's productivity, because an important article is likely to advance engineering more than several of minor significance.

Some universities have taken steps to avoid the short cut of counting publications to evaluate an investigator's research by asking their faculty member to submit a certain number of articles (presumably their most significant articles) for consideration for promotion and tenure. Promotion and tenure committees can then better judge whether the candidate is publishing *significant* research.

The example of discouraging the fragmentation of research illustrates that guidance on responsible research practice, and authorship in particular, rather than addressing every conceivable situation, focuses on those matters that investigators have been *found to overlook*, or situations in which investigators may be *tempted* to act in a way that frustrates the progress of knowledge.

*What is the point of writing reports that contain a full report of one's research (for a given audience); that is of refraining from fragmenting one's research into a larger number of articles?*

### If You Republish Your Previously Published Work, Cite It

What is the point of requiring that any content from your previous publications that is repeated in a new publication be fully referenced in that new publication?

Another obligation when publishing in the technical literature (where one has ample opportunity to make citations) is to cite any place where one has previously reported substantially the same research. In some disciplines this obligation is described under the heading "duplicate publication" (or even under the *self-contradictory* name of "self-plagiarism"[13]). In engineering fields, it is quite

---

[12] This is #2 in the list of obligations of authors in the ACS guidelines.

[13] This term was coined by some editors who were incensed by the publication of articles in their prestigious journals that were word-for-word duplicates. Perhaps the idea was to make the act of duplicate publication sound more like a form of research misconduct. Although the duplicate publication would have been a clear copyright violation assuming that the authors had assigned the copyright to at least one of the journals prior to publication, plagiarism is not a matter of

common and acceptable to present research in several venues. It is not republication *per se* that is forbidden, although some societies (such as the ACM) specify a minimum percentage of *new* content that the second publication must contain. What is forbidden is failure to disclose that the material has been previously presented and where it was published. Thus, the IEEE in its *Publication Services and Products Board (PSPB) Operations Manual* (amended November 16, 2007) says:

> It is common in technical publishing for material to be presented at various stages of its evolution. As one example, this can take the form of publishing early ideas in a workshop, more developed work in a conference and fully developed contributions as journal or transactions papers. The IEEE recognizes the importance of this evolutionary publication process as a significant means of scientific communication and fully supports this publishing paradigm. At the same time the IEEE requires that this evolutionary process be fully referenced.

Editors of journals and other peer-reviewed publications may be unwilling to publish work that *simply duplicates* a previous publication, however. Informing editors of the similarity gives them the opportunity to make an informed decision. The guides to the ethical publication of research that are issued by some engineering societies, such as the ASME, say further that:[14]

1. [A]n author should inform the editor of related manuscripts that the author has under editorial consideration or in press.
2. Copies of these manuscripts should be supplied to the editor, and the relationships of such manuscripts to the one submitted should be indicated.
3. It is unethical for an author to submit for review more than one paper describing essentially the same research or project to more than one journal of primary publication.

*What is the point of requiring that any content from your previous publications that is repeated in a new publication be fully referenced in that new publication?*

### Make Available Any Special Research Materials Used in Reported Research

Why are investigators expected to make available to other investigators any specialized research materials used in their own research? What, if any, specialized research materials are commonly used in research in your field? How difficult is it to make or otherwise obtain them?

To advance research authors are encouraged to make available to others any special research materials needed to replicate or extend the research they report. A fee may be charged to cover the cost of making or providing such materials and material transfer agreements may restrict the use of the materials (to protect the legitimate interests of the author). Given the earlier discussion about the necessary conditions for authorship, it is not ethical for an investigator to demand authorship on research that issues from the use of the materials that she has transferred to others *unless that field generally recognizes supplying such materials*

ownership of intellectual property but of credit. People deserve credit both for their own ideas and for their forms of expression and cannot steal either from themselves.

[14] ASME, 1999.

*as substantial participation in the research.* (The topic of field differences will be revisited later in this chapter.)

> *Why are investigators expected to make available to other investigators any specialized research materials used in their own research? What, if any, specialized research materials are commonly used in research in your field? How difficult is it to make or otherwise obtain them?*

### Disclose Any Financial Conflicts of Interest

What is the point of an author disclosing financial interests that may be affected by publication of some research?

**Financial conflict of interest**, that is, a situation in which a person is in a position of trust that requires her to exercise judgment on behalf of others and has specifically *financial interests* of the sort that might interfere with the exercise of judgment in that position of trust, has received much attention in recent years. Indeed, policies requiring disclosure of financial interests that might conflict with judgment as a researcher or as a public official are very commonly called "conflict of interest policies," as though financial conflict of interest were the definitive or perhaps the only conflict of interest. For example, the latest ACS *Ethical Guidelines to Publication of Chemical Research* (2010) expanded the twelfth of its list of obligations of authors, which deals with this point. As we saw, in addition to specifying what it considers a "significant financial interest," it also says:

> Sources of funding of the research reported should be clearly stated. In addition, all authors should declare . . . any employment or other relationship (within the past three years) with entities that have a financial or other interest in the results of the manuscript (to include paid consulting, expert testimony, honoraria, and membership of advisory boards or committees of the entity). The authors should advise the editor in writing either that there is no conflict of interest to declare, or should disclose potential conflict of interests that will be acknowledged in the published article, whether by insertion of a footnote, or incorporation of a sentence or paragraph in the "acknowledgments" section, or by other format of disclosure to the reader as specified by the journal.[15]

Journals now commonly require authors to disclose to the editor and the readership of the journal any significant financial interest that might be affected by the publication of the manuscript that the author(s) submits. At present, the threshold for significance is usually an interest greater than $10,000 or a 5 percent equity interest (i.e., ownership), say of a company or the stock of a company. Several engineering (and scientific) societies, including the ACSE,[16] require more: They require authors to state whether they have in the preceding three years had any employment relationship, paid consulting, expert testimony work, received honoraria, or held membership on advisory boards or committees of an entity that has a financial or other interest in the results of the manuscript.

---

[15] ACS, 2010, in item 12 of *Ethical Obligations of Authors.*
[16] ASCE, 2009.

Disclosure enables the editor and the readers to judge for themselves whether the research article is biased.

Legislation in 1993 reauthorizing the operations of the National Institutes of Health (NIH) required new federal regulations to minimize conflicts of interest among scientists supported by the NIH, including those receiving NIH grants. The regulations require disclosure of financial interests of more than $10,000. This legislation and the regulations resolved a five-year controversy in which public support for regulations to minimize financial conflicts of interest especially in the evaluation of a medical product or treatment was met by resistance at several major medical schools. Health policy analyst Diana M. Zuckerman summarizes the issues in the following way:

> Many ways exist to analyze data, and if the results are not dramatic one way or the other, a scientist could be motivated to find a significant result where none really exists, to omit some potentially relevant information, or even to unintentionally skew the findings.
>
> That kind of bias is probably inevitable and thus has to be considered acceptable, for example, when scientists working for a particular company are in the early stages of product development. However, such bias becomes unacceptable when the public, not the company, pays for the research, especially when the public could be put at risk as a result.[17]

*What is the point of an author disclosing financial interests that may be affected by publication of some research?*

### Warn Subsequent Investigators of Any Hazards in Conducting the Research You Report

Suppose an investigator has successfully completed experiments despite some special hazards in conducting them. What responsibility does that investigator have for the safety of other investigators who might repeat those experiments?

The AIChE guidelines, following the ACS, state that authors have an obligation to identify any unusual hazards inherent in the chemicals, equipment, or procedures described in their research reports.[18] Subsequent investigators need to know of such dangers, and authors are uniquely in a position to know of such dangers and to inform other investigators, so the stipulation is reasonable. The reason that this subsidiary obligation appears only in the ACS and AIChE enumeration of author obligations is presumably because nonobvious experimental hazards are uncommon in areas of research other than chemistry and chemical engineering.

*Suppose an investigator has successfully completed experiments despite some special hazards in conducting them. What responsibility does that investigator have for the safety of other investigators who might repeat those experiments?*

*What "goods" or values in addition to research integrity and fairness in apportioning credit are ones that authors of research report should pursue?*

---

[17]"Conflict of Interest and Science," *The Chronicle of Higher Education,* October 13, 1993, B1.
[18]The *Ethical Guidelines for AIChE Publications* borrow heavily from the ACS's *Ethical Guidelines to Publication of Chemical Research.*

## Disciplinary or Field Differences in Conventions for Authorship

What justification is there for field differences in conventions for authorship?

We have seen some of the different conventions for assigning credit in different fields. Notice that although conventions for indicating one's role in a research collaboration and the resulting credit vary from one field to the next, those are not ethical variations. They are only conventions for *indicating* credit. Ethically speaking, *it does not matter what convention one uses*, so long as all parties understand the convention and use it consistently. What matters ethically is that credit be apportioned fairly and in a manner consistent with the conventions recognized by readers of the journals and other publications in which the research is reported.

One significant difference (because it affects who deserves to be an author) is the difference in whether someone who only supplies some special material necessary for the research deserves to be a coauthor. In most fields, acknowledgment is more appropriate than authorship for those who create research materials, but there are exceptions, which are presumably justified by the extent of the intellectual contribution represented by creation of the material. In condensed matter physics, for example, it is customary for the investigator who grows the crystals used in an experiment to be a coauthor of the resulting research report.

Similarly, although the contributions of technicians are usually recognized with an acknowledgment rather than with authorship, in high-energy physics the common practice is to include technicians in the famously long author lists.

Some field differences in conventions may be fully justified by the nature of research in that field. Other differences, such as differences in the conventions for designating the team leader on an article (e.g., by showing her as either the last author or as the corresponding author) are simply cultural differences, analogous to the difference between driving on the right side of the road rather than on the left. Differences in authorship conventions pose a special challenge for interdisciplinary work involving fields with different conventions.

> A commonly accepted practice is not necessarily an ethically justified one.

Disciplinary and field differences must be distinguished from common patterns of abuse within a field. In other words, a commonly accepted practice is not necessarily one that is ethically justified. Therefore, if **gift authorship** were particularly common in some field, that would not be a legitimate field difference, but merely a common pattern of abuse.

*What sorts of justification are there for field differences in conventions for authorship?*

## Crediting Others When Publishing outside of the Technical Literature

How is publishing in the popular press different from publishing in engineering and scientific journals in ways that affect one's opportunities to credit others for their research contributions? How can one be fair to others under those conditions?

When publishing in the popular press, giving a popular lecture, giving a press conference, or being interviewed by the press, it is not usually possible to follow the same crediting rules used for publishing in the technical literature. Popular venues do not generally afford the investigator the opportunity to supply footnotes or a bibliography, for example. Opportunities for citation of published work, if available at all, are strictly limited, and acknowledgment may be possible, if at all, only within the main narrative. In these circumstances, it is especially important to avoid revealing others' research results without crediting them or in ways that might prevent the subsequent publication of their work in the technical literature (on the grounds that those results had already appeared in print).

### Ethical Guidelines for Conference Organizers

The ASME has added a useful section to its ethical guidelines for publication, "Ethical Obligations of Conference Organizers." Although that section repeats themes and methods contained in the section in the ACS document on the obligations of journal editors, it is notable for placing on the engineering conference organizers some obligation to watch for unethical behavior.[a]

---

[a]ASME, 1999. They say: Conference Organizers should be alert to possible cases of plagiarism, duplication of previous published work, falsified data, misappropriation of intellectual property, duplicate submission of papers, and inappropriate attribution or incorrect coauthor listing. The organizer may deal directly with such ethical lapses, or, if deemed necessary, may forward them to the ASME Publishing Department. Adapted from "ASME Ethical Standards," 1996–2011, http://www.asme.org/kb/proceedings/proceedings/ethical-standards.

The ACS guidelines give further directives in a section on publishing outside of the scientific literature added in 1994, including:

> Inasmuch as laymen may not understand scientific terminology, the scientist may find it necessary to use common words of lesser precision to increase public comprehension. In view of the importance of scientists' communicating with the general public, some loss of accuracy in that sense can be condoned. The scientist should, however, strive to keep public writing, remarks and interviews as accurate as possible consistent with effective communication.[19]

That section of the ACS guidelines also makes it clear that announcing a discovery to the press is no substitute for peer-reviewed publication in the technical literature and may interfere with publishing in a peer-reviewed venue. They counsel authors to pursue peer-reviewed publication even if they have announced their findings in the popular press.

*How is publishing in the popular press different from publishing in engineering and scientific journals in ways that affect one's opportunities to credit others for their research contributions? How can one be fair to others under those conditions?*

## Responsibilities of Editors and Reviewers That Authors Should Know

In what respects is the trustworthiness of journal editors and reviewers relevant to an author's effort to disseminate knowledge via publication of research?

The publication of reports of research is essential to the growth of knowledge. To do their part in writing and submitting for publication reports of their research, authors must trust editors and reviewers in certain respects. Here we will consider

---

[19]ACS, 2006, item 2 in *Ethical Obligations of Scientists Publishing outside the Scientific Literature*.

the standards of trustworthiness for reviewers and editors that have the greatest importance for authors.

Authors need to be able to trust that editors and reviewers will not take unfair advantage of seeing their unpublished work. It is not surprising that existing guidelines for journal editors and reviewers emphasize that content of their unpublished manuscripts should be neither used nor disclosed by reviewers or editors. (The previously mentioned ACS *Ethical Guidelines to Publication of Chemical Research* provide two reasonable exceptions to the rule against use or disclosure. The identification of these exceptions show how thoroughly the ACS committee writing the guidelines have thought through the issues: They say that a reviewer who learns from an unpublished manuscript that "some of the reviewer's work is unlikely to be profitable, the reviewer, however, could ethically discontinue the work." [In the section on the ethical obligations of editors, they make a similar allowance for an editor to use information to the same limited extent.] They also provide an exception to the rule against disclosure "for an editor who solicits, or otherwise arranges beforehand, the submission of manuscripts," allowing that such an editor "may need to disclose to a prospective author the fact that a relevant manuscript by another author has been received or is in preparation."[20]

Authors rely on editors and reviewers for an unbiased assessment of their work, so editors and reviewers have a responsibility to give unbiased assessments (or recuse themselves from making an assessment). For example, the very first obligation stated in the latest version of the ACS *Ethical Guidelines to Publication of Chemical Research* is:

> 1. An editor should give unbiased consideration to all manuscripts offered for publication, judging each on its merits without regard to race, religion, nationality, sex, seniority, or institutional affiliation of the author(s). An editor may, however, take into account relationships of a manuscript immediately under consideration to others previously or concurrently offered by the same author(s).

The same guidelines list the following as obligations 3, 4, and 5 under "Ethical Obligations of Reviewers of Manuscripts":

> 3. A reviewer (or referee) of a manuscript should judge objectively the quality of the manuscript, of its experimental and theoretical work, of its interpretations and its exposition, with due regard to the maintenance of high scientific and literary standards . . .

> 4. A reviewer should be sensitive to the appearance of a conflict of interest when the manuscript under review is closely related to the reviewer's work in progress or published. If in doubt, the reviewer should return the manuscript promptly without review, advising the editor of the conflict of interest or bias. Alternatively, the reviewer may wish to furnish a signed review stating the reviewer's interest in the work, with the understanding that it may, at the editor's discretion, be transmitted to the author.

> 5. A reviewer should not evaluate a manuscript authored or co-authored by a person with whom the reviewer has a personal or professional connection if the relationship would bias judgment of the manuscript.

[20] *Ethical Obligations of Editors of Scientific Journals*, 2009 edition, pp. 1, 2, 4, available from the American Chemical Society Web site, http://pubs.acs.org/userimages/ContentEditor/1218054468605/ethics.pdf.

The fourth obligation of reviewers also gives guidance on how to ethically manage a conflict of interest situation.

Authors, like other people, are entitled to be treated professionally and with respect. Therefore, personal attacks are disapproved. Such disapproval is reflected in item 10 in the ACS *Ethical Guidelines'* section on ethical obligation of authors, which reads:

> 10. An experimental or theoretical study may sometimes justify criticism, even severe criticism, of the work of another scientist. When appropriate, such criticism may be offered in published papers. However, in no case is personal criticism considered to be appropriate.

Authors deserve to hear reasons for judgments about their research or research reports. Thus, the ACS *Ethical Guidelines*, in the section on the ethical obligations of reviewers of manuscripts, say:

> 7. Reviewers should explain and support their judgments adequately so that editors and authors may understand the basis of their comments. Any statement that an observation, derivation or argument had been previously reported should be accompanied by the relevant citation. Unsupported assertions by reviewers (or by authors in rebuttal) are of little value and should be avoided.

> *In what respects is the trustworthiness of journal editors and reviewers relevant to an author's effort to disseminate knowledge via publication of research?*

# PART 4

# THE FUTURE OF ENGINEERING

# 10 Responsibility for the Environment

---

## Suspected Hazardous Waste[a]

You are an engineering student employed for the summer by a consulting environmental engineering firm. R.J., the engineer who supervises you, directs you to sample the contents of drums located on the property of a client. The look and smell of the drums and samples lead you to believe that analysis of the sample will show it to be hazardous waste. You know that if the material contains hazardous waste, legal restrictions on the transport and disposal of the drums will apply and that federal and state authorities must be notified.

When you inform R.J. of the likely contents of the drums, R.J. proposes to "do the client a favor" – document only that samples have been taken and not proceed with the analysis. R.J. further proposes to tell the client where the drums are located, that they contain "questionable material," and to suggest that they be removed.

Why does R.J. think that incomplete information will be "a favor" to the client?

Is giving incomplete information responsible engineering practice?

Does the law in your state require engineering firms to report any release of hazardous waste to the state's department of environmental protection?

What, if anything, could you do, as a student and a summer hire?

---

[a] Adapted from Case 92–6 of the Board of Ethical Review of the NSPE.

---

## Complying with Poorly Written Environmental Laws

You are an environmental engineer working for a manufacturing company that makes computer components. In the process your plant creates toxic wastes, primarily as heavy metals. Part of your job is to oversee the testing of the effluent from your plant, signing the test results to attest to their accuracy and supplying them to the city.

The allowable levels of heavy metals in the effluent are intended to be several times as stringent as federal law allows, because of recreational use, including swimming and fishing, downstream. However, the law was poorly written. It limits the *concentration* of toxic material in the effluent rather than the *quantity* discharged in a given period. Therefore, the requirement can always be met by diluting the discharge.

Although you are complying with the law, you are concerned that the increased amount of heavy metals you have begun putting in the river may pose a health hazard, especially to some of the residents who regularly eat catfish caught downstream.

What can and should you do and how should you go about it?[a]

---

[a] *Gilbane Gold* is a 24-minute videotaped dramatization produced by the National Institute for Engineering Ethics, National Society of Professional Engineers, 1989. That dramatization raises many ethical issues, including how to cope with such a flaw in environmental legislation.

## The Rise of Ecology and New Ways of Thinking about the Environment

What new concepts and questions about the environment have arisen since the 1960s?

Engineers' and applied scientists' responsibility for the environment in some respects resembles the responsibility for safety, but new ways of thinking about the environment show the matter to be quite complex. This chapter discusses those new ways of thinking, their origins, and the concepts that embody them.

Before 1970, "the **environment**" meant simply the surroundings, the assemblage of stuff nearby. Today, it also has the meaning of the combinations of factors, external to organisms, that influence the flourishing or withering of those organisms. The idea of the environment as an integrated system has only been in wide use since the 1960s, although some argue that it resembles notions of nature or the Earth found in many cultures originating outside Western Europe or notions of nature that predate modern science.[1] The present view of the environment arose with a new scientific discipline, ecology – the study of the relationship between organisms and their environment.

In English, the terms "environment," "ecology," and "ecosystem" have come into common usage to convey the ideas of integrated systems in nature. "Ecology" names a field of study, namely, the science of the relationships among organisms and their environments that was sufficiently obscure in 1971 that the term was not even included in that year's edition of the authoritative *Oxford English Dictionary*. The emergence of the science of ecology has given rise to many new areas of engineering theory and practice – especially related to chemical engineering and civil engineering – that are called "environmental engineering."

The conservation movement has a longer history. During his presidency, Theodore Roosevelt popularized it. **Conservation** efforts bore some resemblance to today's efforts to protect or improve the environment, but they were directed *toward the preservation of specific entities*: recreational areas or natural resources of evident economic significance, such as forests, fish stocks, and navigable waterways. Conservation efforts proceeded without a comprehensive understanding of the relationships between organisms and their environments.

*What new concepts and questions about the environment have arisen since the 1960s? Describe some of the concepts created to express those concerns and questions.*

### Rachel Carson

What was the goal of Rachel Carson's efforts in writing books such as Silent Spring and what results did she achieve?

The twentieth-century figure who did the most to change thinking about what we now call "the environment" was Rachel Carson, a marine biologist. Besides being a meticulous scientist, she had won recognition as a science writer even before her famous book on the effects of pesticides, *Silent Spring*. (Carson had published *Under Sea Wind* in 1941, and in 1951 she won the National Book Award for *The Sea around Us*.) Concerned by the growing evidence of major

[1] Merchant, Carolyn. 1980. *The Death of Nature: Women, Ecology, and the Scientific Revolution*. San Francisco: Harper & Row.

damage to fish, birds, and other animal life because of new mass applications of pesticides to large areas of wilderness, Carson sought someone to write a book about the subject. She finally took on this task herself in *Silent Spring*. This work, published in 1962, changed the consciousness both of the public and of policy makers about the effect of pesticides on the environment.[2] The book influenced the highest levels of government, and these decision makers began to listen to the idea that the ecological process was vital to life and to human well-being.[3]

Rachel Carson in her efforts to bring the danger of pesticides to public attention demonstrated some characteristics shown by Roger Boisjoly in his attempt to bring attention to safety. Among the virtues common to them both was the courage to tell the truth that no one wants to hear (even when it means being harshly criticized), concern for fairness to everyone, and a concern for the safety and well-being of others.

Carson's goal was rather different from that of Boisjoly and from that of LeMessurier, as were her circumstances. Boisjoly's purpose was to alert the decision makers in his company or at NASA to the danger of a fatal explosion. LeMessurier needed to find a way to remedy a flaw in the huge Citicorp Tower while safeguarding the public, a goal that required enlisting the aid of many individuals and organizations. Carson's goal was to reverse a major trend in social policy regarding the use of chemical pesticides. It would not have been sufficient for Carson to have her employer in the 1940s, the U.S. Fish and Wildlife Service, appreciate the problem. Many people in that service had in fact already become aware of the unexpectedly severe harm to wildlife caused by new pesticide use.

### The Enthusiasm for Pesticides like DDT

The enthusiasm for insecticides in the postwar period was evidenced in the award of the 1948 Nobel Prize in Physiology to Paul Hermann Müller of Switzerland for his 1942 discovery of the insecticidal properties of dichlorodiphenyltri-chloroethane, later known as DDT. (DDT had first been synthesized in 1874.)

After World War II, pesticides and the goal of "eradicating" – as contrasted with simply controlling – insect pests gained wide popularity. By the late 1950s this led to massive use of chemical pesticides, including government spraying of vast tracts of land without consent of residents.

To accomplish her goal Rachel Carson needed to counter the enthusiasm for pesticides that had grown out of their use in World War II to combat insect-borne diseases. DDT in particular had proved very effective in controlling lice that had formerly spread disease such as typhus among troops in wartime, killing more members of the armed forces than battle wounds.[4]

In 1958, when Carson began writing *Silent Spring*, almost 200 million dollars' worth of pesticides was sold. Four years later, when the book appeared, that amount had more than doubled.[5] Carson realized the reexamination of the effects of chemical pesticides required first making the hazard posed by insecticides generally understandable to both the voting public and governmental policy makers. Accomplishing this goal

[2] Briggs, Shirley A. 1987. "Rachel Carson: Her Vision and Her Legacy." In *Silent Spring Revisited*, edited by Marco, Gino J., Robert M. Hollingworth, and William Durham. Washington, DC: American Chemical Society, 4–5.

[3] *Ibid.*, 5.

[4] PBS, The American Experience. 1992. *Rachel Carson's Silent Spring*, produced by Neil Goodwin. PBS VIDEO and *Encyclopedia Britannica* online, http://www.eb.com/.

[5] Graham, Frank, Jr. 1970. *Since Silent Spring*. Boston: Houghton Mifflin Company.

required the integration of a diffuse and variable body of data into a clear and compelling story.[6] It also required that she not be dependent on keeping her job, because, if effective, her book was sure to raise a storm of protest. The chemical industry was sure to be especially resistant to her message.

Carson contacted other biologists whose works showed the damage done by pesticides. The letters that she received from those biologists requested anonymity because the biologists were afraid of losing their jobs. When *Silent Spring* did appear in 1962 it received great attention. A condensed version of it appeared in the *New Yorker* in three parts, starting on June 16. Congressman John V. Lindsay inserted the last paragraphs of the first third of that condensed version into the *Congressional Record*. President Kennedy ordered the Science Advisory Committee to study the effects of pesticides. The Book-of-the-Month Club bought it and *Silent Spring* had advance sales of 40,000, although industry critics redoubled their efforts to discredit the book. Some of those attacks only added to the book's publicity, however, and as we shall see, the tide of public opinion had turned and set the stage for landmark environmental legislation of the next two decades.

*What was the goal of Rachel Carson's efforts in writing books such as* Silent Spring *and what results did she achieve?*

### Key U.S. Environmental Legislation, 1969–1986
What changes in thinking about the environment are reflected in the landmark federal laws that were passed in the decades following publication of *Silent Spring*?

Before the late 1960s there was little significant national or state legislation to protect the environment.[7] Individuals could bring lawsuits, but rarely was a single party so harmed by pollution as to take this route.

Below is a listing of major U.S. environmental legislation through 1986. Note the areas of concern reflected in each law. Each measure had its own history, which space does not allow us to examine here, but taken together these acts represent the dawn of the realization of how the environment affects human welfare.[8]

- 1969 – National Environmental Policy Act requires environmental impact statements for actions by federal agencies that affect the environment. Congress then created the Environmental Protection Agency (EPA) to enforce compliance with this law.
- 1970 – Occupational Safety and Health Act establishes the National Institute of Occupational Safety and Health (NIOSH) within the Department of Health and Human Services (DHHS). (NIOSH develops mandatory health and safety standards for business, conducts research on occupational health problems, and produces criteria identifying toxic substances and safe exposure levels for them.)

---

[6] PBS, The American Experience, *op. cit.*

[7] The Delaney Clause in section 409 of the Federal Food, Drug and Cosmetic Act (FFDCA) of 1958 was the only national legislation in place prior to the publication of *Silent Spring*. That clause pertains to "food additives" and specifies that no amount of carcinogenic pesticide residues is to be tolerated in processed foods.

[8] The text of these and other environmental protection laws is available on the EPA Web site at http://www.epa.gov/lawsregs/laws/index.html#env (last updated on September 27, 2010).

- 1970 – Clean Air Act, amended in 1977 and 1990 with National Emission Standards for Hazardous Air Pollutants (NESHAPS), regulates air emissions.
- 1972 – Clean Water Act, amended in 1972, 1977, 1986, and 1995, seeks to "reduce direct pollutant discharges into waterways, finance municipal wastewater treatment facilities, and manage polluted runoff."[9]
- 1976 – Resource Conservation and Recovery Act (RCRA) provides regulation for the on-site handling of toxic chemicals, that is, handling of toxic chemicals at one's facility.
- 1976 – Toxic Substances Control Act (TSCA) provides regulation to protect the public against toxic substances in consumer and industrial products.
- 1980 – Comprehensive Emergency Response, Compensation and Liability Act (CERCLA), popularly known as "the Superfund Act," establishes the Agency for Toxic Substances and Disease Registry (ATSDR) within the Public Health Service (PHS). The PHS is an agency of the DHHS.
- 1986 – Superfund Amendments and Reauthorization Act (SARA) add to the duties of the ATSDR the responsibility for developing and updating a list of toxic substances that pose the most significant threat to human health.[10]

*What changes in thinking about the environment are reflected in the landmark federal laws that were passed in the decades following publication of* Silent Spring?

### The Concept of an Ecosystem

What is an ecosystem and how does attention to it and its health contrast with attention to the health of creatures with moral standing?

The word "ecosystem" has entered the general English vocabulary from the technical vocabulary of ecology. An **ecosystem** is a group of organisms that interact with each other and with their physical environment in ways that affect the population of those organisms.

Thinking in terms of ecosystems – or in terms of systems generally – directs attention away from particular individuals or species to the way their interactions sustain and are sustained by the whole system, even if one's interest is primarily in that individual or species.

An example of a simple ecosystem is that of kelp, sea urchins, and sea otters. Kelp is a commercially valuable ocean plant used in foods, paints, and cosmetics. The kelp forests along the Pacific coasts grow in long streamers attached to the ocean floor. In recent years the kelp suddenly began disappearing. Concurrently, sea urchins had increased because the population of sea otters, a major predator of sea urchins, had fallen off drastically. The sea urchins feeding on the kelp weakened its attachment to the ocean floor, causing it to float away. Understanding the kelp forest as one part of an ecosystem showed that the simplest way to protect the kelp forests was to protect the sea otter. When the population of sea otters

---

[9] See http://www.epa.gov/watertrain/cwa/ (last updated on Friday, September 12, 2008).

[10] I have placed this list in the Online Ethics Center as supplementary material to the story of Rachel Carson. There the list is linked to further information about the acts and agencies in charge of them.

flourished, the sea otter held down the population of sea urchins, and the kelp forests were restored.[11]

Ecosystems vary in their fragility. An ecosystem is more resilient if several component species perform the same function, or "fill the same niche" in the system. For example, the kelp–sea urchin–sea otter system would have been more resilient if sea urchins had other natural predators in addition to sea otters. Then when the otter population diminished, the other predators would have fed on the more plentiful sea urchins, perhaps before they threatened destruction of the kelp.

The systems approach to understanding biological phenomena transforms the ways in which we think of those phenomena. Systems may be naturally occurring, like an ecosystem or the digestive system of an organism; or they can be the product of human endeavors, like transportation systems, political systems, or educational systems. Systems comprise sets of components that work together to perform a function, yet the system is more than the sum of its components. The continued functioning of the whole system is the central concern, rather than the flourishing of any single component or components.

As we saw in the introduction, for a being to have moral standing means that such an individual's well-being must be considered for its own sake. Ecological thinking considers the good of one individual or species *only* in relation to the whole ecosystem. Therefore, concern for the environment does not come down to consideration of the moral standing of individuals or of species. Systems thinking complicates consideration of harms and benefits, and the risk or probability of their occurrence.

*What is an ecosystem and how does attention to it and its health contrast with attention to the health of creatures with moral standing?*

### Hazards and Risks to the Environment

When raising a concern about a hazard to the environment is there a danger that one may be doing something that is outside of one's competence (i.e., for which one is not qualified)?

As we saw in Section 4 of the introduction, "risk" has a technical sense in the context of "risk assessment" or "risk management." Risk in the technical sense is the probability or likelihood of some resulting harm, multiplied by the magnitude of the harm. (As we have noted, the use of the technical sense of risk requires that one be able to meaningfully quantify the resulting harms.)

This technical notion is useful for comparing harms that can be quantified or alternatives in which both make one vulnerable to the same harm. For example, one can compare, say, the risk of death for a driver under 5'4" tall from a deploying airbag with the risk of death from being in an automobile accident unprotected by airbags to see if airbags provide a net benefit to shorter drivers. (Shorter drivers are at greater risk of injury from airbags, because they sit closer to the wheel.) The tolerance of *some* environmental risk is implicit in the authorization of the Environmental Protection Agency (EPA) to regulate (only those) chemicals that

---

[11] Vesilind, P. Aarne, J. Jeffrey Peirce, and Ruth Weiner. 1987. *Environmental Engineering*, second edition. Stoneham, MA: Butterworth, 6–7. Discussion of concepts of environmental engineering in this chapter derive principally from Vesilind et al.

pose an *unreasonable* risk of injury to health or to the environment. As we saw in Chapter 3, however, the tolerance for threats to humans varies widely with the nature of the hazard. Food additives are not permitted if they pose *any* known risk to humans, for example, but foods that have some naturally occurring toxins are allowed to be sold.

The professional responsibility of engineers and applied scientists for environmental protection, like their responsibility for ensuring public safety, requires attention to two sorts of risks:

- Hazards that have gone unrecognized, at least by some key decision makers, and that pose a grave or excessive threat to safety or the environment; and
- Hazards that are a recognized feature of the situation but that cannot be completely eliminated and are mitigated only by increasing other risks and costs.

Addressing the two types of hazards raises different issues.

Environmental hazards that have gone unrecognized are analogous to the safety hazards posed by cold temperatures to the performance of the O-ring seals in the *Challenger* booster rockets, increasing the load on the rods supporting the walkways of the Kansas City Hyatt Regency, and the unrecognized toxicity of some new chemical, such as that discussed in "How Dangerous Is This Chemical?" in Chapter 4. The threat of accident is typical of this sort of hazard, but, as the example of toxicity of a new chemical illustrates, accidents are not the *only* source of such threats. What is essential is that the threat is unrecognized or disregarded. The ethical problem is to bring such risks to light and have them addressed.

Engineers and applied scientists, because of their education and training, are in a special position to recognize both environmental hazards and safety hazards. Their specialized knowledge and training are the basis for the growing consensus that engineers and applied scientists have a professional responsibility to bring environmental as well as safety hazards to light. So it is that engineers generally – not only environmental engineers or those who have environmental protection as an assigned (i.e., official) responsibility – are widely acknowledged as having a professional moral responsibility to prevent environmental damage. For example, the IEA states that its members should practice "in accord with sustainability and environmental principles." The IEEE in its most recent code of ethics (1990) states as the first of the ten points to which IEEE members are committed:

> to accept responsibility in making engineering decisions consistent with the safety, health and welfare of the public, and to disclose promptly factors that might endanger the public or the *environment*. (Italics added.)

It is notable that the IEEE has been in the forefront on this question because electrical engineers are less likely than civil engineers or chemical engineers to be environmental engineers.

Under the second of its listed professional obligations (namely: "Engineers shall at all times strive to serve the public interest") the 2006 revision of the NSPE Code of Ethics states,

> Engineers shall strive to adhere to the principles of sustainable development in order to protect the environment for future generations.

The 2003 revision of the AIChE Code of Ethics[12] states as its first ethical requirement on its members that they

> Hold paramount the safety, health and welfare of the public *and protect the environment* in performance of their professional duties. (Italics added.)

Recognized and unrecognized hazards differ in the extent to which engineers may be prepared or assigned to consider such hazards. Techniques for coping with specific recognized hazards are a part of the subject matter of many engineering disciplines. A specific engineer may or may not be proficient in some specific technique. A professional responsibility not to take work beyond the limits of one's competence is widely acknowledged in engineering and is explicitly stated in the codes of ethics of several engineering societies. For example, the second of seven "fundamental canons" in the Code of Ethics of the American Society of Mechanical Engineers states:

> Engineers shall perform services only in the areas of their competence.

One can undertake the task of checking for overlooked factors that may cause an accident. We saw that structured techniques like fault tree analysis are used in estimating the probability of accidents. Engineers should take on such work only if qualified to use them. It does not make sense to object that some engineer is unqualified to raise a concern about some previously undetected hazard, however. It is possible that the concern about a hazard to the environment will, after more expert assessment, prove groundless, but the engineer would not have been presuming to provide services beyond her competence simply by raising the issue.

The situation with unrecognized hazards is rather different from assessment of the risks from recognized hazards. There is not the same issue of warning about recognized hazards. Further, if they persist even after being recognized, they are usually ones that cannot be eliminated or mitigated without incurring other significant risks and costs. In such cases the task is to evaluate the risk, find ways of mitigating it, and evaluate the risks and costs of actions that would mitigate it. Here the issue of performing services only within limits of one's competence *does* become a significant factor.

> *When raising a concern about a hazard to the environment is there a danger that one may be doing something that is outside of one's competence (i.e., for which one is not qualified)?*

### Illustration from the Exxon Valdez *Oil Spill Case*

Consider Exxon's use of single-hull tankers like the *Exxon Valdez* to transport oil. Was this likely to have been an explicit decision to use single-hulled, rather than double-hulled vessels? How prudent was the decision? How responsible? Give reasons for your judgment.

The contrast between the moral problem posed by recognized hazards and those posed by unrecognized hazards may be further illustrated by considering two of the factors that contributed to the catastrophic oil spill when the Exxon oil tanker *Valdez* went aground on a reef in 1989. First, a seaman had previously

---

[12] Available at http://www.aiche.org/About/Code.aspx (revised January 17, 2003).

warned about alcohol abuse by the captain of the *Valdez*. (Exxon subsequently fired that complainant who at the time of the *Valdez* spill was suing in federal court for wrongful termination. This court case made information about the captain's pattern of alcohol abuse immediately available to the media when the spill occurred.) The observation that someone who is abusing alcohol is not an appropriate commander of a large oil tanker did not require any special expertise or quantification of the harm that he might cause.

In contrast, in all likelihood some sort of cost-benefit calculation underlay Exxon's decision that it was not worth the expense to use double-hulled oil tankers, tankers that would have been less prone to spills. (The cost to Exxon for the cleanup alone was $2.1 billion. Exxon [now known as "Exxon Mobil"] also agreed to pay $900 million for the spill and promised an additional $100 million if any unanticipated damages became evident by 2006. Whether there are such unanticipated damages is the subject of a bitter scientific dispute in which Exxon has invested "some millions" more. This brings the expense to Exxon to more than $3 billion with another $100 million in question, so their calculation was a shortsighted financial decision as well as one negligent of environmental harm.[13])

On the twentieth anniversary of the *Exxon Valdez* spill (August 2009) a report was issued on the spill and its consequences. That report, which is available at http://www.evostc.state.ak.us/, states that twenty years after the spill "*Exxon Valdez* oil persists in the environment and in places, is nearly as toxic as it was the first few weeks after the spill." The April 2010 explosion and sinking of BP's Deepwater Horizon, a semisubmersible drilling rig, led to an oil spill that will be more catastrophic for the environment. (The damage to the environment is not predictable from the volume of oil released, however.[14])

> *Consider Exxon's use of single-hull tankers like the* Exxon Valdez *to transport oil. Was this likely to have been an explicit decision to use single-hulled, rather than double-hulled vessels? How prudent was the decision? How responsible? Give reasons for your judgment.*

### Responsible Behavior in Assessing Risk

In assessing risk, falsifying information would be one clear departure from honest representation. What more is required to give an honest representation?

Cost-benefit and risk-benefit calculations are a frequent component of environmental impact statements. Although one might wish to completely eliminate accidents such as major oil spills, doing so might require ceasing to drill for oil or ceasing to transport it. The lesson from the catastrophic effects of spills such as the *Valdez* spill for those who conduct cost-benefit or risk-benefit calculations with environmental implications is to consider what makes for a morally responsible and far-sighted use of such techniques.

---

[13] Guterman, Lila. 2004. "Slippery Science, 15 Years after the *Exxon Valdez* Oil Spill, Researchers Debate Its Lingering Effects with $100-million on the Line," *Chronicle of Higher Education,* September 24 (last accessed at http://chronicle.com/weekly/v51/i05/05a01201.htm).

[14] A list of oil spills that have occurred since 1967 is available on the Web at http://www.infoplease.com/ipa/A0001451.html.

Analysis of environmental risks is often the official responsibility of some designated engineer or engineers, as well as something that any engineer may find necessary to fulfill her professional moral responsibility for safety. Such analyses are a regular part of the assigned work of many environmental engineers and chemical engineers. As we saw in the introduction on concepts, official obligations and responsibilities differ from moral ones in both their ethical significance and their functioning. There is no guarantee that what an official obligation requires one to do is even ethically permissible, for example. (If it is ethically permissible and one freely takes on the job or assignment, then there is some moral obligation to keep one's implicit promise to perform the job or task.) Given an ethically permissible job assignment or official responsibility, the ethical question is how to carry it out *in an ethically responsible way*. Carrying out moral responsibilities or official responsibilities in a morally responsible manner has both ethical and technical aspects.

The minimal requirements for morally responsible behavior in engineering and science are competence and honesty. Honesty requires more than refraining from falsifying information, of course; it also requires a balanced assessment and presentation of the situation. The requirement for a balanced presentation by engineers and scientists contrasts with the expectation for members of the legal profession who act in an adversarial setting as an advocate for one side or the other. Stephen Unger emphasizes this point in the case of Morris Baslow. Baslow was a marine biologist who investigated the effects of once-through use of waters of New York's Hudson River for cooling of electrical power plants. Baslow found a pattern of fish kill and inhibited growth rate of fish due to increased water temperature. His employer, LMS engineers, in deference to the interests of its client, Consolidated Edison, tried to suppress his results and continued to present a contrary view at hearings held by the EPA and the Federal Energy Commission. Baslow then told them that if they would not release his data, he would. He sent his letter to the EPA and was fired on the same day in 1979. He had insisted on a full and balanced reporting of the data, but LMS had wanted to release only that information that was in the interests of its client.[15]

The requirement for a full and balanced reporting of the facts needs emphasis in the context of environmental analyses, because of the cynical perception that environmental assessments are usually one-sided. There has been some clear dishonesty in these matters. For several years a colleague of mine who works in the area of risk assessment had posted on his office door a particularly outrageous advertisement from a firm doing environmental assessments. It promised to do an environmental study "favorable to you," a clear offer to skew research to suit the client's interests.

Some sources of bias in research cannot be removed. One of these is disciplinary bias, which we examined in the introduction. Researcher investigators use the tools of their discipline and not of another. Frequently, they cannot be expected even to know all the methods from other disciplines that might be used in the situation.

[15] Unger, Stephen H. 1994. *Controlling Technology: Ethics and the Responsible Engineer*, second edition. New York: John Wiley & Sons, Inc., 194–198.

A special challenge to maintaining competence in rapidly developing fields is to keep up with new knowledge and techniques. In areas like environmental engineering or biomedical engineering that have immediate implications for human well-being, the need for an overview of alternative methods is especially pressing.

*In assessing risk, falsifying information would be one clear departure from honest representation. What more is required to give an honest representation?*

### Ecological Thinking and the Question of Who/What Counts

Is resulting harm to people (or some other beings with moral standing) a necessary condition for environmental damage to be ethically objectionable, or do professionals with relevant knowledge have a moral responsibility to seek to prevent it?

What is the justification for the growing opinion, noted earlier in recent revisions of engineering codes of ethics, that engineers have a professional responsibility to protect the environment? Why should anyone, engineer or otherwise, be said to have an ethical responsibility for the environment? Most of the environmental problems for which engineering solutions have been sought involve threats to human health and safety. These are also some of the most widely discussed threats to the environment; indeed, the threat they pose to human health may be the reason that they receive the attention they do. When environmental damage threatens human health and safety, the responsibility to protect the environment against that damage follows directly from the professional responsibility for public health and safety.

Environmental damage that does not directly threaten human health and safety often threatens some other aspect of human well-being. For example, the global warming caused by increases in "greenhouse gases" threatens agriculture and hence economic well-being, as well as produces a threat of physical injury to people from droughts, coastal flooding, and storms. Such examples suggest that concern for the environment may derive solely from a concern about the effect that environmental destruction might have *on humans*.

Some threats to the environment pose no clear threat of injury or disease to people but do threaten some endangered species or a fragile ecosystem. Even in these cases, the possible loss of benefits to humans is often mentioned. For example, the extinction of species caused by the destruction of the Amazonian rain forest is hypothesized to eliminate potential sources of medical remedies. If some nonhuman species are worthy of moral consideration (in the sense discussed in Section 4 of the introduction) then harm to them caused by environmental degradation might be a basis for concern for the environment. Concern for the environment is often discussed as something over and above concern for people or for any specific species, however. The relationship between such a concern and familiar moral categories remains to be spelled out.

*Is resulting harm to people (or some other beings with moral standing) a necessary condition for environmental damage to be ethically objectionable, or do professionals with relevant knowledge have a moral responsibility to seek to prevent it?*

## Moral Standing and the Environment

> Does a concern for the environment stem from a concern for the individuals or species that make it up? Does the environment have moral standing in itself?

Those who hold that people's use of the environment has limits other than what is needed to prevent harm to humans often argue by analogy with the plight of previously disenfranchised people. They argue that in times past, noncitizens, enslaved people, women, and children lacked some legal rights that we regard as basic and as matched by moral rights, many of them human rights. They argue that we do not now think of the "natural environment" as having legal rights only because of the immaturity of present-day ethical reflection. To understand and evaluate such arguments, we need to understand what sorts of objects are held to have moral standing, what that moral standing entails – for example, does it entail having moral "rights"? – and why we should think that those sorts of objects do have that moral standing.

Some of the most quoted arguments are rather unclear about the relationship between having moral standing and having moral rights. For example, in a frequently quoted essay, "Should Trees Have Standing?" Christopher Stone (1972) proposes that legal rights be accorded trees, forests, oceans, rivers, and other "natural objects" – the natural environment as a whole.[16] He equates "having standing" with "having rights" and ascribes rights, although not *the same* rights as humans, to all of these beings. As we saw in the introduction, some people object to ascribing moral and legal rights to beings that cannot choose when they wish to exercise them. To say that such beings have moral standing or are "morally considerable" is not open to the same objections. Having moral standing also morally constrains actions toward such beings.

If we focus attention on individual organisms or even species, the concern for the environment is likely to come out as the implausible assertion that either *every organism* or *every species* has moral standing. Does a bacterium or at least types of bacteria have moral standing?

What is believed to be the last remaining sample of the smallpox virus – a virus that normally cannot survive without a living host – has been artificially maintained. This sample has been preserved, not out of concern for the smallpox virus or for diversity of life forms, but because the sample might provide useful information to people some time in the future.

To ask whether all types of virus or all ticks or all species of tick have moral standing is to bypass the ecological perspective. That perspective focuses on ecosystems, rather than individual organisms or even individual species. The resilience of an ecosystem depends on the multiplicity of species that fill each "niche" – that is, the *unimportance* of an individual species (much less individual organisms) to the survival of the ecosystem.

Species diversity makes an ecosystem more resilient because assaults to one species, say disease, are more likely to be compensated by other species. However, this does not imply that each species – or indeed each feature of the soil or

---

[16] Similar features are found in more recent influential arguments, as well. See for example, Mary Midgley, 1992, "Is the Biosphere a Luxury?" *Hastings Center Report* 2(2): 7–12.

atmosphere – within each ecosystem has a claim to ethical consideration. If each species *is* entitled to such consideration, it must be for other reasons.

## Can We Differentiate Organisms?

It may be difficult to differentiate between one organism and another, let alone distinguish the good of one from that of other organisms. For example, we humans dwell in intimate interdependence with the bacteria that live within us. According to biologist Lynn Margulis the normal adult human is about 15 percent bacteria, dry weight.

When asked about the environment the question of moral standing is more difficult to understand and answer than was the question about the moral standing of individuals or species. The systems thinking characteristic of environmentalism fits poorly with debates about the moral standing of individuals. Because systems are more than the sum of their components, ecosystems may have moral standing that elevates their preservation or well-being above that of the individuals in them. If the flourishing of one group of individuals threatens to throw the system dangerously out of homeostasis, concern for the ecosystem would dictate that those individuals or species would better be sacrificed. If the flourishing of the species is considered primary, the suffering or survival of individual members of that species becomes less important. This tension may help explain why, as Aarne Vesilind has noted, there is not much affinity between the ecology movement and the animal rights movement.

Saying that ecosystems do not have moral standing also leads to paradoxes, however. If ecosystems lack *moral standing*, how can there be any reason to preserve them except to benefit humans?[17]

*Does a concern for the environment stem from a concern for the individuals or species that make it up? Does the environment have moral standing in itself?*

## Some Illustrative Cases

### The Costs of Environmental Protection: The Case of Timbering and the Northern Spotted Owl

Must we, as a society, choose between job loss and environmental degradation? Why or why not?

The northern spotted owl has for hundreds of years lived in the "old-growth" forests of the Pacific Northwest. It feeds on the plant and animal life created by decaying timber. Its habitat has dwindled over the last 150 years as a result of heavy logging in the area, much of it on public land. An estimated 2,000 pairs of spotted owls are all that survive today.

In 1986, environmentalists petitioned the U.S. Fish and Wildlife Service to include the spotted owl among "endangered species." The petition was met with staunch resistance from the timber industry in the region, because it would bar them from clearing the forests that are the owl's habitat. In 1990 the owl was finally declared a "threatened species." Timber companies were required to leave at least 40 percent of the old-growth forests within a radius of 1.3 miles of any spotted owl nest or site of activity. The timber industry claimed that this

[17] Refer to the discussion of moral standing in Section 2 of the introduction on concepts.

requirement would throw thousands of loggers and mill workers in the area out of work.[18]

The controversy over protection of the spotted owl is frequently used to illustrate supposed conflicts between, on the one hand, the ecosystem for which the spotted owl is an indicator, and on the other, almost 30,000 jobs that would be lost in the short run. (In thirty years, if no protection of the owl is undertaken, at the expected rate of timbering, the old-growth forests would be gone, forcing the mills to close, so the only disagreement is over what should be the priorities in the *short run*.) Environmentalists point to both the aesthetic value of this ecosystem and its scientific value, that it has taken millennia to create and would not be restored by reforestation. Those concerned with immediate job loss point to the host of social ills, from domestic violence to suicide, that regularly attend job loss. The question of what, if any, protection of the northern spotted owl should be taken is frequently presented as a debate between two sides, one advocating environmental protection and the other advocating (short-term) job preservation.

As portrayed, the spotted owl case may not be representative of the underlying tradeoffs involved in environmental protection, because in most cases there may be no need to trade off environmental preservation and job creation, even in the short run. A recent study showed that the Endangered Species Act has not had a negative economic impact at the state level. In fact, states with booming economies were found to be the ones that also had the largest number of federally listed species, contradicting the impression that the Endangered Species Act is creating major economic harm. The underlying mechanism for the observed findings seems to be that population growth goes with economic boom and tends to put greater pressure on the environment, leading to new listings of endangered species. The study does not deny the economic effects of environmental preservation, but it finds these effects to be highly localized, such that they are "lost in the noise of background economic fluctuations." The study cites for comparison the recent series of military base closings, and finds the economic effects of those closings to be hundreds of times greater than the combined effects of all the listings under the 20-year history of the Endangered Species Act.[19] The human problems of job loss and dislocation are real, but environmental protection is not a principal source of these difficulties.

As we saw in the first chapter, responsible behavior requires careful consideration of the effects of one's actions. In thinking about the protection of the environment we are still developing ways of understanding our situation and responsibilities in it.

*Must we, as a society, choose between (short-term) job loss and environmental degradation? Why or why not?*

### The 1995 Supreme Court Decision on "Taking" of a Threatened Species
What counts (legally speaking) as a "taking" under the Endangered Species Act of 1973?

[18] The account to this point is based on "Ethics and the Spotted Owl Controversy," *Issues in Ethics*, vol. IV, no. 1 (Winter/Spring), 1991, 1, 6.

[19] Working paper "Endangered Species and State Economic Performance" by Stephen M. Meyer, published by the Project on Environmental Politics and Policy at MIT, reported in "Study finds small economic effects from endangered species protection," by Robert C. Di Iorio, *MIT Tech Talk* 39(26) (April 12, 1995): 1, 8.

As we consider what actions are responsible, quite specific limits have to be set on what individuals and groups may or may not do. The Supreme Court in its decision on *Babbitt, Secretary of Interior, et al. v. Sweet Home Chapter of Communities for a Great Oregon et al.* on June 29, 1995, addressed this very question.

The Endangered Species Act of 1973 makes it unlawful for any person to take endangered or threatened species and defines "take" to mean to "harass, harm, pursue," "wound," or "kill." Secretary of the Interior James Babbitt brought this suit. He interpreted "taking" to include actions that so significantly modified wildlife habitat as to kill or injure protected species. The respondents claimed that Congress did not intend the word "take" to include habitat modification. The District Court found for Babbitt, but the Court of Appeals ultimately reversed the District Court decision. The Court of Appeals held that "harm," like the other words in the definition of "take," should be read as applying only to the perpetrator's direct application of force against the animal taken.

In a 6–3 decision, the Supreme Court held that Babbitt's interpretation of Congress's intent was reasonable.[20]

For the reasons discussed earlier, we will set aside the question of whether the prohibition against "taking" members of endangered species finds its ultimate justification in the moral standing of a species, in the public interest in diversity of species, or in something else. Now we shall turn to another instance in which environmental damage came from an unsuspecting source. This is another case that has taught us that some species and ecosystems are extremely fragile; human actions that disturb the environment may have many far-reaching effects that we are just beginning to appreciate. It is not so much that we know our actions will have certain consequences that will threaten or damage other beings and that they along with ourselves have moral standing. Rather, disturbing the environment is now recognized to frequently have major unpredictable consequences. The threat of those consequences gives us reason to question whether it is wise, prudent, or responsible to go forward. There is increasing appreciation of the danger that unintended consequences may do major and irreversible harm before they are detected or well understood.

An excellent example is the threat of global warming due to an increase in greenhouse gases. The year 1995 was the warmest one on record, according to both the British Meteorologic Service and the NASA Goddard Institute for Space Studies. The second warmest year was 1990, and according to British figures, the period 1991–1995 was warmer than any other five-year period, including 1980–1984 and 1985–1989, even though the 1980s had been the hottest decade on record.[21] It is possible that this warming is due to climate variations that have nothing to do with the increase in greenhouse gases, as those who have warned about this phenomenon acknowledge. However, their argument is that we cannot wait until we are certain that greenhouse gases are a principal cause, because at that point it would be too late to forestall devastating effects.

[20]This account is based on the summary ("syllabus") prepared by the Supreme Court's Reporter of Decisions and distributed by email on the list liibulletin@fatty.law.cornell.edu.

[21]Stevens, William K. 1996. "1995 the Hottest Year Recorded on Earth," *New York Times,* January 4, p. 1.

*What counts (legally speaking) as a "taking" under the Endangered Species Act of 1973?*

### Acid Rain and Unforeseen Consequences of Human Action

What sorts of contributions to addressing climate and environmental damage can engineering (and applied science) make?

An ironic example of environmental damage from unforeseen consequences was the worsening of the problem of acid rain that resulted from a provision of the 1970 Clean Air Act. The burning of fossil fuels, principally by coal- and oil-burning power plants and by automobiles, puts sulfur dioxide and nitrogen oxides into the air. To reduce local air pollution in conformity with the requirements of the Clean Air Act, utilities and smelters built taller smoke stacks. What policy makers had not foreseen was that taller stacks would allow particles and gases to be carried farther and the resulting sulfur dioxide and nitrogen oxides in the atmosphere would produce acid rain. The result of the taller smokestacks was to worsen the problem of acid rain in New England, the Adirondacks, Appalachia, and Canada.

Acid rain can damage forests and soil, degrade ecosystems, and kill fish, as well as damage buildings and statues. This acid rain has made about 1,000 lakes in the United States chronically acidic and has made more than 14,000 Canadian lakes so acidic that fish can no longer live in them.

The Clean Air Act was amended in 1990 and established the Acid Rain Program, which first sought to reduce levels of $SO_2$. The result was a significant reduction in acid rain in the eastern United States by the end of the century.[22] The reduction in $SO_2$ was matched by an increase in the pH of the groundwater in only some of the worst hit locations, however.[23]

As we shall see, a similarly paradoxical result seems to follow from the replacement of chlorofluorocarbons.

The example of the negative effects of an act intended to protect the environment illustrates that environmental protection is not a simple matter of preventing acts that are obviously wrong, such as dumping substances that are known to be toxic. Science and engineering knowledge is essential to the prevention, or even the surveillance and early detection, of damaging consequences of *well-intended and seemingly innocent* actions. The need for new expertise to understand and address environmental issues and deal with threats to the environment gave rise to the new disciplines of environmental engineering and environmental science; increased attention to emergency planning and new techniques for waste reduction, waste separation, and waste management; and cleanup of hazardous chemicals and of radioactive wastes.

The environment is sometimes discussed with a romanticism that makes it seem that what is central to preventing environmental degradation is adopting a different view of or attitude toward nature. (Science and engineering are even

---

[22] One source, "Clean Air Act Reduces Acid Rain in Eastern United States," *Science News,* September 28, 1998, available at http://www.sciencedaily.com/releases/1998/09/980928072644. htm, claimed a 20 percent reduction in $SO_2$ by 1998. The Ecological Society of America, in a 1999 report (available at http://www.esa.org/education_diversity/pdfDocs/acidrainrevisited.pdf) claims a 25 percent reduction in $SO_2$ occurred by 1995.

[23] The Ecological Society of America, 1999.

occasionally represented as the enemies of nature.) More balanced accounts of threats to the environment acknowledge that social practices and innovations of many types caused environmental damage well before the age of science. It was damage to common land from the grazing of agrarian livestock that gave rise to the expression "**the tragedy of the commons**." (This expression refers to the damage to the common good that results when individuals seek their narrow self-interest, regardless of the damage that their combined actions will produce.)

Important as it may be to recognize environmental responsibility as a responsibility sanctioned by religious and cultural traditions, viewing nature as sacred did not prevent the pre-scientific societies from hunting some species to extinction. The fulfillment of responsibility requires knowledge, as well as concern. Just as one can endanger the public safety through ignorance as much as through evil intent or recklessness, so people can and do endanger either the environment or human well-being by acting beyond their competence. Much of the knowledge required for addressing complex environmental questions is engineering knowledge and scientific knowledge.

*What sorts of contributions to addressing climate and environmental damage can engineering (and applied science) make?*

### The Discovery of the Effects of Chlorofluorocarbons on the Ozone Layer

Why was it difficult to recognize that chlorofluorocarbons erode the ozone layer?

Another dramatic example of a major unintended consequence of human action was the depletion of the ozone layer by the use of chlorofluorocarbons. Ozone ($O_3$) is a powerful oxidizer. It has a dual effect on the environment. As a pollutant in the lower atmosphere it has greatly increased in recent decades and is a health hazard to humans and is harmful to crops. However, the ozone *layer* in the stratosphere nine to thirty-one miles above the Earth protects humans and other living organisms from the ultraviolet radiation of the sun. There ozone, together with oxygen, blocks a major part of that radiation. The actual number of ozone molecules comprising the layer is not very large; if the ozone layer were at standard temperature and pressure it would be as thin as a piece of cardboard. Therefore, a relatively modest amount of chemical pollutants can have a significant effect on the ozone layer.[24]

Ozone ($O_3$) is made in the upper atmosphere by the splitting of $O_2$ molecules into atoms of oxygen. These combine with other $O_2$ molecules in the presence of other air molecules to form ozone. As was demonstrated in part by Paul Crutzen in 1970, the decomposition of ozone into oxygen is enhanced by the presence of hydrogen radicals OH and HO2, nitrogen oxides NO and NO2, and free chlorine atoms.[25]

Mario Molina was a post-doctoral fellow in the laboratory of Sherwood Rowland at the University of California at Irvine in 1974 when he and Rowland

---

[24] See the Nobel announcement on the 1995 prize for chemistry at http://nobelprize.org/nobel_prizes/chemistry/laureates/1995/.

[25] *Ibid.*

published an article in *Nature* on the threat to the ozone layer from chlorofluorocarbon (CFC) gases, or "freons." They built on the work of James Lovelock, who had developed the electron capture detector, a device that was able to measure extremely low organic gas contents in the atmosphere. With this device, Lovelock had shown that CFC gases had already spread globally throughout the atmosphere. Molina and Rowland argued that CFCs could be transported up to the ozone layer where the intensive ultraviolet light would cause chemical decomposition, releasing chlorine atoms that would deplete the ozone layer.

Molina and Rowland calculated that continued use of CFC gases would deplete the ozone layer markedly in a few decades. Their calculations drew much criticism, but also much concern. Their work has now proven to be essentially right, and even to have somewhat underestimated the risk.

## Unintended Consequences of HCFCs

The replacement of chlorofluorocarbons (CFCs) in spray cans, air conditioning, and refrigeration with hydrochlorofluorocarbons (HCFCs) also produced unintended consequences, however. HCFCs do not damage the ozone layer, but they act like very potent greenhouse gases, 4,500 times more potent than carbon dioxide. Furthermore, HCFCs may break down to form oxalic acid, which contributes to acid rain.[a]

The example of the negative effects of yet another innovation intended to protect the environment reinforces the point that environmental protection is not a simple matter of preventing clear wrongdoing. Science and engineering knowledge is essential to the prevention, or even the surveillance and early detection, of damaging consequences of well-intended and seemingly innocent actions.

[a]"Chemicals That Eased One Environmental Problem May Worsen Another," *Science News* (March 16, 2010) available at http://www.sciencedaily.com/releases/2010/03/100303114001.htm.

The use of chlorofluorocarbons started out quite innocently. The stability and nontoxicity of these manufactured chemicals had made them seem safe and environmentally benign. They found many uses as refrigerants, industrial cleaning agents, propellants in aerosol sprays, and blowing agents in plastic foams; it was extremely unwelcome news that the use of CFCs was causing major environmental destruction. (Chemical erosion of the ozone layer has already resulted in dramatically increased rates of skin cancer, particularly in places in the southern hemisphere.)

Paul Crutzen, Mario Molina, and Sherwood Rowland jointly received the 1995 Nobel Prize in chemistry for their pioneering work that explained how ozone is formed and decomposes through chemical processes in the atmosphere. These three researchers were honored for what the Nobel Foundation cited as contributing "to our salvation from a global environmental problem that could have catastrophic consequences."[26] This was the first Nobel Prize to be given for environmental science.

The CFCs were phased out after they were found to cause a hole in the ozone layer. Initially they were replaced with hydrochlorofluorocarbons (HCFCs). The HCFCs do not damage the ozone layer, but have been found to act like very potent greenhouse gases, 4,500 times more potent than carbon dioxide.[27]

*Why was it difficult to recognize that chlorofluorocarbons erode the ozone layer?*

[26]*Ibid.*
[27]"Chemicals That Eased One Environmental Problem May Worsen Another," *Science News* (March 16, 2010) available at http://www.sciencedaily.com/releases/2010/03/100303114001. htm.

## Superfund Sites and the Monitoring of Communities for Toxic Contamination

What factors contribute to delay in cleanup of contamination?

Although mistakes by well-intentioned people are the source of some of the current problem of toxic contamination, negligent, reckless, or even deceitful actions have also contributed to it. Extensive contamination with both radioactive and toxic substances has occurred at military installations. The cost of the cleanup of plutonium contamination alone is conservatively estimated at $200 *billion* and the cleanup of all contaminants may cost $400 billion. That is more than the cost of the entire U.S. interstate highway system. The contamination was allowed to increase over many decades. Initially, the Pentagon, the Energy Department, NASA, and the Coast Guard regarded pollution on their property to be none of the public's business. Leaders feared both embarrassment from disclosure and that cleanup would distract them from what they considered more important problems that they were facing. Although from the 1970s environmental groups, state agencies, and the federal EPA complained about toxic contamination, their warnings mostly went unheeded. The cleanup program that is now necessary has been described as "one of the biggest engineering projects ever undertaken." The Hanford Nuclear Reservation in Richland, Washington, is the most extensively contaminated with both radioactive wastes (including plutonium and other radioactive elements) and chemical contaminants. These substances have contaminated the groundwater and soil and are seeping into the huge Columbia River that forms the western part of the border between Oregon and Washington.[28]

Like the Hanford, Chernobyl, and Three Mile Island sites in the United States and the USSR, nuclear facilities elsewhere in the world have also been the site of accidents, and falsification of records. For example, numerous accidents and malfeasance have plagued the nuclear facility in Sellafield in Cumbria, UK.[29] The Sellafield facility manufactures plutonium for nuclear weapons and, beginning in 1956, became the home of the world's first nuclear reactor to generate commercial electricity. In October 2006, it was fined £500,000 for mistakes that led to leakage of highly radioactive liquid within the plant.[30]

## Love Canal

One of the most famous cases of toxic contamination by a corporation is that of Love Canal. In the 1890s, William T. Love had dug a trench as part of a plan to provide hydroelectric power for a model city he proposed to construct near the city of Niagara Falls, New York. Soon a financial depression left Love with no

---

[28] Schneider, Keith. 1991. "Military Has New Strategic Goal in Cleanup of Vast Toxic Waste," *New York Times*, August 5, A1, D3. Extensive information about the Hanford site is available at http://www.halcyon.com/tomcgap/www/hanford.html.

[29] "Sellafield" at the *Global Security.org* Web site, http://www.globalsecurity.org/wmd/world/uk/sellafield.htm, last modified: April 28, 2005 by "Zulu."

[30] *RTE News*. 2006. "British Nuclear Group Fined over Leaks," October 16, accessed at http://www.rte.ie/news/2006/1016/sellafield.html.

investors and he abandoned his project. For the early decades of the twentieth century the trench was used as a swimming hole.

Beginning in the 1930s, chemical companies that had moved into the area began using the canal as a waste dump. In 1942, Hooker Chemical and Plastics Corporation negotiated an agreement to dump wastes at the site and eventually purchased the land for that purpose. It lined the site with cement to keep the chemicals from leaking. Over the next eleven years it put an estimated 352 million pounds of chemical waste in the canal.[31] In 1953, the canal became full with chemical wastes and the municipal wastes that the city of Niagara Falls also dumped at the site.

The Niagara Falls Board of Education became interested in obtaining the land for a new school. Hooker Chemical took samples to demonstrate to the Board of Education the presence of the chemicals at the site, but the board was not dissuaded. Hooker finally sold the land to the board for the nominal price of one dollar.

The Board of Education constructed a grade school and playground near the center of the parcel, adjacent to the landfill. In doing so it partially removed the cap that Hooker had placed on the site. Later the city punctured the trench's cement liner when it installed new sewers and storm drains in the area. As demand for housing increased, the Board of Education sold the remaining land to developers who subdivided it for single-family homes. During the next two decades the waste migrated through the soil, contaminating storm sewers and basements, or surfacing to evaporate and contaminate the air. Beginning in 1958, residents complained to the city about foul odors, oily liquids in their basements, and rashes on children who attended school or played at the site.

Environmental monitoring at the Love Canal was first conducted in 1976. Data collected by the New York Department of Environmental Conservation (NYDEC) and by Calspan Corporation, a private firm hired by the city, revealed extensive contamination of the groundwater, soil, and air. After the local newspaper published these results, the frustrated citizenry took the matter to their congressperson, who involved the federal Environmental Protection Agency (EPA).

The New York State Health Commissioner received a report on the site from the EPA in March 1978. Three months later the State Department of Health conducted a house-to-house health survey. The health department drew blood samples from those living in the ninety-seven homes immediately adjacent to the canal. Two days later Governor Carey declared a state health emergency.

Various citizens' groups emerged in response to the crisis. One such group, under the leadership of Lois Gibbs, later evolved into the Citizens' Clearinghouse for Hazardous Wastes, a citizens group formed to help other communities in similar circumstances.[32]

This case was discussed extensively in 1979 congressional hearings that led to the passage of the Comprehensive Emergency Response, Compensation and

---

[31] Worobec, Mary. 1980. "An Analysis of the Resource Conservation and Recovery Act." *BNA Government Reporter*, Special Report (August 22).

[32] Gibbs, Lois M. 1985. *Centers for Disease Control: Cover-Up, Deceit and Confusion*. Arlington, VA: Citizens' Clearinghouse for Hazardous Wastes; and 1982, *Love Canal: My Story*. Albany: State University of New York Press.

Liability Act (CERCLA), popularly known as "the Superfund Act." Health studies and litigation continued. A settlement was reached between former residents and Occidental Chemical, the parent company of Hooker, in 1983. A new clay cap was installed over the canal in 1984. A consent agreement was reached between the United States and Occidental Chemical in 1989.[33]

Although many sites of toxic contamination exist, the story of Love Canal was a landmark case and one of the best documented, because it was the first to receive national attention. An update on Love Canal and the general situation of schoolchildren's exposure to toxic chemicals may be found at http://www.cnn.com/2008/HEALTH/08/22/toxic.schools/index.html.[34] It should be recognized, however, that there are worse sites in the United States. Furthermore, the dangers resulting from toxic sites are distributed in a way that burdens minority populations. A 1987 study by the United Church of Christ Commission for Racial Justice found that three out of five African Americans and Hispanic Americans live in communities with uncontrolled hazardous waste sites. As Robert Bullard documents in *Dumping in Dixie: Race, Class, and Environmental Quality*, many of the worst sites that have received less attention are in the South and primarily affect the health of people of color.[35]

The threat of toxic contamination will be with us for the foreseeable future in view of the large and increasing number of known toxic waste sites and the frequency of acute chemical "releases" (spills and the like). More than 30,000 hazardous waste sites are known to exist in the United States. Of these, more than 1,200 are large enough or affect large enough populations to be listed on the Environmental Protection Agency's National Priorities List of Superfund cleanup sites. The inventory of potential Superfund sites has been growing each year.[36] The inventory of *potential* Superfund sites is growing at the rate of 2,000 or 3,000 annually.[37]

Along with toxic exposure from dumps, toxic exposures due to chemical releases are frequent. There are about 1,200 acute chemical releases in a typical year. In 1986, accidental releases resulted in 210 deaths, 6,490 injuries, and 533

[33] The account here is drawn from a longer account in *Monitoring the Community for Exposure and Disease: Scientific, Legal, and Ethical Considerations* by Nicholas A. Ashford, Carla Bregman, Dale B. Hattis, Abyd Karmali, Christine Schabacker, Linda-Jo Schierow, and Caroline Whitbeck, a report supported by the Agency for Toxic Substance and Disease Registry (ATSDR) and the National Institute for Occupational Safety and Health (NIOSH), 1991. That account makes extensive use of the transcript of the hearings held by the U.S. House of Representatives (U.S. Congress 1979a) as well as the specific sources cited earlier. For a valuable insight into the thinking of managers within the Hooker Corporation, see chapter 2.4, "The Hooker Memos," in Alastair Gunn and P. Aarne Vesilind, 1986, *Environmental Ethics for Engineers*. Chelsea, MI: Lewis Publishers.
[34] "Despite Love Canal's lessons, schoolchildren are still at risk," *CNNHealth.com*, last updated 9:02 a.m. EDT, Friday, August 22, 2008.
[35] Bullard, R. D. 1990. *Dumping in Dixie: Race, Class, and Environmental Quality*. Boulder, CO: Westview Press.
[36] Commentary by Charles Xintaras at the 1993 Hazardous Waste Conference on Education in Environmental Health, at http://atsdr1.atsdr.cdc.gov:8080/cx12a.html.
[37] Zuras, A. D., F. J. Prinznar, and C. S. Parrish. 1985. "The National Priorities List Process." In AIChE, *Management of Uncontrolled Hazardous Waste Sites*. New York: AIChE, 1–3.

evacuations.[38] Engineers and scientists are necessarily involved in investigating these sites, releases, and accidents, attempting to prevent or mitigate the effects of future accidents and dumping and cleaning up what has already occurred.

The great expense of the needed cleanup makes it a responsibility that citizens and public officials are tempted to ignore. A case in point is the cleanup of the previously mentioned plutonium contamination at military installations, which will cost hundreds of billions of dollars. Ignoring such contamination leaves the toxic and radioactive materials to continue to seep into the soil and poison groundwater.

*What factors contribute to delay in cleanup of toxic and radioactive contamination?*

## Environmental Norms in U.S. Corporations

What might one learn about a company's policies to reduce pollution before accepting employment there?

The attitudes of U.S. corporations toward the environment vary widely although environmentalism in corporate America has increased significantly in the last two decades. Because of the difference in corporate attitudes, engineers will find a variety of levels of support for fulfilling their newly recognized responsibility for the environment.

We have already considered how engineering societies have recently made additions to their codes of ethics about protection of the environment. Attitudes toward polluting practices are changing in U.S. corporations as well. Because many engineers and scientists work in corporate environments, understanding corporate attitudes and the variety among them is especially important for engineers and scientists, especially those at the beginning of their careers. How much has changed in the last ten to fifteen years is highlighted by a study of corporate attitudes conducted by Joseph M. Petulla from 1982 to 1985.[39] Even during the three years of his study, Petulla noted increasing environmental concern. He placed companies in one of three categories based on their management practices regarding environmental pollution. Contrary to his expectation, Petulla did not find a direct correlation between these practices and company size. He did find that the "corporate culture" as determined by the CEO, and, in some cases, the senior corporate attorney, was a major factor in predicting a company's management strategy.

The most environmentally concerned companies demonstrated what Petulla characterized as "supportive compliance" with environmental laws. Compliance was endorsed by the CEO, and carried through by well-trained personnel

---

[38] Binder, S. and S. Bonzo. 1989. "Letter to the Editor." *American Journal of Public Health* 79(12): 1681.

[39] Petulla, Joseph M. 1989. "Environmental Management in Industry." In *Ethics and Risk Management in Engineering,* edited by Albert Flores (Lanham, MD: University Press of America). This study was brought to my attention by reading Charles E. Harris', "Manufacturers and the Environment: Three Alternative Views" in *Environmentally Conscious Managing: Recent Advances,* edited by Mo Jamshidi, Mo Shahinpoor, and J. H. Mullins (Albuquerque, NM: ECM Press, 1991).

using the best pollution control equipment, supported by ongoing research and cooperation with government agencies and community groups. However, Petulla found that only 9 percent of the companies surveyed fell into this category. The middle category, into which 58 percent of his sample fell, demonstrated "cost-oriented environmental management"; that is, they complied with the law but demonstrated no general commitment to preventing environmental degradation. The final group, into which 29 percent of Petulla's sample fell, was classified as demonstrating "crisis-oriented environmental management." They had no full-time staff assigned to environmental protection and addressed issues of pollution only when forced to, frequently finding it cheaper to pay fines and lobbying fees rather than to prevent pollution. One representative from this group expressed his reason for this strategy by saying, "Why the hell should we cooperate with the government or anyone else who takes us away from our primary goal [of making money]?"[40]

The best known statement of the position that making money is the primary goal of corporations is the one mentioned at the beginning of Chapter 5 by economist Milton Friedman, who said that the responsibility of managers is to "make as much money as possible while conforming to the basic rules of society, both those embodied in law and those embodied in ethical custom." Friedman specifically said that stockholders' money should not be spent on "avoiding pollution."[41] His view is that it is the *responsibility* of a manager to avoid polluting only to the extent that it is legally required, and even to *legally resist* compliance where doing so will make more money.

Some companies began initiatives in environmental protection somewhat earlier, but many of them only became significant in the 1980s. The aim of such programs is not merely to reduce pollution from waste. They also often aim at changing manufacturing methods so as not to use hazardous substances and not to produce them or pass them on in the product or by-products.

The 1990s saw a rise in the popular support for control of environmental pollution by groups ranging from the Green Party in California to the Religious Right. In this changing climate many corporations now at least wish to appear environmentally concerned. This has been particularly true of chemical companies. It is notable that some chemical companies in the United States dropped "chemical" from their names.

In the decade from 2000 to 2010, some environmental concerns (such as the interest in organically grown food) became popular and many companies sought to change at least their negative image as polluters. To change the reality as well as the public perception, by:

- Reducing toxicity of components and ingredients in their products
- Reducing the exposure of workers to toxic substances in the creation (manufacture, growth, etc.), transport, and sale of their products

---

[40] *Ibid.*, 146.

[41] Friedman, Milton. 1970. "The Social Responsibility of Business Is to Increase Its Profits," *The New York Times Magazine* (September 13) reprinted in *Ethical Issues in Engineering*, edited by Deborah Johnson (Englewood Cliffs, NJ: Prentice-Hall, Inc., 1991), 78–83.

- Encouraging or improving the recycling of their products and/or components at the end of the product life
- Reducing toxicity in their by-products
- Improving the disposal (or, hopefully, finding reuse) of their waste products
- Reducing their demand for raw materials the manufacture or growth of which cause environmental degradation

Even with greater corporate commitment to environmental protection, there is still a range of corporate attitudes. Experienced engineers have described to me great differences even within the same industry. It is important for new engineers to know about a company's policies before joining it, if they are to fulfill responsibilities vis-à-vis the environment without running afoul of management.

*What might one learn about a company's concern to reduce pollution before working there?*

## From "Global Warming" to "Climate Change"

What is meant by "climate change"? How, if at all, is it different from "global warming"?

In the last decades of the twentieth century, increasing attention was given to observations of increases in average temperatures around the world. Some claimed that these increases were largely due to human actions such as the burning of fossil fuels, which contributes to the presence of **greenhouse gases** (i.e., gases that contribute to the **greenhouse effect** on Earth). The greenhouse effect warms the Earth's surface and lower atmosphere by selectively transmitting short-wave radiation to the Earth or reflecting it back to Earth. Others looked to such factors as the cyclical patterns of changing average temperatures as an explanation of the rise in temperature. Because of the initial observations of rising temperatures the phenomenon was first termed "**global warming**." Later, as it became clear that *other* climactic changes were also occurring, the term "global warming" was dropped in favor of "**climate change**."

Many of the changes other than temperature rise are regional, rather than worldwide. They include changes in rainfall, wind patterns, and storm patterns (and therefore coastal erosion), changes that may disrupt agriculture and fishing. Such changes often alter patterns of disease incidence in humans and other species upon which humans depend. However, as the National Research Council pointed out:

global temperature is easier to project than regional changes such as rainfall, storm patterns, and ecosystem impacts.[42, *]

---

[42] National Research Council. 2008. "Understanding and Responding to Climate Change," 2008 edition, p. 10.

*Animations of the shrinking icecap at the North Pole are available on the Web site of the U.S. National Oceanic and Atmospheric Administration (NOAA) at http://www.gfdl.noaa.gov/the-shrinking-arctic-ice-cap-ar4#movies.

Figure 10.1. Iceberg (Photo by Jon Brack; courtesy of the National Science Foundation)

Awareness of environmental degradation has increased markedly even as the estimated costs of remediation have increased. Major abrupt climate change, as illustrated by the Dust Bowl drought in the U.S. Great Plains of the 1930s, is very difficult for people and ecosystems to adapt to. One postulated threat of temperature rise is that melting icebergs, such as the one shown in Figure 10.1, might reduce the salinity of ocean water, disrupting the existing ocean circulation that currently warms Europe and making locations in Europe warmer than locations at comparable latitudes in the United States. (For example, Rome is on about the same latitude as New York City.) The cooling of Europe would greatly disrupt agriculture and fishing.

*What is meant by "climate change"? How, if at all, is it different from "global warming"?*

## Technological Innovation in Response to Environmental Challenges

What environmental challenges today are of greatest interest to you to work on as an engineer?

Since the September 11, 2001 attacks, both terrorism and political instability have received more attention in the United States. Furthermore, environmental degradation in general (and climate change in particular) has been identified as a factor that contributes to political instability. This increases the roster of negative consequences *for humans* that are now attributed to environmental degradation (increased storm damage, coastal erosion, agricultural loss, political instability), whatever the harm to the environment for its own sake.

## Scientific Estimates of Climate Change

In the 1980s, controversy swirled about the causes of recently observed climactic changes. The National Academies' report, "Understanding and Responding to Climate Change,"[a] which laid out the basics of a scientific understanding, and its more comprehensive reports on climate change issued in 2010 have settled many of those questions from a scientific point of view, however. (The headline the Academies issued for the news briefing that accompanied the release of the first three of these reports was "STRONG EVIDENCE ON CLIMATE CHANGE UNDERSCORES NEED FOR ACTIONS TO REDUCE EMISSIONS AND BEGIN ADAPTING TO IMPACTS."[b]) Those reports provide the most thorough scientific assessments to date of climate change, the state of scientific studies of climate change, and expert opinion on how best to cope with those changes that have progressed too far to forestall.

---

[a]"Understanding and Responding to Climate Change," 2008 edition.
[b]"STRONG EVIDENCE ON CLIMATE CHANGE UNDERSCORES NEED FOR ACTIONS TO REDUCE EMISSIONS AND BEGIN ADAPTING TO IMPACTS," news briefing, issued May 19, 2010, available at http://www8.nationalacademies.org/onpinews/newsitem.aspx?RecordID=05192010. At the time of this writing, this page also has links to the three reports that have been issued.

Some of the assessments of the mechanisms and consequences of climate change and most of the measures for coping with it will come from engineering. Some of the best known of these to which engineering has already contributed are the development of alternative energy sources to coal and petroleum, such as wind power, solar power, geothermal mining, and nuclear power. (The development of nuclear power is surrounded with greater controversy, both because of the threat of catastrophic accidents, after the Chernobyl and Three Mile Island accidents, and because of the stubborn problem of disposing of nuclear waste.)

Awareness of environmental degradation has increased markedly even as the estimated threats from such degradation, especially degradation that contributes to climate change, have increased. Unfortunately new environmental disasters regularly appear, which leads me to predict that one or more new ones will be developing as you read this book. (As I write, the latest disaster is a huge oil spill resulting from an explosion of a BP offshore oil rig in the Gulf of Mexico.)

*What environmental challenges today are of greatest interest to you to work on as an engineer?*

## The Concern with Sustainability and Sustainable Development

What does "sustainable development" mean and how does concern to achieve sustainable development relate to the environmental goals that preceded it?

In August 1987, the World Commission on Environment and Development issued its report to the United Nations. (That report is officially titled "Our Common Future" but widely known as "the Brundtland Report," so named for the chairperson of the commission.) That report[43] established the concept of *sustainable*

---

[43]The report is available at http://www.un-documents.net/ocf-12.htm#II.

*development* as a fundamental concept for many writers on issues of human well-being or environmental protection. Some of those writers gave their own definitions of "sustainable development," leading to some confusion. The definition in the Brundtland Report is:

> **Sustainable development** is development that meets the needs of the present without compromising the ability of future generations to meet their own needs.

The report goes on to point out that this definition makes use of two key notions:

> [T]he concept of "needs," in particular the essential needs of the world's poor, to which overriding priority should be given; and the idea of limitations imposed by the state of technology and social organization on the environment's ability to meet present and future needs.

The concern with the world's poor dictates that economic development be a priority. At the same time, the threat of environmental disaster makes sustainability a goal. The two goals are combined in the goal of sustainable development.

## Wants and Needs

One reason that some may find the Brundtland Report definition inadequate is its employment of the concept of "needs." The distinction between wants and needs is notoriously culturally determined. Although there is general agreement about the human need for food, clothing, shelter, and jobs (the four specific needs named in chapter 2 of the Brundtland Report) it is common for people in a culture to place other goals and values, which might receive little emphasis in some other culture, ahead of securing food, clothing, shelter, or jobs. Lack of cross-cultural agreement may stand in the way of getting international cooperation required for institutional and legal change called for in chapter 12 of the Brundtland Report.

As we have seen, the 2006 revision of the NSPE Code of Ethics added the clause:

> Engineers shall strive to adhere to the principles of *sustainable development*. (Italics added.)

The World Federation of Engineering Organizations (WFEO) carries on the front page of its Web site the slogan "Engineers Shape the Sustainable Future."[44] Furthermore, WFEO and FIDIC (International Federation of Consultant Engineers) together with the Union des Associations Techniques Internationales (UATI) created the World Engineering Partnership for *Sustainable Development* (WEPSD) in 1992. These actions reflect engineering's embrace of the concepts of sustainability and sustainable development.

*What does "sustainable development" mean and how does concern to achieve sustainable development relate to the environmental goals that preceded it?*

## Summary and Conclusion

We have seen how a new awareness of environmental degradation and an ecological understanding developed from the early 1960s. New disciplines of environmental engineering and science give powerful evidence of the relevance of these disciplines to environmental protection and sustainability. A variety of reasons have been given for thinking that engineers have a special professional responsibility for the environment. The environmental effects that have received the bulk of attention have been the ones that carry clear implications for human

[44] See http://www.wfeo.org/index.php (January 29, 2010).

health, safety, and well-being. The latter effects fall under the well-recognized responsibility of engineers to ensure the public health, safety, and welfare.

Some environmental damage does not have clear implications for human health and safety, however. Should responsibility for such environmental damage be regarded as analogous to the engineer's responsibility for quality? Is there any reason to think that engineers have any special responsibility to prevent such damage? (Of course, *all* people might have some responsibility, or specific engineers, such as environmental engineers, might have specific responsibilities stemming from their positions of trust.) There is not a clear answer to the foregoing questions, but we may not need one. Ecological thinking leads to the recognition of the interconnectedness of organism and species and the recognition that the good of one is often highly dependent on others or on the whole. Therefore, the question of the moral standing of individuals and species may be of little importance so long as *some* interdependent beings have moral standing.

Recent examples of acid rain and erosion of the ozone layer have taught us that even well-intended actions can have far-reaching negative effects on the environment, effects that clearly endanger human health, safety, and well-being. Engineering and applied science are essential to anticipating the consequences of human action on the environment before the effects become catastrophic.

# 11 A Note on End Use and "Macro" Issues

## The "End-Use Problem"

When people speak of the **"end-use problem"** in engineering ethics, they are speaking about the question of whether or to what extent engineers are responsible for what others do with the technologies that the engineers have helped to design, manufacture, or maintain.

In Chapter 4, we examined an argument that engineers, at least those who worked on medical life-support technology, bear guilt because of the harm resulting from the overuse of life-support technology. That was a stringent argument that engineers are accountable for the end use of their work. I argued that it was too demanding because engineers are not in a position to foresee that the technology would be so misused and in fact there are other interventions, such as the requirement that health care facilities respect living wills and other so-called advance care directives, that directly address the problem of misuse and allow life-support technology to continue to benefit patients with whom it is appropriately used.

When I conducted interviews as part of a study of the responsibilities of biomedical engineers, I found biomedical engineers to be a rather diverse group. Although some come to biomedical engineering through a series of fortuitous occurrences, others came to biomedical engineering because of a concern about the *end use* of their work. In some cases, this took the form of wanting to provide devices that would directly benefit people. In others, it took the form of wanting to avoid military work without leaving their special technical area. I even found one respected biomedical engineer who took early retirement when he discovered that one of his principal inventions proved not to benefit patients.

Engineers are sometimes hired to answer technical questions and solve technical problems that arise in the service and pursuit of goals that they are not told about. Furthermore, even when they do know the larger outcome of their work – say, a consumer product or a government program – that outcome may be used to serve some very different goal, as when sewing needles are used to torture people. There was a time when engineering, like science, claimed to be value-free, although the claim was always less plausible for engineering, precisely because engineering characteristically addresses practical problems (and which problems are judged *important to address* is a value-laden judgment). However, today major challenges have been offered to the view that even science is value-free, as we saw in Chapter 8.

If engineering is not value-free does that mean that any and all engineers involved in the design or manufacture of any goods bear moral responsibility for what people do with them, including the extreme example of a sadistic use of sewing needles? That is too extreme, as we saw in Chapter 1. Just as predictability is crucial in determining whether engineers were responsible for the overuse of the medical devices they helped to create, so engineers are answerable only for those uses of their work that they could predict. Of course, predictability itself is not merely a matter of the nature of some technology and the state of engineering knowledge, but of *what has been learned about previous uses* and misuses of technology. Once a danger is evident there is also the question of what alternatives exist to achieve desirable functions of the technology in question, without allowing for the predictable undesirable use. No good substitute has been found for knives, for example, that is, nothing can do the work of a knife that does not also have the potential to be used as a deadly weapon.

---

## Should You Be Concerned with This Possible Use of Your Work?

You are a graduate student, working in an algorithms research lab. This is your first year there and your advisor, Dr. Dna, has assigned you to a project, designing and benchmarking an algorithm for comparing similar sections of DNA. The funding for this project comes from GeneTech, a private company.

Your friend Jo, an older student who is working on a master's thesis, has repeatedly told you that you should take a break during your first summer and get an internship. Jo claims that it will not set you that far back on your degree progress and that the experience you will gain will make it much easier to find a job when you graduate. You agree and realize that GeneTech may be a good choice. Because of your time on their project, you are familiar with the kind of work they do and enjoy doing it, as well.

Before sending your resumé to GeneTech, you do some research on the company. You find out that the company researches and markets techniques for genetic analysis. However, you also find out that it is a subsidiary of a large health insurance company. On discovering this, you have a sinking feeling. The algorithm that you have been developing is to identify specific differences between similar sections of genetic code. You can envision only two purposes for it: for research on mutations in sections of DNA or for identifying the probability that an organism will express a certain phenotypic trait. What if this health insurance company is funding your research because it wants to lay the groundwork for a genetic testing program to discover which applicants are genetically predisposed to phenotypic expression of chronic diseases, which it could use to deny applicants coverage?

You feel some ethical reservations about continuing work on this project, so you go to Dr. Dna to ask for advice. She tells you it is not your job to ask where the research is going, only to be the first to get it done and publish it. Furthermore, the Genetic Information Nondiscrimination Act (GINA), which took effect in May 2008, prohibits U.S. insurance companies and employers from discriminating on the basis of information derived from genetic tests. She thinks it highly unlikely that the political climate will ever change to the extent that the law will be repealed and genetic testing of applicants for health insurance be allowed. You agree with her that genetic screening by health insurance providers will probably never be legal, but you still feel uneasy about working on the project.

What should you do?

## What Are "Macro" Issues?

The goal of this book is to give you tools to address the moral problems that commonly arise in engineering work and to critically examine arguments about the issues raised. If there are some moral problems that are problems for the engineering *profession*, then if you join that profession, those problems are ones that through your professional organizations you may help address. Certainly one issue for engineering (or any other profession) is maintenance of its ethical standards for members of that profession. We have already seen that trustworthiness of members of a profession is better than regulation both for members of a profession and for the public, and how some professions, such as accounting, have lost the public trust after major scandals.

As we saw at the end of Chapter 1, Bill Wulf when he was president of the NAE spoke of "macro problems" in engineering ethics, by which he meant problems for the *engineering profession* rather than problems that the individual engineers can address (and which are sometimes called "micro issues" in contrast). The example he gave is the problem of complex systems, the complexity of which makes their behavior inherently unpredictable. It remains to be seen whether the engineering profession is willing and able to address such issues. Certainly, the leadership position of the president of the NAE is a position well suited to raise such issues. (This use of the term *macro* is unrelated to other uses such as in the term *macroengineering*.)

We have seen the scope of engineering responsibility expand many times in the course of the twentieth century. There are still limits on what engineers might in principle foresee and hence some limitations on what engineers can accomplish or be responsible for, however. Unfortunately, some nonengineers seek to ignore this limitation and seek to blame engineers for any and all negative effects of technology or even aspects of modern life! Now that ABET has instituted requirements for ethics education in engineering programs, we see some engineering departments farming out that education to humanities and social science departments. This action cannot help but give the message that real engineers know nothing about the ethics of their own profession. Furthermore, those in humanities and social science disciplines, rather than learning about the engineering work world for which engineering ethics education ought to prepare student engineers, succumb to the temptation to operate with distorted stereotypes of engineers and to exhort engineering students to change those negative features of the world that those humanities and social science scholars are at a loss to address themselves. I recommend that you carefully examine whether any purported macro issue in engineering ethics is in fact an issue that the engineering profession is best qualified to solve or whether it is a difficult policy issue regarding technology that scholars wish some person or group would solve.

It is better for everyone if professionals (and other experts) are trustworthy. That requires professions and other bodies of experts to devote attention and effort to developing and transmitting and supporting high standards of professional practice.

## The Use of Human Growth Hormone as an Example of an Issue for the Whole Society

The use of human growth hormone (HGH), somatotropin, has been criticized as a drug treatment because its inappropriate use poses unknown hazards to children. (Its use for weight loss and its use as a performance-enhancing drug by athletes have also been criticized, although it may not be effective for either of these purposes.) The hormone was commercially developed as an effective treatment for a form of human dwarfism that is due to deficiency of the growth hormone. However, the use of the genetically engineered human growth hormone in the United States far exceeds the amount needed to treat the 2,500 cases of growth hormone deficiency found in children each year. It appears that some parents are subjecting their children – especially their male children – to unknown risks of side effects to increase their height, merely because the children would otherwise be of *normally* short stature.

In the human growth hormone case, like the case of life-support technology and unlike the dune buggy case discussed in Chapter 3, the intended use of HGH to treat children who have dwarfism is regarded as too important to forgo. Therefore, there has not been a significant drive to ban the hormone altogether.

Serious arguments have been offered against developing treatments for traits that, although outside the statistical norm, are not painful or seriously debilitating to their bearer. One argument is that making these conditions the object of treatment contributes to stigmatizing them and the people who have them. An argument specifically against the use of HGH as a treatment to increase height is that doing so will simply change the standards for short stature. Its use will create a *new* and slightly taller group of people who will be in the "normally short stature" category.[1]

[1] This scenario is an adaptation of one created by Michael Jolson (Case Western Reserve University [CWRU] 2006.

# Epilog: Making a Life in Engineering

We have examined many aspects of professional responsibility for engineers and scientists. Many aspects of moral life lie outside considerations of professional responsibility. Family responsibility and general civic responsibility are two other major areas of moral responsibility that clearly lie outside the scope of considerations of this book. What about the choice of work in engineering and science? Such decisions are made within the context of many other decisions, including family obligations. For example, family obligations may restrict the geographical region in which you seek work. Practical considerations such as the need to pay back education loans also influence job choices. The choice of work is more intimately connected to professional ethics, because it significantly influences the opportunities for expressing one's values in one's work. Work that fulfills one's aspirations as well as ambitions and need for income is a major element in a meaningful life. How does one find opportunities to do such work? This is a problem, indeed a design problem, that a person addresses many times in a life, if at all. (I say "If at all," because many people in the world today and throughout human history have had little opportunity to pursue many aspirations in their work life beyond providing subsistence to themselves and their families.)

The current range of possibilities for a young adult with talent in engineering or science may itself be daunting, and it would only add to that burden to attempt to catalog the value dimensions of work choice. In any case, I am reluctant to do so, because I have too often seen humanists and social scientists much more ready to instruct engineers and scientists about the goals they should pursue than to consider the social implications of their own work in humanities and social science.

Instead, I will tell you about two of the many engineers it has been my pleasure to know who have acted on their aspirations. The two stories I shall tell are about two engineers at very different stages of their career and making contributions of very different sorts based on quite different concerns and priorities. Your own aspirations may or may not be like either of the two engineers whose stories I tell here.

## Miguel Barrientos, Building a Water Pump for Andean Alpaca Breeders

First is a story of an undergraduate project carried out by Miguel Barrientos (MIT '93) to design and manufacture a human-powered treadle pump to meet the needs of Andean alpaca breeders.[1] Miguel found the project a fulfilling experience. It enabled him to gain valuable experience in the field of appropriate technology – that is, the development of technology suited to the needs of small producers, rural and urban, especially in the developing world. It was personally rewarding because it enabled him to work for the betterment of his country, Bolivia.

As Miguel knew, people who live in rural areas of Bolivia continually face the problem of supplying water for their houses, crops, and animals. This problem became particularly acute immediately preceding his project, because of a drought in the Andean regions of the country that had begun in 1989. The drought had become a major obstacle to most of the development projects that operate in the Bolivian Andes.

From June 8 to September 4, 1992, Miguel worked as a technical consultant for the Alpaca Wool Production and Processing Project (PPPLA). The PPPLA endeavors to improve breeding and veterinarian practices among alpaca breeders in the Andean zones of Bolivia, and receives funding from the United Nations Development Program (UNDP), the United Nations Capital Development Fund (UNCDF), and Appropriate Technology International (ATI). Miguel was asked to identify an appropriate type of water pump and assist with its production. The pump had to be suitable for use with a water table 2 to 5 meters deep, inexpensive to manufacture, and simple to maintain and repair. ATI had previously tested a human-powered treadle pump for use in Africa. The area for the intended use of the pump in Bolivia is cold and dry because of the high altitude of the zone (4,000 to 5,000 meters above sea level).

Miguel began his project by visiting with some of the alpaca breeders for whom the pump was to be built to better understand their needs and conditions of the pumps' use. The design considerations Miguel identified for the treadle water pump were that it be

- versatile – its design should suit it for use to feed livestock, irrigate pastures and crops, and even provide water for households.
- inexpensive – sell for about 80 U.S. dollars (existing pumps sold for at least 140 U.S. dollars).
- repairable by the alpaca breeders themselves.

The project needed skilled artisans in Bolivia to build the pump for the Andes. Miguel located a small machine shop called "Khana Wayra" – which in the Aymara language means "Light of the Wind" – whose owners were experienced in producing wind turbines, manual and wind-powered water pumps, solar heaters, and drilling equipment. With the members of Khana Wayra, Miguel built and tested a prototype treadle pump. By July, the pump was ready for a demonstration trip to the project area.

Miguel modified the original design of the treadle pump in several ways: Cylinders made of PVC pipe replaced the cylinders made of sheet steel to make it easier to maintain. He enlarged the valve box and valve plate to accommodate PVC cylinders, which were slightly larger than the metal cylinders. He redesigned

---

[1] This account is primarily based on the final report on the project "Designing Tools for Developing Countries" written by Miguel Barrientos (MIT '93) and my discussions with him about the project.

the inlet pipe and the treadle support so that they could be detached from the main body to make the pump easier to transport.

Khana Wayra's machinists were experienced in building water pumps, so they completed the prototype pump in four days.

Miguel and Pablo Garay, a member of Khana Wayra, tested the prototype pump briefly in early July, satisfactorily pumping water at floor level. Several days later when they took the pump to a well, they found that the outlet valves were not making a tight seal, allowing air to be sucked into the pump. The leakage was due to the imperfect roundness of the cylinders, and to the poor seal made by the rubber disc they were using. Khana Wayra did not have a lathe where the cylinders could be turned down, so the only way to improve suction was to replace the rubber disc. They molded leather cups to replace the rubber discs, treated them with vegetable oil, and tested the pump again. They successfully pumped water from a depth of 1.5 meters, but suction was too weak to pump from a depth of 3.5 meters. They then replaced the inlet pipe, which had a diameter of 1.5 inches, with a 1-inch diameter pipe, and tested the pump again. The pump functioned well, pumping water at a rate of approximately 1.3 liters per second.

With the machinists of Khana Wayra, Miguel then determined the maximum depth at which the pump could function with the new 1-inch diameter pipe. They pumped water successfully first from 2.5 meters and then from 3 meters, and finally from 4.8 meters. The pump operated flawlessly even in the 4.8-meter well, extracting water at a rate of approximately 0.5 liters per second. It did not fully fill the cylinders in a 6-meter well, but because the theoretical depth limit for a treadle pump at 4,000 meters above sea level (atmospheric pressure = 470 mm Ha) is 6.4 meters, they were satisfied with the performance of the prototype.

Next came field testing. The field testing had several purposes:

1. To demonstrate the characteristics of the treadle pump to the alpaca breeders.
2. To test the pump in the area where it was to be used.
3. To identify flaws of the pump design, both through tests and feedback from the potential users.

At the first site, Miguel first tried the pump in one location where water can be found at a depth of 1.5 meters, but found the water too muddy to pump efficiently. In that location, observers were disappointed in the amount of physical labor required to extract the water. Miguel then tried the pump at a new location, where he found clear water at a depth of 2 meters. One of the spectators volunteered to try it out. The pump operated flawlessly, pumping water at a rate that impressed the people watching the demonstration.

The field testing and comments from local observers at Cosapa and later at Wariscata led Miguel to recommend the following changes before the pump went into mass production:

1. Increased support to the treadles' axle to make the pump structure more rigid.
2. Clear marking on the inlet pipe at the point of attachment of the treadle support pipe, to facilitate assembly.

3. Construction of the treadles of 4-inch by-2-inch hardwood to make the operation of the pump safer.

4. Attachment of the treadle support pipe to the inlet pipe so that the pump operator does not have to get too close to the well.

5. Widening the baseboard to increase the stability of the pump.

6. Increasing the thickness of the pulley's rope because the rope wears rapidly.

7. Use of stainless steel springs in the outlet valves to prevent corrosion, and modification of the valves to eliminate leaks.

8. Gluing the rubber seal of the valve box to the baseboard to avoid deformation caused by suction.

9. Special tooling to accelerate the production rate of treadle pumps. (This tooling was subsequently received from ATI.)

10. Provision of blueprints and instructions to those observers who expressed interest in producing some of the parts themselves.

Along with these recommendations, Miguel gave Khana Wayra new blueprints of the pump incorporating all the modifications to the original pump design, and instructed Khana Wayra's members on the use of the tooling they received from ATI.

## Jim Melcher, Witnessing against Waste and Violence

At his untimely death of cancer at age 54, James R. Melcher was the Julius A. Stratton Professor of Electrical Engineering and Physics and director of MIT's Laboratory for Electromagnetic and Electronic Systems (LEES), one of the large interdisciplinary laboratories at MIT. He was widely known in electrical engineering for his practical applications of continuum electromechanics, a broadly interdisciplinary field that draws on electromagnetics, fluid and solid mechanics, heat transfer, and physical chemistry. His strong interest in the ethical questions raised by engineering work was long evident to his colleagues in the MIT Electrical Engineering and Computer Science Department and around the world.

The son of a Methodist minister, Jim was a kind and modest person, with the courage to look unflinchingly at very difficult problems. He did not shrink from considering any implications of engineering activities. I recall one of the institute professors at MIT who had known Jim since graduate school describing Jim as a "saint."

A key experience for Jim in becoming a vocal critic of militarism and energy dependence had been his sabbatical year at the Cavendish Laboratory at Cambridge University in 1971 where he was working on his text, *Continuum Electromechanics*. He described the effect of seeing the United States from abroad as convincing him to have a stronger influence on the course the United States was taking.[2]

[2] Melcher, James D. 1991. "America's Perestroika, Living a New National Agenda," *Technology Review* (April): 4–11. This article, my memory of conversations with Jim, and the obituary for Jim published in the January 9, 1991, issue of *Tech Talk*, are the principal sources for the present account.

The first Arab oil boycott took place shortly after Jim's return to the United States. The experience of the effects of this boycott together with his growing awareness of U.S. energy vulnerability led him to undertake a striking witness: to give up his second car, and make his daily commute, eighteen miles round trip, by bicycle. Jim had always gotten plenty of exercise, in part to control his diabetes, but with university athletic facilities at his disposal he had many exercise options that were pleasanter than biking eighteen miles a day through Boston winters. Bicycle racing was one of the side benefits that attracted graduate students to work with Jim in the Laboratory for Continuum Electromechanics, which he had founded within LEES.

Jim's research now became more applied: limiting air pollution from diesel exhaust and coal combustors. His students, too, took their theoretical work on topics such as the mass transfer of electric fields in fluidized beds and immediately applied them to making environmentally friendly ways to recycle asphalt concrete.

Jim recounts his realization that "For the sake of oil, our government cast its lot with an obscenely rich dictatorship [in Iran] which was out of step with popular movements." He particularly recalled an interview with the then shah of Iran on a U.S. news program. The shah, whose dictatorship is widely acknowledged to have been supported by the CIA, unblinkingly affirmed he was God. Jim observed that had the U.S. government given direct subsidies to U.S. corporations to purchase oil at even $2/gallon, it would have been cheaper in financial terms than trying to keep oil "cheap" by military force. "Without a hidden military subsidy of foreign oil, domestic oil would be competitive. Embedded in shale, for example, there is more oil in the United States than in the Middle East. Shale oil would be competitive if the price of oil were little higher than it is now – and if that price held steady as crises came and went." He tried unsuccessfully to advance the exploration of these domestic alternatives.

Jim was additionally appalled at what he saw as the overreadiness of the United States to resort to military means or CIA tactics in support of unpopular dictatorships. Jim found the quick resort to violence in support of materialism quite inconsistent with Christianity as he understood it.

When President Regan announced the Strategic Defense Initiative (SDI) or Star Wars in 1983, Jim, like many of his peers, saw this as much as a political initiative as a technical one, and was concerned about what he saw as the Pentagon's efforts to make university pursuit of SDI funds look like an endorsement of the project, which many believed was fundamentally flawed and promised only to increase a growing national deficit. The talents of LEES were ideally suited to SDI research, and Jim made it clear to the faculty in his laboratory that he would not block any proposals for SDI funding from individual faculty in his lab, but he sought to find other sources of support and was quite successful in doing so. (At this time engineering salaries had yet to undergo "hardening" and many universities including Jim's expected their engineering faculty to raise a great deal of grant money and even to cover half of their own salary by this means.)

Personally, Jim wanted to do more, so he became part of a campaign to convince *senior* faculty in engineering and science departments around the country to pledge not to take SDI funding. On May 13, 1986, he joined with three Nobel laureates at a Washington press conference to make public more than 3,700 names of those who had pledged not to take SDI funds.

In his last months, as Jim squarely faced his losing bout with cancer, he turned to writing his experience of making larger sense of things as he made a life in engineering. I leave you with some of his words from that article:

> To really integrate the way you earn your living with your social and even spiritual aspirations, for people in any line of work, is the true test of an education. Your values must become part of your professional thinking, which is best learned "hands on."[3]

[3] *Ibid.*, 11.

# References

Alberts, Bruce and Kenneth Shine. 1994. "Scientists and the Integrity of Research," *Science*, 266: 1660.

Allen, Anita L. 1987. *Uneasy Access: Privacy for Women in a Free Society*. Totowa, NJ: Rowman and Allanheld.

_____. 1995. "Privacy in Health Care." In the second edition of the *Encyclopedia of Bioethics*. New York: Macmillan, pp. 2064–2073.

American Chemical Society. 1994. "Ethical Guidelines to Publication of Chemical Research," *Accounts of Chemical Research*, 27(6): 179–181, revised 2010. The 2010 revision is available as a pdf at http://pubs.acs.org/page/policy/index.html or http://pubs.acs.org/userimages/ContentEditor/1218054468605/ethics.pdf.[1]

American Institute of Aeronautics and Astronautic (AIAA) Publications Committee. "Ethical Guidelines & Procedures [for Publication of Aeronautics and Astronautics Research]," http://www.aiaa.org/content.cfm?pageid=720.[2]

American Institute of Chemical Engineers (AIChE). "Ethical Guidelines for AIChE Publications," http://www.aiche.org/Publications/Resources/Ethics.aspx.[3]

American Society of Civil Engineers (ASCE). 2003, 2006, 2009. "Professional Ethics and Conflict of Interest: Resolution 502," http://www.asce.org/Content.aspx?id=8404.

American Society of Mechanical Engineers (ASME). 1999. "Ethical Obligations of Authors," http://www.asme.org/Publications/ConfProceedings/Author/Ethics.cfm.[4]

---

[1] These guidelines were first issued in 1985. Subsequent revisions have been principally to add materials (such as a section on publishing outside the technical literature) or make minor changes. Note that the ACS has many engineers as members and *Chemical and Engineering News* is one of its publications. These guidelines provided a model for similar guidelines of many other scientific and engineering societies, including the American Optical Society (which often publishes engineering research) and many of the guidelines from other engineering societies listed herein.

[2] These guidelines contain brief guidelines for authors (about plagiarism and six other violations) and a statement of procedures for investigating ethical violations. The opening paragraph makes clear that the AIAA Publications Committee developed these guidelines in response to "an increasing number and variety of ethical violations." The guidelines appear to be original and contain no acknowledgments. Another copy of the same document, titled "Publication Ethical Standards: Guidelines and Procedures," is available at http://www.writetrack.net/aiaa/documents/Ethical_Guidelines.pdf.

[3] This document contains three sections for editors, for authors, and for reviewers. No authorship, date, or acknowledgments are given for these guidelines.

[4] This document contains this acknowledgment: "These ethical standards have been to a large extent compiled from the existing standards of The American Chemical Society and ASME acknowledges its appreciation to ACS for granting permission to quote from the ACS Ethical

Ashford, Nicholas A. 1986. "Medical Screening in the Workplace: Legal and Ethical Considerations." *Seminars in Occupational Medicine*, 1 (1): 67–79.

Ashford, Nicholas A., Carla Bregman, Dale B. Hattis, Abyd Karmali, Christine Schabacker, Linda-Jo Schierow, and Caroline Whitbeck. *Monitoring the Community for Exposure and Disease: Scientific, Legal, and Ethical Considerations*, a report supported by the Agency for Toxic Substance and Disease Registry (ATSDR) and the National Institute for Occupational Safety and Health (NIOSH), 1991.

Association for Computing Machinery (ACM). 2006, revised 2010. "Policy and Procedures on Plagiarism," http://www.acm.org/publications/policies/plagiarism_policy/.[5]

Austin, J. L. 1961. "A Plea for Excuses." In *Philosophical Papers*, edited by J. O. Urmson and G. J. Warnock, pp. 123–152. London: Oxford University Press.

Babbage, Charles. 1830. *Reflections on the Decline of Science in England in Science and Reform: Selected Works of Charles Babbage*. Cambridge; New York: Cambridge University Press, 1989.

Baier, Annette. 1986a. "Extending the Limits of Moral Theory." *Journal of Philosophy*, 77: 538–545.

———. 1986b. "Trust and Antitrust." *Ethics*, 96: 232–260. Reprinted in *Moral Prejudices*, pp. 95–129. Cambridge, MA: Harvard University Press.

———. 1990. "A Naturalist View of Persons." Presidential Address delivered before the Eighty-Seventh Annual Eastern Division Meeting of the American Philosophical Association in Boston, Massachusetts, December 29, 1990. *APA Proceedings*, 65(3): 5–17.

———. 1993. "Claims, Rights, Responsibilities." In *Prospects for A Common Morality*, edited by J. P. Reeder and G. Outka, Princeton University Press, and in a collection of Baier's essays titled *Passions of the Mind*, forthcoming from Harvard University Press.

Barber, Bernard. 1983. *The Logic and Limits of Trust*. New Brunswick, NJ: Rutgers University Press.

Baron, Marcia. 1984. *The Moral Status of Loyalty*. Dubuque, IA: Kendall/Hunt Publishing Co.

Beardsley, Tim. 1996. "Profile: Thereza Imanishi-Kari." *Scientific American*, 275(5) (November): 50–52.

Beauchamp, Tom and James Childress (eds.). 1979. *Principles of Biomedical Ethics*. London: Oxford University Press.

Beecher, Henry K. 1966. "Ethics and Clinical Research." *New England Journal of Medicine*, 274(24) (June 16): 1354–1360.

Guidelines to Publication of Chemical Research, (Chem. Rev. 1995, 95, pp. 11A–13A. Copyright 1985, 1989, 1995, American Chemical Society)." Acknowledgment is also given to ASCE and AGU for drawing on their guidelines in the development of the ASME document.

[5] This document addresses issues only of plagiarism and duplicate publication, which it calls "self-plagiarism," but it cites the ACM Code of Ethics and Professional Conduct, 1992 revision, which is available at http://www.acm.org/about/code-of-ethics, and the "Software Engineering Code of Ethics and Professional Practice," version 5.2 (a product of the ACM/IEEE-CS Joint Task Force on Software Engineering Ethics and Professional Practices), which is available at http://www.acm.org/about/se-code/, for overarching considerations about credit and about intellectual property.

Benner, Patricia. 1984. *From Novice to Expert: Excellence and Power in Clinical Nursing*. Reading, MA: Addison-Wesley.

Benner, Patricia and Judith Wrubel. 1989. *The Primacy of Caring: Stress and Coping in Health and Illness*. Reading, MA: Addison-Wesley.

Binder, S. and S. Bonzo. 1989. "Letter to the Editor." *American Journal of Public Health*, 79(12): 1681.

Bird, Stephanie J. and David E. Housman. 1995. "Trust and the Collection, Selection, Analysis and Interpretation of Data: A Scientist's View." *Science and Engineering Ethics*, 1(4) (October): 371.

Bird, Stephanie J. and Jerome Rothenberg. 1988. "To Screen or Not to Screen: Drugs, DNA, AIDS." Unpublished manuscript.

Block, Peter. 1993. *Stewardship – Choosing Service over Self-Interest*. San Francisco: Berrett-Koehler Publishers.

Board of Ethical Review of the National Society of Professional Engineers. 1976. *Opinions of the Board of Ethical Review*, Volume IV. Alexandria, VA: National Society of Professional Engineers.

_____. 1981. *Opinions of the Board of Ethical Review*, Volume V. Alexandria, VA: National Society of Professional Engineers.

_____. 1989. *Opinions of the Board of Ethical Review*, Volume VI. Alexandria, VA: National Society of Professional Engineers.

_____. 1994. *Opinions of the Board of Ethical Review*, Volume VII. Alexandria, VA: National Society of Professional Engineers.

Bosk, Charles L. 1979. *Forgive and Remember: Managing Medical Failure*. Chicago: University of Chicago Press.

Brandt, Allan M. 1985. *No Magic Bullet: A Social History of Venereal Disease in the United States since 1880*. New York: Oxford University Press.

Briggs, Shirley A. 1987. "Rachel Carson: Her Vision and Her Legacy." In *Silent Spring Revisited*, edited by Gino J. Marco, Robert M. Hollingworth, and William Durham, pp. 3–11. Washington, DC: American Chemical Society.

Broad, William J. and Nicholas Wade. 1982. *Betrayers of the Truth*. New York: Simon and Schuster.

Broome, Taft M. Jr. 1986. "The Slippery Ethics of Engineering." *Washington Post* (December 28): 451.

Buciarelli, Louis L. 1985. "Is Idiot Proof Safe Enough?" *Applied Philosophy*, 2(4): 49–57; reprinted in *Ethics and Risk Management in Engineering*, edited by Albert Flores, pp. 201–209. Landam, New York, and London: University Press of America.

Bullard, R. D. 1990. *Dumping in Dixie: Race, Class and Environmental Quality*. Boulder, CO: Westview Press.

Buzzelli, Donald E. 1993. "The Definition of Misconduct in Science: A View from NSF." *Science*, 259: 584–648.

Callahan, Daniel, Arthur Caplan, and Bruce Jennings (eds.). 1985. *Applying the Humanities*. New York: Plenum.

Caplan, Arthur L. 1992. *When Medicine Went Mad: Bioethics and the Holocaust*. Totowa, NJ: Humana Press.

Cohen, Jon. 1994. "U.S.–French Patent Dispute Heads for a Showdown." *Science*, 265 (July 1): 23–25.

Committee on Academic Responsibility Appointed by the President and Provost of MIT. 1992. *Fostering Academic Integrity*. Boston: Massachusetts Institute of Technology.

Committee on Editorial Policy of Council of Biology Editors (CBE). 1990. *Ethics and Policy in Scientific Publication*. Bethesda, MD: Council of Biology Editors, Inc.

Committee on Engineering Design Theory and Methodology, the Manufacturing Studies Board of the National Research Council. 1991. *Improving Engineering Design: Designing for Competitive Advantage*. Washington, DC: National Academies Press.

Committee on Science, Engineering and Public Policy of the National Academy of Sciences, National Academy of Engineering, and Institute of Medicine. 1995. *On Being a Scientist*, second edition. Washington, DC: National Academies Press.

Culliton, Barbara J. 1988. "Harvard Tackles the Rush to Publication." *Science*, 241 (July 29): 525.

Curd, Martin and Larry May. 1984. *Responsibility for Harmful Actions*. Dubuque, IA: Kendall/Hunt Publishing Co.

Dandekar, Natalie. 1991. "Can Whistleblowing be Fully Legitimated?" *Business and Professional Ethics Journal*, 10(1): 89–108.

Davis, Michael. 1988. "Avoiding the Tragedy of Whistleblowing." *Business and Professional Ethics Journal*, 8(4): 3–19.

———. 1991. "Thinking like on Engineer." *Philosophy and Public Affairs* 20: 2(Spring 1991) 150–167.

———. 1997. "Better Communications between Engineers and Managers: Some Ways to Prevent Many Ethically Hard Choices." *Science and Engineering Ethics*, 3(2): 171–212.

Djerassi, Carl. 1989. *Cantor's Dilemma*. New York: Penguin Books, USA.

———. 1994. *The Bourbaki Gambit*. Athens: University of Georgia Press.

Dowie, Mark. 1977. "Pinto Madness." *Mother Jones* (September/October): 19–32.

Elliston, Frederick Keenan and Lockhart van Schaick. 1985. *Whistleblowing: Managing Dissent in the Workplace*. New York: Praeger Scientific.

Elstein, Arthur S., Lee S. Shulman, and Sarah A. Sprafka. 1978. *Medical Problem Solving: An Analysis of Clinical Reasoning*. Cambridge, MA: Harvard University Press.

Flumerfelt, R. W., C. E. Harris, M. J. Rabins, and C. H. Samson. 1992. *Engineering Ethics* (Texas A&M), final report to the NSF on Grant Number DIR-9012252.

French, Peter. 1982. "What is Hamlet to McDonnell-Douglas or McDonnell-Douglas to Hamlet: DC-10." *Business and Professional Ethics Journal*, 1(2): 1–14.

Friedman, Milton. 1970. "The Social Responsibility of Business Is to Increase Its Profits." *The New York Times Magazine* (September 13); reprinted in *Ethical Issues in Engineering*, 1991, edited by Deborah Johnson, pp. 78–83. Englewood Cliffs, NJ: Prentice-Hall.

Friedman, Paul J. 1992. "The Troublesome Semantics of Conflict of Interest," *Ethics and Behavior*, 2(4): 245–251.

Gibbs, Lois M. 1982. *Love Canal: My Story*. Albany: State University of New York Press.

_____. 1985. *Centers for Disease Control: Cover-up, Deceit and Confusion*. Arlington, VA: Citizens' Clearinghouse for Hazardous Wastes.

Gilligan, Carol. 1982. *In a Different Voice: Psychological Theory and Women's Development*. Cambridge, MA: Harvard University Press.

Goleman, Daniel. 1985. *Vital Lies, Simple Truths*. New York: Simon & Schuster, Inc.

Goodstein, David L. 1994. "Whatever Happened to Cold Fusion?" *Engineering and Science*, Fall: 15–25; reprinted from *the American Scholar*, Autumn 63: 4.

_____. 1995. "Ethics and Peer Review." *Biotechnology*, 13 (June): 618.

Graham, Loren. 1993. *The Ghost of the Executed Engineer: Technology and the Fall of the Soviet Union*. Cambridge, MA: Harvard University Press.

Gunn, Alastair and P. Aarne Vesilind. 1983. "Ethics and Engineering Education." *Journal of Professional Issues in Engineering*, 109, no. 2.

_____. 1986. *Environmental Ethics for Engineers*. Chelsea, MI: Lewis Publishers.

_____. 1990. "Why Can't You Ethicists Tell Me the Right Answers?" *Journal of Professional Issues in Engineering*, 116, no. 1.

Hampshire, Stuart. 1949. "Fallacies in Moral Philosophy." *Mind*, 58: 466–482; reprinted in *Revisions: Changing Perspectives in Moral Philosophy*, edited by Stanley Hauerwas and Alasdair MacIntyre, 1983.

_____. 1989. *Innocence and Experience*. Cambridge, MA: Harvard University Press.

Harris, Charles E., Jr. 1991. "Manufacturers and the Environment: Three Alternative Views." In *Environmentally Conscious Managing: Recent Advances*, edited by Mo Jamshidi, Mo Shahinpoor, and J. H. Mullins. Albuquerque, NM: ECM Press.

Harris, Charles E., Jr., Michael S. Pritchard, and Michael J. Rabins. 1995. *Engineering Ethics*. Belmont, CA: Wadsworth.

Hartley, Diane. 1978. "Implications of a Major Urban Office Complex," Vols. 1 & 2. Senior Thesis, Princeton University.

Hauerwas, Stanley. 1977. *Truthfulness and Tragedy*. Notre Dame, IN: University of Notre Dame Press.

_____. 1983. "Constancy and Forgiveness, The Novel as School for Virtue." *Notre Dame Literary Journal*, 15(3).

Hauerwas, Stanley and Alasdair MacIntyre (eds.). 1983. *Revisions: Changing Perspectives in Moral Philosophy*. Notre Dame, IN: University of Notre Dame Press.

Hoerr, John, with William G. Glaberson, Daniel B. Moskowitz, Vicky Cahan, Michael A. Pollock, and Jonathan Tasini. 1985. "Beyond Unions." *Business Week*, July 8.

HHS Commission on Research Integrity. 1995. "Professional Misconduct Involving Research." *Professional Ethics Report*, VIII, no. 3 (Summer).

Holton, Gerald. 1978. "Subelectrons, Presuppositions, and the Millikan-Ehrenhaft Dispute." In *Historical Studies in the Physical Sciences*, 11: 166–224; reprinted in the collection of Holton's essays, *Scientific Imagination*, Cambridge, MA: Cambridge University Press, 1978, pp. 25–83.

_____. 1994. "On Doing One's Damnedest: The Evolution of Trust in Scientific Findings." Chapter 7 in Holton's *Einstein, History, and Other Passions*. New York: American Institute of Physics.

Hurst, J. Willis and H. Kenneth Walker (eds.). 1972. *The Problem-Oriented System*. Baltimore, MD: Williams and Wilkins Company.

Institute for Electrical and Electronic Engineering (IEEE). 2007. *Publication Guidelines*. Section 8.2, Publication Services and Products Board Operations Manual (November 16), available at http://www.ieee.org/portal/pages/iportals/aboutus/whatis/index.html.

IEEE. 2007. *Introduction to the Guidelines for Handling Plagiarism Complaints*. Available at http://www.ieee.org/web/publications/rights/Plagiarism_Guidelines_Intro.html

Jackall, Robert. 1988. *Moral Mazes*. New York: Oxford University Press.

Jackson, C. Ian. 1986. *Honor in Science*. New Haven, CT: Sigma Xi, the Scientific Research Society.

Janis, Irving. 1982. *Groupthink*, second edition. Boston, MA: Houghton Mifflin.

Jasanoff, Sheila S. 1985. *Controlling Chemicals: The Politics of Regulation in Europe and the U.S.* Ithaca, NY: Cornell University Press.

Jonsen, Albert R. 1977. "Do No Harm: Axiom of Medical Ethics." In *Philosophical Medical Ethics: Its Nature and Significance*, edited by S. F. Spicker and H. T. Engelhardt, Jr. Dordrecht, Holland: D. Reidel Publishing Co.

_____. 1991. "Of Balloons and Bicycles, or the Relationship between Ethical Theory and Practical Judgment." *Hastings Center Report*, 21(5): 14–16.

Jonsen, Albert R. and Stephen Toulmin. 1988. *The Abuse of Casuistry: A History of Moral Reasoning*. Berkeley: University of California Press.

Kaufman, Ron. 1992. "After 5 Years, Heated Controversy Persists in Science Copyright Case." *The Scientist*, September 14: 1, 4, 5, 10.

Kiang, Nelson. 1995. "How are Scientific Corrections Made?" *Science and Engineering Ethics*, 1, 4(October): 347.

Korenman, Stanley G. and Allan C. Shipp with Association of American Medical Colleges ad hoc Committee on Misconduct and Conflict of Interest. 1994. *Teaching the Responsible Conduct of Research through a Case Study Approach*. New York: Association of American Medical Colleges.

Knox, Richard A. 1993. "Care at the End Not as Costly as Assumed." *Boston Globe*, 243(105): A1, April 14.

Kuhn, Thomas S. 1977. *The Essential Tension*. Chicago: University of Chicago Press.

Lachman, Judith A. 1990. *Issues in Management, Law and Ethics*, chapter 22, unpublished manuscript.

Ladd, John. 1970. "Morality and the Ideal of Rationality in Formal Organizations." *The Monist*, 54(4): 488–516.

_____. 1975. "The Ethics of Participation." In *Participation in Politics: NOMOS XVI*, pp. 98–125. New York: Atherton-Leiber.

_____. 1976. "Are Ethics and Science Compatible?" In *Science, Ethics, and Medicine*, pp. 49–78. Hastings-on-Hudson: The Hastings Center.

_____. 1979. "Legalism and Medical Ethics." In *Contemporary Issues in Biomedical Ethics*, pp. 1–35. Clifton, NJ: Humana Press.

_____. 1982a. "The Distinction between Rights and Responsibilities: A Defense." *Linacre Quarterly*, 49: 121–142.

_____. 1982b. "Philosophical Remarks on Professional Responsibility in Organizations." *Applied Philosophy*, 1 (Fall): 58–70.

_____. 1985. "The Quest for a Code of Professional Ethics: An Intellectual and Moral Confusion." In *AAAS Professional Ethics Project: Professional Ethics Activities in the Science and Engineering Societies*, edited by Rosemary Chalk,

Mark S. Frankel, and Sallie B. Chafer. Washington, DC: AAAS Press. Reprinted in *Ethical Issues in Engineering*, edited by Deborah Johnson, pp. 130–136. Englewood Cliffs, NJ: Prentice-Hall, 1990.

Lammers, Stephen and Allen Verhey (eds.). 1998. *On Moral Medicine: Theological Perspective in Medical Ethics* 2/e. Grand Rapids, MI: Wm. B. Eerdmans Pub.

Layton, Edwin T. 1985. "Theory and Application in Science and the Humanities." In *Applying the Humanities*, edited by Daniel Callahan, Arthur L. Caplan, and Bruce Jennings, pp. 57–70. New York: Plenum Press.

Leveson, Nancy G. and Clark S. Turner. 1993. "An Investigation of the Therac-25 Accidents." *Computer* (published by IEEE) (July): 18–41.

Levine, Robert J. 1986. *Ethics and the Regulation of Clinical Research*, second edition. Baltimore, MD: Urban & Schwarzenberg.

_____. 1995. "Informed Consent: Consent in Human Research." In *Encyclopedia of Bioethics*, second edition, pp. 1241–1250. New York: Macmillan.

Long, Thomas A. and James F. Thorpe. 1987. "The Challenger Case: A Flight from Responsibility?" ASEE Annual Conference Proceedings.

Longino, Helen. 1990. *Science as Social Knowledge*. Princeton, NJ: Princeton University Press.

Lucent Technologies Investigation Committee. 2002. *Report of the Investigation Committee on the Possibility of Scientific Misconduct in the Work of Hendrik Schön and Coauthors*. Distributed by the American Physical Society with permission of Lucent Technologies.

Luegenbiehl, Heinz C. 1983. "Codes of Ethics and the Moral Education of Engineers." *Business and Professional Ethics Journal*, 2(4): 41–61. Reprinted in *Ethical Issues in Engineering*, edited by Deborah Johnson, pp. 137–154. Englewood Cliffs, NJ: Prentice-Hall, Inc., 1990.

Lytton, William B. 1996. *Combating Corruption in Foreign Markets. The Evolving Role of Ethics in Business: Conference Report*. New York: Conference Board, Inc.

MacIntyre, Alasdair. 1981. *After Virtue: A Study in Moral Theory*. Notre Dame, IN: University of Notre Dame Press.

_____. 1984a. "Does Applied Ethics Rest on a Mistake?" *The Monist*, 67(4): 499–512.

_____. 1984b. *Three Rival Versions of Moral Inquiry: Encyclopaedia, Genealogy and Tradition*. Notre Dame, IN: University of Notre Dame Press.

_____. 1988. *Whose Justice? Which Rationality?* Notre Dame, IN: University of Notre Dame Press.

Mann, Charles C. 1994. "Radiation: Balancing the Record." *Science*, 263(5146): 470–474.

Marshall, Eliot. 1996. "Fraud Strikes Top Genome Lab." *Science*, 274 (November 8): 908.

Martin, Mike W. and Roland Schinzinger. 1989. *Ethics in Engineering*, second edition. New York: McGraw-Hill Publishing Company.

McConnell, Malcolm. 1987. *Challenger: A Major Malfunction*. Garden City, NY: Doubleday.

Mazur, Allan. 1989. "The Experience of Universities in Handling Allegations of Fraud or Misconduct in Research." In *Project on Scientific Fraud and Misconduct: Report on Workshop Number Two*, edited by Rosemary Chalk, pp. 67–94. Washington, DC: American Association for the Advancement of Science.

Merchant, Carolyn. 1980. *The Death of Nature: Women, Ecology, and the Scientific Revolution*. San Francisco: Harper & Row.

Middleton, William W. 1986. "Ethical Process Enforcement and Sanctions – The Engineering and Physical Science Societies." Delivered at the AAAS-IIT Workshop on Professional Societies and Professional Ethics, May 23, 1986.

Millikan, Robert A. 1913. "On the Elementary Electrical Charge and the Avogadro Constant." *Physical Review*, 2: 109–143.

Mnookin, Seth. 1996. "Department of Defense DNA Registry Raises Legal, Ethical Issues." *Gene Watch*, 10(1) (August): 1, 3, 11.

Morgenstern, Joe. 1995. "Fifty-Nine Story Crisis." *The New Yorker*, May 29, pp. 45–53.

Murray, Thomas. 1983. "Warning: Screening Workers for Genetic Risk." *Hastings Center Report*, 13(1): 5–8.

———. 1992. "The Human Genome Project and Genetic Testing: Ethical Implications." In *The Genome, Ethics, and the Law*, AAAS-ABA National Conference of Lawyers & Scientists. Washington, DC: AAAS.

National Research Council Panel on Scientific Responsibility and the Conduct of Research. 1992. *Responsible Science: Ensuring the Integrity of the Research Process*, Vol. I. Washington, DC: National Academies Press.

Nazario, Sonia L. 1991. "McDonnell Douglas Jet Evacuation Drills Leave 44 Injured." *Wall Street Journal*, Wednesday, October 30, A3–A4.

Newhouse, John. 1982. *The Sporty Game*. New York: Alfred A. Knopf, Inc.

Nissenbaum, Helen. 1996. "Accountability in a Computerized Society." *Science and Engineering Ethics*, 2(1).

Nissenbaum, Helen and Deborah Johnson. 1995. *Computers, Ethics and Social Values*. Englewood Cliffs, NJ: Prentice-Hall.

Norman, Colin. 1988. "Stanford Inquiry Casts Doubt on 11 Papers." *Science*, 242 (November 4): 659–661.

Office of the Inspector General, National Science Foundation. 1992. *Semiannual Report to the Congress*. No. 7: April 1, 1992–September 30, 1992, 22.

———. 1993. *Semiannual Report to the Congress*. No. 9: April 1, 1993–September 30, 1993, 37.

PBS. The American Experience: *Rachel Carson's Silent Spring*.

Perrow, Charles. 1984. *Normal Accidents*. New York: Basic Books.

Petroski, Henry. 1985. *To Engineer Is Human*. New York: St. Martin's Press.

Petulla, Joseph M. 1989. "Environmental Management in Industry." In *Ethics and Risk Management in Engineering*, edited by Albert Flores. Landam, MD: University Press of America.

Pfeifer, Mark P. and Gwendolyn L Snodgrass. 1990. "The Continued Use of Retracted, Invalid Scientific Literature." *Journal of the American Medical Association*, 263(10): 1420–1423.

Pincoffs, Edmund. 1971. "Quandary Ethics." *Mind*, 80: 552–571. Reprinted in *Revisions: Changing Perspectives in Moral Philosophy*, edited by Stanley Hauerwas and Alasdair MacIntyre.

Pooley, Eric. 1996. "Nuclear Warriors." *Time*, March 4, pp. 46–54.

Postel, Sandra. 1988. "Controlling Toxic Chemicals." In *State of the World*, edited by Linda Starke, pp. 118–136. New York: W. W. Norton & Co.

Potter, Elizabeth. 1994. "Locke's Epistemology and Women's Struggles." In *Modern Engendering*, edited by Bat Ami Bar On. New York: SUNY Press.

Powers, Madison. 1993. "The Right of Privacy Reconsidered." Commissioned paper for *Genetic Privacy Collaboration*, April 22, 1993.

Proctor, Robert. 1988. *Racial Hygiene: Medicine under the Nazis*. Cambridge, MA: Harvard University Press.

Racker, Efraim. 1989. "A View of Misconduct in Science." *Nature*, 339 (May): 91–93.

Rapp, Rayna. 1987. "Moral Pioneers: Women, Men and Fetuses on a Frontier of Reproductive Technology." *Women and Health*, 13(1–2): 101–117. Also published in *Women, Embryos, and Women's Rights: Exploring the New Reproductive Technologies*, edited by Elaine Baruch, Amadeo d'Amado, and Joni Seager.

Rawls, John. 1957. "Outline of Decision Procedure for Ethics." *Philosophical Review*, 66: 177–197.

Regan, Tom. 1983. *The Case for Animal Rights*. Berkeley: University of California Press.

Reich, Warren T., editor in chief. 1978. *Encyclopedia of Bioethics*. London: Macmillan.

Rennie, Drummond, Veronica Yank, and Linda Emanuel. 1997. "When Authorship Fails: A Proposal To Make Contributors Accountable," *Journal of the American Medical Association*, 278: 579–585.[6]

Roberts, Leslie. 1991. "Misconduct: Caltech's Trial by Fire." *Science*, 253: 1344–1347.

Roddis, W. M. Kim. 1993. "Structural Failures and Engineering Ethics." *Journal of Structural Engineering ASCE*, (May).

Rorty, Amelie Oksenberg. 1995. "The Many Faces of Morality." *Midwest Studies in Philosophy*, XX: 67–82.

Rothstein, Mark A. 1987. "Drug Testing in the Workplace: The Challenge to Employment Relations and Employment Law." *Chicago-Kent Law Review*, 63: 683–743.

Rowe, Mary P. 1990. "Barriers to Equality: The Power of Subtle Discrimination to Maintain Unequal Opportunity." *Employee Responsibilities and Rights Journal*, 3(2): 153–163.

Rowe, Mary P. and Michael Baker. 1984. "Are You Hearing Enough Employee Concerns?" *Harvard Business Review*, 62(3): 127–135.

Sarasohn, Judy. 1993. *Science on Trial*. New York: St. Martin's Press.

Scanlon, Walter E. 1980. *Alcoholism and Drug Abuse in the Workplace: Employee Assistance Programs*. New York: Prager.

Schneewind, J. B. 1985. "Applied Ethics and the Sociology of the Humanities." In *Applying the Humanities*, edited by Denial Callahan, Arthur L. Caplan, and Bruce Jennings, pp. 57–70. New York: Plenum Press.

———. 1968–69. "Moral Knowledge and Moral Principles." *Knowledge and Necessity, Royal Institute of Philosophy Lectures*, 3: 249–262, reprinted in *Revisions: Changing Perspectives in Moral Philosophy*, edited by Stanley Hauerwas and Alasdair MacIntyre.

[6] A proposal for a policy change to make investigators less likely to seek or accept credit through the mechanism of undeserved authorship.

Schneider, Keith. 1991. "Military Has New Strategic Goal in Cleanup of Vast Toxic Waste." *New York Times*, August 5, A1, D3.

Schwab, Adolf J. 1996. "Engineering Ethics in the U.S. and Germany." *IEEE Institute*, June.

Shore, Eleanor. 1995. "Effectiveness of Research Guidelines in Prevention of Scientific Misconduct." *Science and Engineering Ethics*, 1(4)(October): 383.

Singer, Peter. 1977. *Animal Liberation*. New York: Avon Books.

———. 1985. *In Defense of Animals 1985*. New York: Blackwell.

———. 1990. "The Significance of Animal Suffering." *Behavioral and Brain Sciences*, 13(1): 9–12.

Stone, Christopher. 1972. "Should Trees Have Standing? Towards Legal Rights for Natural Objects." 45 *SO. CAL. L. REV.* 450.

Stewart, Walter and Ned Stewart. 1987. "The Integrity of the Scientific Literature." *Nature*, 325 (January 15): 207–214.

Susskind, Lawrence E. 1990. "A Negotiation Credo for Controversial Siting Disputes." *Negotiation Journal*. October.

Tannenbaum, Jerrold and Andrew Rowan. 1985. "Rethinking the Morality of Animal Research." *Hastings Center Report*, 15 (October): 32–36.

Taubes, Gary. 1993. *Bad Science*. New York: Random House.

Thomson, Judith Jarvis. 1990. *The Realm of Rights*. Cambridge, MA: Harvard University Press.

Thoreau, Henry David. 1849. "Civil Disobedience." Reprinted in *Civil Disobedience in Focus*, edited by Hugo A. Bedau, pp. 28–48. New York and London: Routledge, 1991.

Thorpe, James F. and William H. Middendorf. 1979. *What Every Engineer Should Know about Product Liability*. New York and Basal: Marcel Dekker, Inc.

Toulmin, Stephen. 1981. "The Tyranny of Principles: Regaining the Ethics of Discretion." *Hastings Center Report*, 11(6): 31–39.

Turkington, Richard, George Trubow, and Anita Allen. 1992. *Privacy: Cases and Materials*. Houston, TX: John Marshall Press.

Unger, Stephen H. 1994. *Controlling Technology: Ethics and the Responsible Engineer*, second edition. New York: John Wiley.

Vesilind, P. Aarne, J. Jeffrey Peirce, and Ruth Weiner. 1987. *Environmental Engineering*, second edition. Stoneham, MA: Butterworth.

Wald, Mathew L. 1996. "Two Northeast Utilities Plants Face Shutdown." *The New York Times*, March 9.

Weil, Vivian. 1983a. "Beyond Whistleblowing: Defining Engineers' Responsibilities." *Proceedings of the Second National Conference on Ethics in Engineering*. Center for the Study of Ethics in the Professions, Illinois Institute of Technology, Chicago.

———. 1983b. "The Browns Ferry Case." In *Engineering Professionalism and Ethics*, edited by James H. Schaub and Karl Pavlovic. New York: John Wiley & Sons.

———. 1989. "Military Research, Secrecy, and Ethics." In *Ethical Issues Associated with Scientific and Technological Research for the Military*, edited by Carl Mitcham and Philip Siekevitz, pp. 193–99. *Annals of the New York Academy of Sciences*, vol. 577.

Weil, Vivian and Rachelle Hollander. 1990. "Sharing Scientific Data II: Normative Issues." *IRB*, 12: 7–8.

Weil, Vivian and John W. Snapper, (eds.). 1989. *Owning Scientific and Technical Information: Value and Ethical Issues*. New Brunswick, NJ: Rutgers University Press.

Weinstein, Milton C. and Harvey V. Fineberg. 1980. *Clinical Decision Analysis*. Philadelphia: W. B. Saunders Company.

Weiss, Philip. 1989. "Conduct Unbecoming?" *New York Times Magazine*, October 29, pp. 40–41, 68.

Weisskoph, Steven. 1989. "The Aberdeen Mess." *Washington Post Magazine*, January 15, p. 55.

Westin, Alan F. 1988. *Resolving Employment Disputes without Litigation*. Washington, DC: Bureau of National Affairs.

Westrum, Ron. 1991. *Technologies and Society*. Belmont, CA: Wadsworth.

Wheeler, David L. 1992. "U.S. Attorney: Leave 'Baltimore Case' to the Scientists." *The Chronicle of Higher Education*, July 22, A7.

———. 1995. "Making Amends to Radiation Victims." *The Chronicle of Higher Education*, October 13.

Whitbeck, Caroline. 1985. "Why the Attention to Paternalism in Medical Ethics?" *Journal of Health Politics, Policy and Law*, 10(1): 181–187.

———. 1991. "Ethical Issues Raised by New Medical Technologies." In *The New Reproductive Technologies*, pp. 49–64. Hillsdale, NJ: Lawrence Erlbaum Publishers.

———. 1992a. "Ethical Considerations in Biomonitoring Communities for Toxic Contamination." *Technology and Discovery: Proceedings of the Sixth International Conference of Society for Philosophy and Technology*, edited by Joe Pitt and Elana Lugo and slightly adapted from the author's contribution to *Monitoring the Community for Exposure and Disease*, a report to the Agency for Toxic Substances and Disease Registry. Nicholas Ashford, Principal Investigator.

———. 1992b. "The Trouble with Dilemmas: Rethinking Applied Ethics." *Professional Ethics*, 1(1 & 2): 119–142.

———. 1995. "Trust." In *Encyclopedia of Bioethics*, second edition. New York: Macmillan, pp. 2499–2504.

———. 2001. *Responsible Authorship, Group Mentoring in Responsible Research Conduct*, Online Ethics Center, available at http://www.onlineethics.org/Resources/TeachingTools/20357/19237/auth.aspx.[7]

———. 2004. "Trust and the Future of Research," *Physics Today*, 48–53. Available at http://scitation.aip.org/journals/doc/PHTOAD-ft/vol_57/iss_11/48_1.shtml.

Williams, Bernard. 1985. *Ethics and the Limits of Philosophy*. Cambridge, MA: Harvard University Press.

---

[7]The module contains a background discussion, cases/scenarios, and a method for using discussion of these scenarios or student interviews of faculty research supervisors to engage the members of laboratories or departments in the group mentoring of trainees in the responsible conduct of research.

———. 1988. "Formal Structures and Social Reality." In *Trust: Making and Breaking Cooperative Relations*. edited by Diego Gambetta, pp. 3–13. Oxford: Basil Blackwell.

———. 1993. *Shame and Necessity*. Berkeley: University of California Press.

Winslow, Ron. 1994. "FDA Halts Tests on Device That Shows Promise for Victims of Cardiac Arrest." *Wall Street Journal*, May 11, p. B8.

Woolf, Patricia K. 1986. "Pressure to Publish and Fraud in Science." *Annals of Internal Medicine*, 104(2): 254–256.

Worobec, M. 1980. "An Analysis of the Resource Conservation and Recovery Act." *BNA Government Reporter*, Special Report (August 22).

Young, Iris. 1990. *Justice and the Politics of Difference*. Princeton, NJ: Princeton University Press.

Zuckerman, Diana M. 1993. "Conflict of Interest and Science." *The Chronicle of Higher Education*, October 13, B1.

Zuras, A. D., F. J. Prinznar, and C. S. Parrish. 1985. "The National Priorities List Process." In *Management of Uncontrolled Hazardous Waste Sites*, edited by AIChE, pp. 1–3. New York: AIChE.

# Index